Lecture Notes in Computer Scie

T0238619

Commenced Publication in 1973
Founding and Former Series Editors:
Gerhard Goos, Juris Hartmanis, and Jan van Leeuwen

Narciso Martí-Oliet   Miguel Palomino (Eds.)

Recent Trends in Algebraic Development Techniques

21st International Workshop, WADT 2012
Salamanca, Spain, June 7-10, 2012
Revised Selected Papers

 Springer

Volume Editors

Narciso Martí-Oliet
Miguel Palomino
Universidad Complutense de Madrid, Facultad de Informática
Departamento de Sistemas Informáticos y Computación
28040 Madrid, Spain
E-mail: {narciso, miguelpt}@ucm.es

ISSN 0302-9743   e-ISSN 1611-3349
ISBN 978-3-642-37634-4   e-ISBN 978-3-642-37635-1
DOI 10.1007/978-3-642-37635-1
Springer Heidelberg Dordrecht London New York

Library of Congress Control Number: 2013934733

CR Subject Classification (1998): F.3, D.2.4, D.3.1, F.4, I.1, C.2.4

LNCS Sublibrary: SL 1 – Theoretical Computer Science and General Issues

Typesetting: Camera-ready by author, data conversion by Scientific Publishing Services, Chennai, India

Printed on acid-free paper

Springer is part of Springer Science+Business Media (www.springer.com)

Preface

The 21st International Workshop on Algebraic Development Techniques (WADT 2012) was held in Salamanca, Spain, during June 7–10 2012.

The algebraic approach to system specification encompasses many aspects of the formal design of software systems. Originally born as a formal method for reasoning about abstract data types, it now covers new specification frameworks and programming paradigms (such as object-oriented, aspect-oriented, agent-oriented, logic, and higher-order functional programming) as well as a wide range of application areas (including information systems, concurrent, distributed, and mobile systems).

The WADT workshop series focuses on the algebraic approach to the specification and development of systems and aims at providing a platform for presenting recent and ongoing work, to meet colleagues, and to discuss new ideas and future trends.Typical, but not exclusive, topics of interest are:

- Foundations of algebraic specification
- Other approaches to formal specification, including process calculi and models of concurrent, distributed, and mobile computing
- Specification languages, methods, and environments
- Semantics of conceptual modeling methods and techniques
- Model-driven development
  Graph transformations, term rewriting, and proof systems
- Integration of formal specification techniques
- Formal testing and quality assurance
- Validation and verification

During the workshop, 36 abstracts were presented and compiled in a technical report. In addition to the presentations of ongoing research results, the program included three invited lectures by Roberto Bruni (Università di Pisa, Italy), Francisco Durán (Universidad de Málaga, Spain), and Kim G. Larsen (Aalborg University, Denmark).

As for previous WADT workshops, after the meeting authors were invited to submit full papers for the refereed proceedings. There were 25 submissions from which 16 were selected for these final proceedings after being reviewed each by at least three referees.

Even a small workshop like WADT requires the effort of many people to make it happen. We first want to thank all the authors who submitted papers and showed interest in the subject. The members of the Program Committee and the external referees appointed by them worked hard to satisfactorily complete the reviewing process on time, for which we are most grateful. Special thanks also to Gustavo Santos, whose help in the organization of this event was invaluable.

The workshop took place under the auspices of IFIP WG 1.3, and it was organized by the Departamento de Sistemas Informáticos y Computación at

Universidad Complutense de Madrid. We gratefully acknowledge the sponsorship by the Spanish Ministerio de Economía y Competitividad, IFIP TC1, Facultad de Informática of Universidad Complutense de Madrid, Caja España–Duero Obra Social, Universidad de Salamanca, and IMDEA Software Institute.

We also thank EasyChair for making the life of the organizers so much easier.

February 2013   Narciso Martí-Oliet

                                                      Miguel Palomino

Organization

Steering Committee

Michel Bidoit	CNRS, France
Andrea Corradini	Università di Pisa, Italy
José Luiz Fiadeiro	Royal Holloway, University of London, UK
Rolf Hennicker	Ludwig-Maximilians-Universität München, Germany
Hans-Jörg Kreowski	Universität Bremen, Germany
Till Mossakowski (Chair)	German Research Center for Artificial Intelligence, Germany
Fernando Orejas	Universitat Politécnica de Catalunya, Spain
Francesco Parisi-Presicce	Università di Roma, Italy
Grigore Roşu	University of Illinois at Urbana-Champaign, USA
Andrzej Tarlecki	Warsaw University, Poland

Program Committee

The Program Committee is composed of the members of the Steering Committee plus:

Fabio Gadducci	Università di Pisa, Italy
Narciso Martí-Oliet (Co-chair)	Universidad Complutense de Madrid, Spain
Tom Maibaum	McMaster University, Canada
Miguel Palomino (Co-chair)	Universidad Complutense de Madrid, Spain
Markus Roggenbach	Swansea University, UK
Martin Wirsing	Ludwig-Maximilians-Universität München, Germany

Organizing Committee

Narciso Martí-Oliet	Universidad Complutense de Madrid, Spain
Miguel Palomino	Universidad Complutense de Madrid, Spain
Gustavo Santos	Universidad de Salamanca, Spain
Ignacio Fábregas	Universidad Complutense de Madrid, Spain
Isabel Pita	Universidad Complutense de Madrid, Spain
Adrián Riesco	Universidad Complutense de Madrid, Spain
David Romero	Universidad Complutense de Madrid, Spain
Fernando Rosa	Universidad Complutense de Madrid, Spain

Additional Reviewers

Bocchi, Laura
Borzyszkowski, Tomasz
Bruni, Roberto
Castro, Pablo
Chrząszcz, Jacek
Ciancia, Vincenzo
Ciobaca, Stefan
Codescu, Mihai
David de Frutos Escrig
Demasi, Ramiro
Dietrich, Dominik
Ermler, Marcus
Gorla, Daniele
Hildebrandt, Thomas
James, Phillip
Klarl, Annabelle
Klin, Bartek
Knapp, Alexander

Kuske, Sabine
Lluch Lafuente, Alberto
Lopez Pombo, Carlos Gustavo
Luettgen, Gerald
Meredith, Patrick
Moore, Brandon
Palamidessi, Catuscia
Pawłowski, Wiesław
Riesco, Adrian
Roggenbach, Markus
Schubert, Aleksy
Seisenberger, Monika
Tribastone, Mirco
Tronci, Enrico
Tuosto, Emilio
Tutu, Ionut
Vandin, Andrea

Table of Contents

Open Multiparty Interaction*

Chiara Bodei[1], Linda Brodo[2], and Roberto Bruni[1]

[1] Dipartimento di Informatica, Università di Pisa, Italy
[2] Dipartimento di Scienze Politiche, Scienze della Comunicazione e Ingegneria dell'Informazione, Università di Sassari, Italy

Abstract. We present the `link`-calculus, a process calculus based on interactions that are *multiparty*, i.e., that may involve more than two processes and are *open*, i.c., the number of involved processes is not fixed or known a priori. Communications are seen as chains of links, that record the source and the target ends of each hop of interactions. The semantics of our calculus mildly extends the one of CCS in the version without message passing, and the one of π-calculus in the full version. Cardelli and Gordon's Mobile Ambients, whose movement interactions we show to be inherently open multi-party, is encoded in our calculus in a natural way, thus providing an illustrative example of its expressiveness.

Introduction

An *interaction* is an action by which communicating processes can influence each other. Interactions in the time of the Web are something more than input and output between two entities. Actually, the word itself can be misleading, by suggesting a reciprocal or mutual kind of actions. Instead, interactions more and more often involve many parties and actions are difficult to classify under output and input primitives. We can imagine an interaction as a sort of puzzle in which many pieces have to be suitably combined together in order to work.

As networks have become part of the critical infrastructure of our daily activities (for business, social, health, government, etc.) and a large variety of loosely coupled processes have been offered over global networks, as services, more sophisticated forms of interactions have emerged, for which convenient formal abstractions are under study. For example, one important trend in networking is moving towards architectures where the infrastructure itself can be manipulated by the software, like in the Software Defined Networking approach, where the control plane is remotely accessible and modifiable by software clients, using open protocols such as OpenFlow, making it possible to decouple the network control from the network topology and to provide Infrastructure as a Service over data-centers and cloud systems. Another example is that of complex biological interactions as the ones emerging in bio-computing and membrane systems.

As a consequence, from a foundational point of view, it is strategic to provide the convenient formal abstractions and models to naturally capture these new

---

* Research partially supported by the EU through the FP7-ICT Integrated Project 257414 ASCEns (Autonomic Service-Component Ensembles).

N. Martí-Oliet and M. Palomino (Eds.): WADT 2012, LNCS 7841, pp. 1–23, 2013.

communication patterns, by going beyond the ordinary binary form of communication. These models should be sufficiently expressive to faithfully describe the complex phenomena, but they have also to provide a basis for the formal analysis of such systems, by offering sufficient mathematical structure and tractability. We present here a process calculus, called `link`-calculus, which takes interaction as its basic ingredient. The described interactions are *multiparty*, i.e., they may involve more than two processes and are *open*, i.e., the number of involved processes is not known *a priori*. Communication actions are given in terms of links, that record the source and the target ends of each hop of interactions. Links can be indeed combined in link chains that route information across processes from a source to a destination. Despite the inherent complexity of representing more sophisticated forms of interaction, we show that the underlying synchronisation algebra and name handling primitives are quite simple and a straight generalisation of dyadic ones. This is witnessed by the operational semantics rules of our calculus, that in the simpler version (i.e., without message passing) resemble the rules of CCS [21], while in the full one they resemble the rules of π-calculus [22].

Finally, we address a more technical issue, by providing a natural encoding of Cardelli and Gordon's (pure) Mobile Ambients (MA) [9] in the `link`-calculus. We have chosen Mobile Ambients as one of the most representative examples of process calculi for compartmentalisation based on the principles of location mobility and location awareness, from which many other models originated as abstractions for global computing, for systems biology and membrane systems. Our encoding highlights the multi-party nature of interactions in MA and shows that the spatial aspects due to the ambient nesting can be dealt with in passing to a flat calculus such as our `link`-calculus. We prove a tight correspondence at the level of reduction semantics and we provide a new bisimilarity semantics for MA as a side result. We are confident that analogous results can be obtained for many descendants of MA, as e.g., the Brane Calculus [8].

The paper, where we assume the reader has some familiarity with process calculi, is organised as follows. In Sections 1 and 2 we define the `link`-calculus, starting from introducing its fragment without name mobility, called *Core Network Algebra*. In Section 3, we recall the basics of MA. In Section 4, we define the encoding from MA to the `link`-calculus. In Section 5, we draw some final remarks, outline some future research avenues and point to some related works.

1   A Core Network Algebra

We start by defining a network-aware extension of CCS, called *Core Network Algebra* (CNA for short), whose communication actions are given in terms of *links*. A link is a pair that record the source and the target ends of a communication, meaning that the input available at the source end can be forwarded to the target one. Links are combined in *link chains* to describe how information can be routed across ends. Link chains also allow seamless realisation of multiparty synchronisations. While in this section we focus on the basic ideas of link interaction, we shall enhance the model with name mobility in the next section.

Links and Link Chains. Let $\mathcal{C}$ be the set of channels, ranged over by $a, b, c, \ldots$. Let $\mathcal{C} \cup \{\tau\} \cup \{*\}$ be the set of actions, ranged over by $\alpha, \beta, \gamma, \ldots$, where the symbol τ denotes a *silent* action, and the symbol $*$ denotes a non-specified action.

A *link* is a pair $\ell = {}^{\alpha}\backslash_{\beta}$; it can be read as forwarding the input available on α to β, and we call α the *source end* of ℓ and β the *target end* of ℓ. A link $\ell = {}^{\alpha}\backslash_{\beta}$ is *valid* if either $\alpha, \beta \neq *$ or $\ell = {}^{*}\backslash_{*}$. In the first case, the link is called *solid*. The link ${}^{*}\backslash_{*}$ is called *virtual*. We let $\mathcal{L}$ be the set of valid links. Examples of non valid links are ${}^{\tau}\backslash_{*}$ and ${}^{*}\backslash_{a}$.

As it will be shortly explained, a virtual link is a sort of "missing link" inside a link chain, a non specified part that can be supplied, as a solid link, by another link chain, via a suitable composition operation.

A *link chain* is a finite sequence $s = \ell_1 \ldots \ell_n$ of (valid) links $\ell_i = {}^{\alpha_i}\backslash_{\beta_i}$ s.t.:

1. for any $i \in [1, n-1]$, $\begin{cases} \beta_i, \alpha_{i+1} \in \mathcal{C} & \text{implies } \beta_i = \alpha_{i+1} \\ \beta_i = \tau & \text{iff } \alpha_{i+1} = \tau \end{cases}$
2. if $\forall i \in [1, n].\alpha_i, \beta_i \in \{\tau, *\}$, then $\forall i \in [1, n].\alpha_i = \beta_i = \tau$.

The first condition says that any two adjacent solid links must agree on their ends: it also imposes that τ cannot be matched by $*$. The second condition disallows chains made of virtual links only.

The empty link chain is denoted by ϵ. A non-empty link chain is *solid* if all its links are so. A link chain is *simple* if it includes exactly one solid link (and one, none or many virtual links). For ℓ a solid link and s a simple link chain, we write $\ell \in s$ if ℓ is the *only* solid link occurring in s. We write $|s|$ to denote the *length* of s, i.e., the number of links in s.

We say that an action a is *matched* in s if: 1) $\alpha_1, \beta_n \neq a$, and 2) for any $i \in [1, n-1]$, either $\beta_i = \alpha_{i+1} = a$ or $\beta_i, \alpha_{i+1} \neq a$. Otherwise, we say that a is *unmatched* (or *pending*) in s. For instance, a is matched in the sequence ${}^{\tau}\backslash_a^a\backslash_{\tau}$, while it is not matched in the sequences ${}^{\tau}\backslash_a^*\backslash_*$ and in ${}^a\backslash_a^a\backslash_a^a\backslash_a$.

We can rephrase usual communication primitives of process algebras as links.

- The output action $\bar{a}$ (resp. the input action a) of CCS can be seen as the link ${}^{\tau}\backslash_a$ (resp. ${}^a\backslash_{\tau}$) and the solid link chain ${}^{\tau}\backslash_a^a\backslash_{\tau}$ as a CCS-like communication.
- The action a of CSP can be seen as the link ${}^a\backslash_a$ and the solid link chain ${}^a\backslash_a^a\backslash_a^a\backslash_a$ as a CSP-like communication among three peers over a.

The following basic operations over links and link chains are partial and strict, i.e., they may issue $\bot$ (undefined) and the result is $\bot$ if any argument is $\bot$. To keep the notation short: if one of the sub-expressions in the righthand side (RHS) of any defining equation is undefined, then we assume the result is $\bot$; if none of the conditions in the RHS of any defining equation is met, then the result is $\bot$.

Merge. Two link chains can be merged if they are to some extent "complementary", i.e., if they have the same length, each provides links that are not specified in the other and together they form a (valid) link chain. The virtual links in a chain can be seen as the not yet specified part in the chain, and possibly provided by another link chain after a merge. If there is a position where both link chains carry solid links, then there is a clash and the merge is not possible (undefined). If the merge would result in a non valid sequence, then the merge is not possible. Formally, for $s = \ell_1...\ell_n$ and $s' = \ell'_1...\ell'_n$, with $\ell_i = {}^{\alpha_i}\backslash_{\beta_i}$ and $\ell'_i = {}^{\alpha'_i}\backslash_{\beta'_i}$ for any $i \in [1, n]$, we define $s \bullet s'$ by letting:[1]

$$s \bullet s' \triangleq (\ell_1 \bullet \ell'_1)...(\ell_n \bullet \ell'_n) \qquad \alpha \bullet \beta \triangleq \begin{cases} \alpha & \text{if } \beta = * \\ \beta & \text{if } \alpha = * \end{cases}$$
$${}^{\alpha}\backslash_\beta \bullet {}^{\alpha'}\backslash_{\beta'} \triangleq {}^{(\alpha \bullet \alpha')}\backslash_{(\beta \bullet \beta')}$$

Roughly, the merge is defined element-wise on the actions of a link chain, by ensuring that whenever two actions are merged, (at least) one of them is $*$ and that the result of the merge is still a link chain. Intuitively, we can imagine that s and s' are two parts of the same puzzle separately assembled, where solid links are the pieces of the puzzle and virtual links are the holes in the puzzle and their merge $s \bullet s'$ puts the two parts together, without piece overlaps.

Example 1. Let $s_1 = {}^{\tau}\backslash_a^*\backslash_*^*\backslash_*$, $s_2 = {}^*\backslash_*^a\backslash_b^*\backslash_*$, and $s_3 = {}^*\backslash_*^*\backslash_b^*\backslash_\tau$ three link chains of the same length. Then s_1 and s_2 can be merged to obtain $s = s_1 \bullet s_2 = ({}^{\tau}\backslash_a^\bullet\backslash_*^*)({}^*\backslash_*^\bullet\backslash_b^a)({}^*\backslash_*^\bullet\backslash_*^*) = ({}^{\tau\bullet*}\backslash_{a\bullet*})({}^{*\bullet a}\backslash_{*\bullet b})({}^{*\bullet*}\backslash_{*\bullet*}) = {}^{\tau}\backslash_a^a\backslash_b^*\backslash_*$. Similarly, s and s_3 can then be merged to obtain: $s \bullet s_3 = {}^{\tau}\backslash_a^a\backslash_b^b\backslash_\tau$.

Lemma 1. *(i) If s is solid, then for any s' we have $s \bullet s' = \bot$.*
(ii) The merge of link chains is a commutative and associative operation.
(iii) For any ℓ, ℓ': $\ell \bullet \ell' = {}^*\backslash_*$ if and only if $\ell = \ell' = {}^*\backslash_*$.

Restriction. Certain actions of the link chain can be hidden by restricting the channel where they take place. Of course, restriction is possible only if this process does not introduce any unmatched communication. Formally, for $s = \ell_1...\ell_n$, with $\ell_i = {}^{\alpha_i}\backslash_{\beta_i}$ with $i \in [1, n]$, we define the restriction operation $(\nu a)s$ by letting

$$(\nu a)s \triangleq {}^{\alpha_1}\backslash(\nu a)({}^{\alpha_2}_{\beta_1})\backslash...\backslash(\nu a)({}^{\alpha_n}_{\beta_{n-1}})\backslash_{\beta_n} \text{ if } \alpha_1, \beta_n \neq a \qquad (\nu a)({}^{\alpha}_{\beta}) \triangleq \begin{cases} {}^{\tau}_{\tau} & \text{if } \alpha = \beta = a \\ {}^{\alpha}_{\beta} & \text{if } \alpha, \beta \neq a \end{cases}$$

Lemma 2. *(i) For any a, ℓ: $(\nu a)\ell = {}^*\backslash_*$ if and only if $\ell = {}^*\backslash_*$.*
(ii) For any a, s, s' such that a does not appear in s: $(\nu a)(s \bullet s') = s \bullet (\nu a)s'$.
(iii) For any a, b, s: $(\nu a)(\nu b)s = (\nu b)(\nu a)s$.

Example 2. Let $s = {}^{\tau}\backslash_a^a\backslash_b^*\backslash_*$ and $s' = {}^*\backslash_*^*\backslash_b^b\backslash_\tau$. Then, $(\nu a)s = {}^{\tau}\backslash(\nu a)({}^a_a)\backslash(\nu a)({}^*_b)\backslash_* = {}^{\tau}\backslash_\tau^*\backslash_b^*\backslash_*$. Similarly, $(\nu a)(s \bullet s') = {}^{\tau}\backslash_\tau^*\backslash_b^b\backslash_\tau = ((\nu a)s) \bullet s'$.

---

[1] As anticipated, we remark that in the defining equations for merge it is implicitly understood that: if $\ell_i \bullet \ell'_i = \bot$ for some i, then $s \bullet s' = \bot$; if the sequence $(\ell_1 \bullet \ell'_1)...(\ell_n \bullet \ell'_n)$ is not a link chain, then $s \bullet s' = \bot$; if $\alpha \bullet \alpha' = \bot$ or $\beta \bullet \beta' = \bot$, then ${}^{\alpha}\backslash_\beta \bullet {}^{\alpha'}\backslash_{\beta'} = \bot$; if $\alpha, \beta \neq *$, then $\alpha \bullet \beta = \bot$.

Process Syntax. The CNA processes are generated by the following grammar (for brevity, we omit the CCS-like renaming operator $P[\Phi]$ because it will be later subsumed by the syntax in Section 2):

$$P, Q ::= \mathbf{0} \mid X \mid \ell.P \mid P + Q \mid P|Q \mid (\nu a)P \mid \mathsf{rec}X.P$$

where ℓ is a solid link (i.e., $\ell = {}^{\alpha}\backslash_{\beta}$ with $\alpha, \beta \neq *$).

Roughly, processes are built over a CCS-like syntax (with nil process $\mathbf{0}$, prefix $\ell.P$, sum $P + Q$, parallel $P|Q$, restriction $(\nu a)P$ and recursion $\mathsf{rec}X.P$, for X a process variable), but where the underlying synchronisation algebra is based on link chains. This is made evident by the operational semantics, presented below.

As usual, $(\nu a)P$ binds the occurrences of a in P, the sets of free and of bound names of a process P are defined in the obvious way and denoted, respectively, by $fn(P)$ and $bn(P)$, processes are taken up to alpha-conversion of bound names, and we shall often omit trailing $\mathbf{0}$, e.g., by writing ${}^{a}\backslash_{b}$ instead of ${}^{a}\backslash_{b}.\mathbf{0}$

Operational Semantics. The idea is that communication can be routed across several processes by combining the links they make available to form a link chain. Since the length of the link chain is not fixed a priori, an open multiparty synchronisation is realised.

The operational semantics is defined in terms of a Labelled Transition System (LTS) whose states are CNA processes, whose labels are link chains and whose transitions are generated by the SOS rules in Fig. 1. Notice that the SOS rules are very similar to the CCS ones, apart from the labels that record the link chains involved in the transitions: moving from dyadic to *linked* interaction does not introduce any complexity burden. We comment in details the rules *(Act)*, *(Res)*, and *(Com)*. In rules *(Res)* and *(Com)* we leave implicit the side conditions $(\nu a)s \neq \perp$ and $s \bullet s' \neq \perp$, respectively (they can be easily recovered by noting that otherwise the label of the transition in the conclusion would be undefined).

The rule *(Act)* states that $\ell.P \xrightarrow{s} P$ for any simple link chain s whose unique solid link is ℓ. Intuitively, $\ell.P$ can take part in any interaction, in any (admissible) position. To join in a communication, $\ell.P$ should suitably enlarge its link ℓ to a simple link chain s including it, whose length is the same for all participants in order to proceed with the merge operation. Following the early style, the suitable length can be inferred at the time of deducing the input transition. Note that if one end of ℓ is τ, then ℓ can only appear at one extreme of s.

In the *(Res)* rule, the operator (νa) applied to s, can serve different aims: *floating*, if a does not appear in s, then $(\nu a)s = s$; *hiding*, if a is matched in s (i.e., a appears as ends already matched by adjacent links), then all occurrences of a in s are renamed to τ in $(\nu a)s$; *blocking*, if a is pending in s (i.e., there is a 'non-matched' occurrence of a in s), then $(\nu a)s = \perp$ and the rule cannot be applied.

In the *(Com)* rule the link chains recorded on both the premises' transitions are merged in the conclusion's transition. This is possible only if s and s' are to some extent "complementary". Contrary to CCS, the rule *(Com)* can be applied several times to prove that a transition is possible, because $s \bullet s'$ can still contain virtual links (if s and s' had a virtual link in the same position). However, when $s \bullet s'$ is solid, no further synchronisation is possible (by Lemma 1 (ii)).

$$\frac{\ell \in s \quad (\ell \ \textit{only}\ \text{solid link in } s)}{\ell.P \xrightarrow{s} P}\ \text{(Act)} \qquad \frac{P \xrightarrow{s} P'}{P+Q \xrightarrow{s} P'}\ \text{(Lsum)} \qquad \frac{P \xrightarrow{s} P'}{(\nu a)P \xrightarrow{(\nu a)s} (\nu a)P'}\ \text{(Res)}$$

$$\frac{P \xrightarrow{s} P'}{P|Q \xrightarrow{s} P'|Q}\ \text{(Lpar)} \qquad \frac{P \xrightarrow{s} P' \quad Q \xrightarrow{s'} Q'}{P|Q \xrightarrow{s \bullet s'} P'|Q'}\ \text{(Com)} \qquad \frac{P[X \mapsto \mathsf{rec}X.P] \xrightarrow{s} P'}{\mathsf{rec}X.P \xrightarrow{s} P'}\ \text{(Rec)}$$

Fig. 1. SOS semantics of the CNA (rules (*(Rsum)*) and (*Rpar*) omitted)

Example 3. Let $P = {}^\tau\backslash_a.P_1|(\nu b)Q$ and $Q = {}^b\backslash_\tau.P_2|{}^a\backslash_b$. The process ${}^\tau\backslash_a.P_1$ can output on a, while ${}^b\backslash_\tau.P_2$ can input from b; the process ${}^a\backslash_b$ provides a one-shot link forwarder from a to b. Together, they can synchronise by agreeing to form a solid link chain of length 3, as follows, where $(\nu b)^*\backslash^a_*\backslash^b_b\backslash_\tau = {}^*\backslash^a_*\backslash^\tau_\tau\backslash_\tau.$

$$\cfrac{\cfrac{}{{}^\tau\backslash_a.P_1 \xrightarrow{{}^\tau\backslash^*_a\backslash^*_*} P_1}\ \text{(Act)} \qquad \cfrac{\cfrac{\cfrac{}{{}^b\backslash_\tau.P_2 \xrightarrow{{}^*\backslash^*_*\backslash^b_\tau} P_2}\ \text{(Act)} \quad \cfrac{}{{}^a\backslash_b.0 \xrightarrow{{}^*\backslash^a_*\backslash^*_b} 0}\ \text{(Act)}}{Q \xrightarrow{{}^*\backslash^a_*\backslash^b_\tau} P_2|0}\ \text{(Com)}}{(\nu b)Q \xrightarrow{{}^*\backslash^a_\tau\backslash^\tau_\tau} (\nu b)(P_2|0)}\ \text{(Res)}}{P \xrightarrow{{}^\tau\backslash^a_a\backslash^\tau_\tau} P_1|(\nu b)(P_2|0)}\ \text{(Com)}$$

The following lemma, whose proof goes by rule induction, shows that labels behave like an accordion. Any label s in a transition is interchangeable or replaceable with any chain having one or more ${}^*\backslash_*$ either on the left or on the right or on both sides of s. It is also replaceable with any composition where each ${}^*\backslash_*$ inside s is replaced by one or more ${}^*\backslash_*$. This fact can be exploited later, when the abstract semantics is given.

Lemma 3

(i) If $P \xrightarrow{s} P'$ and $s^*\backslash_*$ (resp. ${}^*\backslash_*s$) is valid, then $P \xrightarrow{s^*\backslash_*} P'$ (resp. $P \xrightarrow{{}^*\backslash_*s} P'$). Vice versa, if $P \xrightarrow{s'} P'$ and $s' = s^*\backslash_*$ or $s' = {}^*\backslash_*s$, then $P \xrightarrow{s} P'$.

(ii) If $P \xrightarrow{s_1{}^*\backslash_*s_2} P'$ then $P \xrightarrow{s_1{}^*\backslash^*_*\backslash_*s_2} P'$. Vice versa, if $P \xrightarrow{s_1{}^*\backslash^*_*\backslash_*s_2} P'$ then $P \xrightarrow{s_1{}^*\backslash_*s_2} P'$.

(iii) If $P \xrightarrow{s_1{}^\alpha\backslash^a_a\backslash_\beta s_2} P'$ then $P \xrightarrow{s_1{}^\alpha\backslash^*_a\backslash^a_*\backslash_\beta s_2} P'$. Vice versa, if $P \xrightarrow{s_1{}^\alpha\backslash^*_a\backslash^a_*\backslash_\beta s_2} P'$ then $P \xrightarrow{s=s_1{}^\alpha\backslash^a_a\backslash_\beta s_2} P'$.

Note that if $P \xrightarrow{{}^a\backslash^\tau_\tau\backslash_b} P'$, then it is not the case that a virtual link can be inserted in the middle of the chain, because the τ's may represent a communication on a restricted channel and ${}^a\backslash^*_\tau\backslash^\tau_\tau\backslash_b$ is not a valid link chain anyway.

Routing examples. We give a few examples to show how flexible is CNA for defining "routing" policies. We have already seen a one-shot, one-hop forwarder from a to b, that can be written as ${}^a\backslash_b.0$. Its persistent version is just written as $P^a_b \triangleq \mathsf{rec}X.{}^a\backslash_b.X$. Moreover, with $P^a_b|P^b_a$ we obtain a sort of name fusion between a and b, i.e., a and b can be interchangeably used. An alternating forwarder from a to b first and then to c can be defined as $A^a_{b,c} \triangleq \mathrm{rec}X.\,({}^a\backslash_b.\,{}^a\backslash_c.X)$. A persistent non-deterministic forwarder, from a to $c_1...c_n$ can be written, e.g., as $P^a_{c_1...c_n} \triangleq \mathrm{rec}X.\,({}^a\backslash_{c_1}.X + \cdots + {}^a\backslash_{c_n}.X)$. Similarly, $P^{b_1...b_m}_a \triangleq \mathrm{rec}X.\,({}^{b_1}\backslash_a.X + \cdots + {}^{b_m}\backslash_a.X)$ is a persistent non-deterministic forwarder, from $b_1...b_m$ to a. By combining the two processes as $(\nu\,a)(P^{b_1...b_m}_a \,|\, P^a_{c_1...c_n})$, then we obtain a persistent forwarder from any of the b_i's to any of the c_j's.

Abstract semantics. As usual, we can use the LTS semantics to define suitable behavioural equivalences over processes. We are interested in bisimilarity. However, when comparing two labels we abstract away from the number and identities of hops performed and from the size of the sequences of virtual links.

Definition 1. *We let $\bowtie$ be the least equivalence relation over link chains closed under the following axioms (remind that $\alpha, \beta, \gamma \in \mathcal{C} \cup \{\tau, *\}$):*

$$ {}^*\backslash_* s \bowtie s \qquad s_1{}^\alpha\backslash^\gamma_\gamma\backslash_\beta s_2 \bowtie s_1{}^\alpha\backslash_\beta s_2 \qquad s^*\backslash_* \bowtie s $$

A link chain is *essential* if it is composed by alternating solid and virtual links, and has solid links at its ends. An essential link chain has minimal length with respect to equivalent link chains and concisely represents the missing "paths" of interactions between those already completed. For example, we have that ${}^*\backslash^a_\tau\backslash^*_b\backslash^*_*\backslash^c_d\backslash^*_a\backslash_*$ is equivalent to the essential chain ${}^a\backslash^*_b\backslash^c_*\backslash_a$.

Lemma 4
(i) Let s and s' be two essential link chains s.t. $s \bowtie s'$, then $s = s'$.
(ii) Let s be an essential link chain. For any $s' \bowtie s$ we have $|s'| \geq |s|$.

It is immediate to check that by orienting the axioms in Def. 1 from left to right we have a procedure to transform any link chain s to a unique essential link chain s' such that $s \bowtie s'$. We write $\mathsf{e}(s)$ to denote such unique representative, which enjoys the following nice properties, useful in the proof of the following Proposition 1. The first part of the lemma says that $\mathsf{e}(\cdot)$ induces an equivalence over link chains that is a congruence with respect to juxtaposition of link chains (when it is well-defined). The second part says that taken two link chains s_1 and s_2 in the same equivalence class, and given any sequence s that can be merged with s_1, then it is possible to find a link chain s', in the same equivalence class as s, such that it can be merged with s_2 and the result is equivalent to $s \bullet s_1$.

Lemma 5
(i) If $\mathsf{e}(s_1) = \mathsf{e}(s_2)$, then for any s such that $s_1 s$ (resp. $s s_1$) is a link chain we have that also $s_2 s$ (resp. $s s_2$) is a link chain and $\mathsf{e}(s_1 s) = \mathsf{e}(s_2 s)$ (resp. $\mathsf{e}(s s_1) = \mathsf{e}(s s_2)$).
(ii) If $\mathsf{e}(s_1) = \mathsf{e}(s_2)$, then for any s such that $s \bullet s_1 \neq \bot$ there exists a link chain $s' \bowtie s$ such that $s' \bullet s_2 \neq \bot$ and $\mathsf{e}(s \bullet s_1) = \mathsf{e}(s' \bullet s_2)$.

In the following, we write $P \to P'$ when $P \xrightarrow{s} P'$ for some s s.t. $\mathsf{e}(s) = {}^\tau\backslash_\tau$.

Definition 2. *A* network bisimulation **R** *is a binary relation over CNA processes such that, if* $P\,\mathbf{R}\,Q$ *then:*

- *if* $P \overset{s}{\to} P'$, *then* $\exists\ s',\ Q'$ *such that* $\mathsf{e}(s) = \mathsf{e}(s')$, $Q \overset{s'}{\to} Q'$, *and* $P'\,\mathbf{R}\,Q'$;
- *if* $Q \overset{s}{\to} Q'$, *then* $\exists\ s',\ P'$ *such that* $\mathsf{e}(s) = \mathsf{e}(s')$, $P \overset{s'}{\to} P'$, *and* $P'\,\mathbf{R}\,Q'$.

We let $\sim_n$ denote the largest network bisimulation and we say that P is *network bisimilar* to Q if $P \sim_n Q$.

Example 4. Consider the two processes $P \triangleq \mathsf{rec}X.\,{}^a\backslash_b.X$ and $Q \triangleq \mathsf{rec}X.\,(\nu c)$ $({}^a\backslash_c \mid {}^c\backslash_b.X)$. We have that whenever $P \overset{s}{\to} P'$, then $P' = P$ and $\mathsf{e}(s) = {}^a\backslash_b$. Similarly, whenever $Q \overset{s}{\to} Q'$, then $Q' = (\nu c)(\mathbf{0}|Q)$ and $\mathsf{e}(s) = {}^a\backslash_b$. Then we prove that $P \sim_n Q$ by showing that the relation **R** below:

$$\mathbf{R} \triangleq \{(P, R) \mid \exists n.R = C^n[Q]\}$$

is a network bisimulation, where $C^n[Q]$ is inductively defined by letting $C^0[Q] \triangleq Q$ and $C^{n+1}[Q] \triangleq C[C^n[Q]]$ for $C[\cdot]$ the context $(\nu c)(\mathbf{0}|\cdot)$

Proposition 1. *Network bisimilarity is a congruence.*

Proof. The proof uses standard arguments. The only non-trivial case is that of parallel composition. We want to prove that if $P \sim_n Q$ then for any R we have $P|R \sim_n Q|R$. We define the relation $\mathbf{R}_| \triangleq \{(P|R, Q|R) \mid P \sim_n Q\}$ and show that $\mathbf{R}_|$ is a bisimulation. Suppose $P \sim_n Q$ and $P|R \overset{s}{\to} T$. We want to prove that $Q|R \overset{s}{\to} T'$ with $T\,\mathbf{R}_|\,T'$. There are three cases to be considered, depending on the last SOS rule used to prove $P|R \overset{s}{\to} T$. If the used rule is

- *(Rpar)*, then it means that $R \overset{s}{\to} R'$ for some R' with $T = P|R'$. But then, by using *(Rpar)* we have $Q|R \overset{s}{\to} Q|R'$ and $P|R'\,\mathbf{R}_|\,Q|R'$ by definition of $\mathbf{R}_|$.
- *(Lpar)*, then it means that $P \overset{s}{\to} P'$ for some P' with $T = P'|R$. By assumption we know that $P \sim_n Q$ and therefore there exists $s',\ Q'$ s.t. $Q \overset{s'}{\to} Q'$ with $\mathsf{e}(s) = \mathsf{e}(s')$ and $P' \sim_n Q'$. By applying the rule *(Lpar)* we have that $Q|R \overset{s'}{\to} Q'|R$ and we have done because $P'|R\,\mathbf{R}_|\,Q'|R$ by definition of $\mathbf{R}_|$.
- *(Com)*, then it means that $P \overset{s_1}{\to} P'$, $R \overset{s_2}{\to} R'$, for some s_1, s_2, P', R' with $s = s_1 \bullet s_2$ and $T = P'|R'$. By assumption we know that $P \sim_n Q$ and therefore there exists $s_1',\ Q'$ s.t. $Q \overset{s_1'}{\to} Q'$ with $\mathsf{e}(s_1) = \mathsf{e}(s_1')$ and $P' \sim_n Q'$. Now it may be the case that $s_1' \bullet s_2$ is not defined, but by the previous technical lemmata, we know that s_1' and s_2 can be stretched resp. to s_1'' and s_2' by inserting enough virtual links to have that $s_1'' \bullet s_2'$ is defined, $\mathsf{e}(s_1 \bullet s_2) = \mathsf{e}(s_1'' \bullet s_2')$, $Q \overset{s_1''}{\to} Q'$ and $R \overset{s_2'}{\to} R'$. We conclude by applying rule *(Com)*: $Q|R \overset{s'}{\to} Q'|R'$ with $P'|R'\,\mathbf{R}_|\,Q'|R'$ by definition of $\mathbf{R}_|$.

Analogously to CCS, it is immediate to check that several useful axioms over processes hold up to network bisimilarity, like the commutative monoidal laws for | and +, the idempotence of + and the usual laws about restriction.

2   The Calculus of Linked Interactions

We can now enhance the algebra to deal with name passing and call it the *calculus of linked interactions* (link-*calculus* for short), by extending, in the syntax, link prefixes with tuples of arguments, where each link in the whole chain simply carries the same list of arguments, but with different (send/receive) capabilities.

$$P, Q ::= \cdots \mid \ell t.P$$

In this way, we keep apart the interaction mechanism from the name-passing one that eventually fit together in synchronisations. We have just borrowed from π-calculus the name handling machinery (and liberated it from dyadic interaction legacy), still having input, output and extrusion mechanisms.

In the tuple $t = \langle \vec{w} \rangle$, names can be used either as values or as variables. To be distinguished, variables are underlined. During communication, variables are instantiated by values, while values are used for matching arguments, as in [1].

Example 5. Consider e.g., the two tuples in two "complementary" prefixes like $^{\tau}\backslash_a \langle id, n, \underline{x} \rangle.P$ and $^a \backslash_\tau \langle id, \underline{y}, m \rangle.Q$, where $\underline{x}$ is an input for P, $\underline{y}$ is an input for Q, and id is a name known by both processes: the two links can be merged, the first parameters must match exactly, n is assigned to y, while m is assigned to x.

This mechanism allows, e.g., a form of multi-way communication, where all peers involved in the chain link can express arguments to be matched and provide actual arguments to replace the formal ones of other peers. For a tuple t, we let $vals(t)$ and $vars(t)$ denote the set of values and the set of variables of t, respectively. We say that a tuple t is *ground* if $vars(t) = \emptyset$.

We assume action names are admissible values, i.e., as in π-calculus we have the possibility to communicate (names of) means of communication. Similarly to $(\nu\, a)$, the prefix $\ell t.P$ binds the occurrences of the variables $vars(t)$ in P (and the notions of free names $fn(P)$, bound names $bn(P)$ and alpha-conversion are updated accordingly). In the following, given two sets of names S and T, we write $S\#T$ as a shorthand for $S \cap T = \emptyset$.

Operational semantics. The operational semantics is defined in terms of an LTS whose states are link-calculus processes, whose labels are pairs sg of link chains and tuples and whose transitions are generated by the SOS rules in Fig. 2.

In a label of the form sg, with s solid, g must be the empty tuple $\langle \rangle$, i.e., it is not possible to observe the arguments of a completed communication. We let s abbreviate $s\langle \rangle$.

We denote by $\sigma = [x_1 \mapsto v_1, ..., x_n \mapsto v_n]$ the substitution that replaces each x_i with v_i, and is the identity otherwise, and set $vars(\sigma) = \{x_1, ..., x_n\}$. For a tuple $t = \langle w_1, ..., w_n \rangle$, a link $\ell = {}^\alpha\backslash_\beta$, and a substitution $\sigma = [x_1 \mapsto v_1, ..., x_n \mapsto v_n]$ we define $t\sigma$ and $\ell\sigma$ element-wise as:

$$t\sigma \triangleq \langle w_1\sigma, ..., w_n\sigma \rangle \qquad a\sigma \triangleq \begin{cases} v_i & \text{if } a = x_i \text{ for some } i \in [1, n] \\ a & \text{otherwise} \end{cases}$$

$$\ell\sigma \triangleq {}^{\alpha\sigma}\backslash_{\beta\sigma} \qquad \underline{a}\sigma \triangleq \begin{cases} v_i & \text{if } a = x_i \text{ for some } i \in [1, n] \\ \underline{a} & \text{otherwise} \end{cases}$$

The application of σ on processes is defined as below (note that, as processes are taken up to alpha-conversion of bound names, it is always possible to find suitable representatives of $\ell t.P$ and $(\nu a)P$, such that $(\ell t.P)\sigma$ and $((\nu a)P)\sigma$ are well-defined):

$$\mathbf{0}\sigma \triangleq \mathbf{0} \qquad\qquad (\ell t.P)\sigma \triangleq (\ell\sigma)(t\sigma).(P\sigma), \text{if } vars(t)\#(vars(\sigma) \cup vals(\sigma))$$
$$(P+Q)\sigma \triangleq P\sigma + Q\sigma$$
$$(P|Q)\sigma \triangleq P\sigma|Q\sigma \qquad ((\nu a)P)\sigma \triangleq (\nu a)P\sigma, \text{if } \{a\}\#(vars(\sigma) \cup vals(\sigma))$$
$$X\sigma \triangleq X \qquad\qquad (\mathsf{rec}X.P)\sigma \triangleq \mathsf{rec}X.(P\sigma)$$

We say that g is a *full instance* of t and write $g \preceq_\sigma t$ if $vars(\sigma) = vars(t) \wedge g = t\sigma$.

Like in the π-calculus, names in the tuple can be extruded during the communication. In the labels of transitions, we need to annotate positions in the tuple to distinguish between arguments that are taken in input (i.e., they are guessed instances), or that are extruded. We underline the former and overline the latter. A name can be extruded when it is not already annotated; after the extrusion, it will be overlined. Formally, given a (annotated) tuple g, we define $(\nu a)sg$ and $(\nu a)g$ as follows:

$$(\nu a)(sg) \triangleq ((\nu a)s)((\nu a)g) \qquad\qquad (\nu a)w \triangleq \begin{cases} w & \text{if } w \neq a, \overline{a}, \underline{a} \\ \overline{a} & \text{if } w = a \end{cases}$$
$$(\nu a)\langle w_1, ..., w_n\rangle \triangleq \langle(\nu a)w_1, ..., (\nu a)w_n\rangle$$

We let $ex(g)$ denote the set of extruded (i.e., overlined) names appearing in g. We write $a \in g$ if the name a appears in the tuple g (with or without annotation).

Lemma 6. *If $(\nu a)g \neq \perp$, then $vars((\nu a)g) = vars(g)$.*

Two annotated tuples can be merged when they list exactly the same values in the same order, and if the values in matching positions are annotated in some compatible way. Formally, if $\langle \vec{w} \rangle = \langle w_1, ..., w_n \rangle$ and $\langle \vec{w'} \rangle = \langle w_1', ..., w_n' \rangle$:

$$sg \bullet s'g' \triangleq (s \bullet s')(g \bullet g') \qquad \langle \vec{w} \rangle \bullet \langle \vec{w'} \rangle \triangleq \langle w_1 \bullet w_1', ..., w_n \bullet w_n' \rangle$$

$$w \bullet w' \triangleq \begin{cases} w & \text{if } (w = w' = v) \vee (w = w' = \underline{v}) \\ v & \text{if } (w = v \wedge w' = \underline{v}) \vee (w = \underline{v} \wedge w' = v) \\ \overline{v} & \text{if } (w = \overline{v} \wedge w' = \underline{v}) \vee (w = \underline{v} \wedge w' = \overline{v}) \end{cases}$$

Example 6. Back to Ex. 5, $^\tau\backslash_a^*\backslash_*\langle id, n, \underline{m}\rangle \bullet {}^*\backslash_a^*\backslash_\tau\langle id, \underline{n}, m\rangle = (^\tau\backslash_a^*\backslash_* \bullet {}^*\backslash_a^*\backslash_\tau)$ $(\langle id, n, \underline{m}\rangle \bullet \langle id, \underline{n}, m\rangle) = {}^\tau\backslash_a^*\backslash_\tau\langle id \bullet id, n \bullet \underline{n}, \underline{m} \bullet m\rangle = {}^\tau\backslash_a^*\backslash_\tau\langle id, n, m\rangle$. Recall that in the early-style (Act) rule the values to be received are guessed and so the prefix variables $\underline{y}$ and $\underline{x}$ are replaced by $\underline{n}$ and $\underline{m}$, respectively.

Lemma 7. *If $g \bullet g' \neq \perp$, then $vars(g \bullet g') \subseteq vars(g) \cap vars(g')$.*

A close look at the SOS rules in Fig. 2 shows that they resemble the early semantic rules of π-calculus. The main difference is that we are dealing with a multi-party form of communication, hence the "close" rule must be applied when the communication has been completed. Let us briefly comment on the rules.

Rule *(Act)* allows the process $\ell t.P$ to offer the tuple t in a communication on the link ℓ. More precisely, following an early style, the actual tuple to be communicated

$$\frac{\ell \in s \quad (\ell \ only \ \text{solid link in} \ s)}{\ell t.P \xrightarrow{sg} P\sigma} \ (Act) \qquad \frac{g \preceq_\sigma t}{} \qquad \frac{P \xrightarrow{sg} P'}{P + Q \xrightarrow{sg} P'} \ (Lsum)$$

$$\frac{P[X \mapsto \mathsf{rec}X.\,P] \xrightarrow{sg} P'}{\mathsf{rec}X.\,P \xrightarrow{sg} P'} \ (Rec) \qquad \frac{P \xrightarrow{sg} P' \quad ex(g)\#fn(Q)}{P|Q \xrightarrow{sg} P'|Q} \ (Lpar)$$

$$\frac{P \xrightarrow{sg} P' \quad a \notin g}{(\nu\,a)P \xrightarrow{(\nu\,a)sg} (\nu\,a)P'} \ (Res) \qquad \frac{P \xrightarrow{sg} P' \quad a \in g}{(\nu\,a)P \xrightarrow{(\nu\,a)sg} P'} \ (Open)$$

$$\frac{P \xrightarrow{sg} P' \quad Q \xrightarrow{s'g'} Q' \quad \begin{array}{c} ex(g)\#fn(Q) \\ ex(g')\#fn(P) \end{array} \quad s \bullet s' \ \text{is not solid}}{P|Q \xrightarrow{sg \bullet s'g'} P'|Q'} \ (Com)$$

$$\frac{P \xrightarrow{sg} P' \quad Q \xrightarrow{s'g'} Q' \quad \begin{array}{c} ex(g)\#fn(Q) \\ ex(g')\#fn(P) \end{array} \quad \begin{array}{c} s \bullet s' \ \text{is solid} \\ g \bullet g' \ \text{is ground} \end{array}}{P|Q \xrightarrow{s \bullet s'} (\nu\, ex(g \bullet g'))(P'|Q')} \ (Close)$$

Fig. 2. SOS semantics of the link-calculus (rules *(Rsum)* and *(Rpar)* omitted)

must be a full instance of t (see the condition $g \preceq_\sigma t$): the communication is that of sg, where variables appearing in t are replaced in $g = t\sigma$ by actual parameters. The substitution σ is also applied to the continuation, after the transition ($P\sigma$).

In rules *(Res)* and *(Open)* we leave implicit the side condition $(\nu\,a)sg \neq \bot$. Rule *(Res)* is applicable whenever $a \notin g$, in which case $(\nu\,a)g = g \neq \bot$ and thus $(\nu\,a)(sg) = ((\nu\,a)s)g$. Rule *(Open)* models the extrusion of a. Note that, since $(\nu\,a)sg \neq \bot$ and $a \in g$, we have that the only possibility for a to appear in g is without annotations (otherwise $(\nu\,a)g = \bot$). Then, by definition of $(\nu\,a)g$, all occurrences of a within g are overlined in $(\nu\,a)g$ (to denote the name extrusion).

In rule *(Com)*, the annotated tuples are "complementary" and can be merged, by merging the link chains and the two tuples. Note that we leave implicit the side condition $sg \bullet s'g' \neq \bot$, because the $sg \bullet s'g'$ annotates the label of the conclusion transition. In addition, rule *(Com)* checks that the extruded names of one process do not clash with the free names of the other process (like in ordinary π-calculus) and finally that the communication is not completed yet ($s \bullet s'$ not solid).

Rule *(Close)* premises differ from *(Com)* one, because *(Close)* is applicable only when the communication has been fully completed and cannot be further extended ($s \bullet s'$ is solid), in which case we must close, i.e., put back the restriction of all names extruded in the communication ($\nu\, ex(g \bullet g'')$). Still, we need to make sure that $g \bullet g'$ has no unresolved input, i.e., that all requested values have been issued ($g \bullet g'$ is ground). Moreover, the observed label is just $s \bullet s'$, as explained before. This is similar to the π-calculus mechanism, according to which the synchronisation of e.g., $a\langle x \rangle$ and $\bar{a}\langle x \rangle$ yields τ and not $\tau \langle x \rangle$.

While rules *(Lsum)*, *(Rsum)* and *(Rec)* are straightforward, rules *(Lpar)* and *(Rpar)* need just to check that extruded names of one process do not clash with free names of the other process (like in ordinary π-calculus).

Note that if the only used tuples are the empty ones $\langle\rangle$, then the semantics rules coincide with the ones of CNA.

Example 7. Consider the following restricted process, built on the processes in Ex. 5, $S \triangleq (\nu\, a)(^{\tau}\backslash_a\langle id, n, \underline{x}\rangle.P|(\nu\, m)^a\backslash_\tau\langle id, \underline{y}, m\rangle.Q)$. In one step, S can reduce to $S' \triangleq (\nu\, a)(\nu\, m)(P[x \mapsto m]|Q[y \mapsto n])$, via the communication $^{\tau}\backslash^{\tau}_{\tau}\langle id, n, \overline{m}\rangle$, where the restriction on a hides the matched occurrences of a in $^{\tau}\backslash^a_a\backslash_\tau$ and the restriction on m causes the extrusion of the name m (to process P). Moreover, since the chain link $^{\tau}\backslash^{\tau}_{\tau}$ is solid (i.e., it cannot be extended further) and the tuple $\langle id, n, \overline{m}\rangle$ is ground, we remove the tuple from the observed label, i.e., $S \xrightarrow{^{\tau}\backslash^{\tau}_{\tau}} S'$. No other interaction is possible.

Abstract semantics. By analogy with the early bisimilarity of the π-calculus, we extend the notion of network bisimilarity to consider the tuples of names.

Definition 3. *A* linked bisimulation **R** *is a binary relation over* link-*calculus processes such that, if P **R** Q then:*

- *if $P \xrightarrow{sg} P'$ with $ex(g)\#fn(P)$, then there exists s' and Q' such that $\mathsf{e}(s) = \mathsf{e}(s')$, $Q \xrightarrow{s'g} Q'$, and P' **R** Q';*
- *if $Q \xrightarrow{sg} Q'$ with $ex(g)\#fn(Q)$, then there exists s' and P' such that $\mathsf{e}(s) = \mathsf{e}(s')$, $P \xrightarrow{s'g} P'$, and P' **R** Q'.*

We let $\sim_l$ denote the largest linked bisimulation and we say that P is *linked bisimilar* to Q if $P \sim_l Q$.

The following result may look surprising, since early bisimilarity is not a congruence in the case of π-calculus, due to the input prefix context. The classic counterexample is $P = x|\overline{y}$ being bisimilar to $Q = x.\overline{y} + \overline{y}.x$ but $a(y).P$ being not bisimilar to $a(y).Q$ (when the name received on y is x, a τ move is available for $P[y \mapsto x]$ but not for $Q[y \mapsto x]$). The fact is that the SOS semantics of the link-calculus can collect the ready set of prefixes concurrently available to be executed (e.g., by separating them through virtual links); thus the above P and Q are not considered as bisimilar. In other words, virtual links allow to establish an interaction between different ends, as if a substitution was available to rename one end into the other, i.e., the semantics is already substitution closed.

Proposition 2. *Linked bisimilarity is a congruence.*

3   Background on Mobile Ambients

In this section, we briefly recall Mobile Ambients (MA) syntax and semantics, in order to show the encoding of MA in the link-calculus, in the next section.

$$\frac{}{n[\,\text{in}\,m.P\,|\,Q\,]\,|\,m[\,R\,] \to m[\,n[\,P\,|\,Q\,]\,|\,R\,]}\ \text{(In)}$$

$$\frac{}{m[\,n[\,\text{out}\,m.P\,|\,Q\,]\,|\,R\,] \to n[\,P\,|\,Q\,]\,|\,m[\,R\,]}\ \text{(Out)}$$

$$\frac{}{\text{open}\,n.P\,|\,n[\,Q\,] \to P\,|\,Q}\ \text{(Open)} \qquad \frac{P \to Q}{(\nu\,n)P \to (\nu\,n)Q}\ \text{(Res)} \qquad \frac{P \to Q}{n[\,P\,] \to n[\,Q\,]}\ \text{(Amb)}$$

$$\frac{P \to Q}{P\,|\,R \to Q\,|\,R}\ \text{(Par)} \qquad \frac{P' \equiv P \quad P \to Q \quad Q \equiv Q'}{P' \to Q'}\ \text{(Cong)}$$

$P \equiv P$	$Q \equiv P \Rightarrow P \equiv Q$	$P \equiv Q, Q \equiv R \Rightarrow P \equiv R$
$P\,|\,\mathbf{0} \equiv P$	$P\,|\,Q \equiv Q\,|\,P$	$(P\,|\,Q)\,|\,R \equiv P\,|\,(Q\,|\,R)$
$(\nu\,n)\mathbf{0} \equiv \mathbf{0}$	$(\nu\,n)(\nu\,m)P \equiv (\nu\,m)(\nu\,n)P$	$P \equiv Q \Rightarrow P|R \equiv Q|R$
	$(\nu\,n)(P\,|\,Q) \equiv P\,|\,(\nu\,n)Q,\ \text{if}\ n \notin \mathit{fn}(P)$	$P \equiv Q \Rightarrow (\nu\,n)P \equiv (\nu\,n)Q$
$!P \equiv P\,|\,!P$	$(\nu\,n)(m[\,P\,]) \equiv m[\,(\nu\,n)P\,],\ \text{if}\ n \neq m$	$P \equiv Q \Rightarrow n[\,P\,] \equiv n[\,Q\,]$

Fig. 3. Reduction and Structural Congruence Rules for the Mobile Ambients

Mobile Ambients (MA) [9] is a calculus for *mobility* that includes both mobile agents and mobile computational ambients in which agents interact. Ambients are *bounded* locations, such as a web page or a virtual address space, that can be moved as a whole. Each ambient has a name and can include sub-ambients to form a hierarchical structure, that can be dynamically modified by agents.

We focus here on the so-called *pure* MA, i.e., disregarding communication primitives and variables for brevity. However our results in Section 4 can be easily extended to the more general case.

Let n range over the numerable set of names $\mathcal{N}$. The set of mobile ambient processes $\mathcal{P}_{MA}$ (with metavariable P) and the set of capabilities $\mathcal{C}ap$ (with metavariable M) are defined below:

$$P ::= \mathbf{0} \mid (\nu n)P \mid P|Q \mid !P \mid n[\,P\,] \mid M.P$$
$$M ::= \text{in}\,n \mid \text{out}\,n \mid \text{open}\,n$$

The first four constructs are quite standard in process calculi, while the other ones are specific to ambients: $n[\,P\,]$ denotes the ambient n in which the process P runs; $M.P$ executes an action depending on the capability M and then behaves as P. There are three kinds of capabilities: one for entering, one for exiting and one for opening up an ambient. In the process in $m.P$ the entry capability allows the immediately surrounding ambient n (if any) to find a sibling ambient named m where to enter (i.e., to become a child of m). Similarly, in the process out $m.P$ the exit capability allows the immediately surrounding ambient n to exit its parent ambient (if named m) and to become a sibling of m. Finally, in open $m.P$ the open capability allows the boundary of an ambient m located in parallel to be dissolved. Each interaction involves more than two parties.

The only binder is $(\nu\, n)$ and the sets of free names and of bound names of a process P are defined in the obvious way and denoted, respectively, by $fn(P)$ and $bn(P)$. As usual, the restriction of a sequence of names $\vec{m} = \{m_1, ..., m_k\}$ for a process P is denoted as $(\nu\vec{m})P$ and stands for $(\nu m_1)...(\nu m_k)P$. We shall denote by $\sigma = [n \mapsto m]$ the (capture-avoiding) substitution that replaces n by m, and by $P\sigma$ the process obtained by applying σ to P to replace all free occurrences of n in P by m. Processes are taken up to alpha-conversion of restricted names, i.e., $(\nu\, n)P$ denotes the same process as $(\nu\, m)(P[n \mapsto m])$ whenever $m \notin fn(P)$.

The semantics of the MA is given by the reduction and the structural congruence rules in Fig. 3. Besides the one-step reduction rules for movement capabilities $((In)$, (Out) and $(Open))$ and the usual reduction rule for congruence $(Cong)$, the other rules propagate reductions across scopes (Res), ambient nesting (Amb) and parallel composition (Par).

Example 8. Consider the process $P \triangleq m[\,s[\,\mathsf{in}\, n.R\,] \mid T\,] \mid \mathsf{open}\, m.Q \mid n[G]$ (for suitable R, T, Q, G) that can execute an $\mathsf{open}\, m$ followed by an $\mathsf{in}\, n$:

$$P \to s[\,\mathsf{in}\, n.R\,] \mid T \mid Q \mid n[G] \to T \mid Q \mid n[\,s[\,R\,] \mid G\,]$$

We write $P \downarrow_n$, and say that P *has barb* n, if $P \equiv (\nu\vec{m})(n[P_1]\|P_2)$ for some names $\vec{m}$ and processes P_1 and P_2 with $n \notin \vec{m}$. We say that P *has weak barb* n, written $P \Downarrow_n$ if there exists P' such that $P \to^* P'$ and $P' \downarrow_n$, for $\to^*$ the reflexive and transitive closure of the (immediate) reduction relation $\to$. Let $\mathbf{R}$ a binary relation on processes. We say that:

- $\mathbf{R}$ is *preserved by contexts* if $P\, \mathbf{R}\, Q$ implies $C[P]\, \mathbf{R}\, C[Q]$ for any context $C[\cdot]$;
- $\mathbf{R}$ is *closed under reductions* (or *reduction closed*) if whenever $P\, \mathbf{R}\, Q$ and $P \to P'$ then there exists Q' such that $Q \to^* Q'$ and $P'\, \mathbf{R}\, Q'$;
- $\mathbf{R}$ is *barb preserving* if $P\, \mathbf{R}\, Q$ and $P \downarrow_n$ implies $Q \Downarrow_n$.

Definition 4. *Reduction barbed congruence, written $\cong$, is the largest relation over processes, which is reduction closed, barb preserving, and preserved by all contexts.*

4   Encoding Mobile Ambients

We are now ready to show our encoding of MA in the link-calculus, that follows the idea developed in [5]. We pick MA as an interesting case because ambient interactions are inherently multi-party, even when they apparently involve only two parties. Each movement involves an ambient and the process exercising the corresponding capability, but the resulting reconfiguration of the hierarchical structure also impacts on the ambients and processes of the context.

Rule *(In)* requires a three-party interaction (at least), involving: 1) the process in $m.P$ with the capability to enter the ambient m; 2) its parent ambient $n[\cdot]$ to be moved; 3) the ambient $m[\cdot]$ to be entered. The rule can be successfully applied only when all the three entities are available. Any encoding based on binary interactions must deal with atomicity issues in the completion of the move, with conflicting, concurrent operations on the same ambient and with the possibility of retract/roll-back in case only two peers out of three are available.

Similarly, rule *(Out)* requires a three-party interaction (at least), involving: 1) the process out $m.P$ with the capability to leave the ambient m; 2) its parent ambient $n[\cdot]$ to be moved; 3) the "grand-parent" ambient $m[\cdot]$ (where $n[\cdot]$ is enclosed) to be exited.

Rule *(Open)* apparently requires a two-party interaction only, but as a matter of fact it is more complex than the other two rules, as it introduces the need of multi-party interactions with an unbounded number of peers. This is because when the ambient $n[Q]$ is dissolved, its content Q must be relocated, which may consist of an unbounded (and not known a priori) number of parallel processes: they all participate to the interaction! We adopt here a syntactic solution with no semantic impact on the rest: we replace $n[\cdot]$ with some sort of blind forwarder that leaves Q unaware of the deletion of $n[\cdot]$. However, the presence of forwarders complicates the interactions needed by rules *(In)* and *(Out)*, because the three parties can now be connected via chains of forwarders of arbitrary length. Our forwarders are reminiscent of forwarders in π-calculus [15], inspired by [28].

Roughly, the idea is to define an encoding assigning to any MA process P a corresponding link-calculus process $[\![P]\!]_{\tilde{a}}$ (the role of names $\tilde{a}$ will be made clear later) such that:

 – for any reduction $P \to P'$ there is a step $[\![P]\!]_{\tilde{a}} \to [\![P']\!]_{\tilde{a}}$;
 – and vice versa, for any silent step $[\![P]\!]_{\tilde{a}} \to Q$ there is an MA process P' such that $Q = [\![P']\!]_{\tilde{a}}$ and $P \to P'$.

Unfortunately, a direct encoding has to deal with the presence of forwarders, so that in general:

 – for any reduction $P \to P'$ we can find a step $[\![P]\!]_{\tilde{a}} \to Q$ but Q can differ from $[\![P']\!]_{\tilde{a}}$ because of the presence of forwarders;
 – and vice versa, for any silent step $[\![P]\!]_{\tilde{a}} \to Q$ we can find an MA process P' such that $P \to P'$ but, again, Q can differ from $[\![P']\!]_{\tilde{a}}$ because of the presence of forwarders.

One possible turnaround is to show that the correspondence holds up to some suitable abstract equivalence instead of strict equality. Instead, we provide a tighter correspondence that introduces forwarders in the syntax of MA, with no effect whatsoever on the semantics and expressiveness, and that allows us to recover the stronger correspondence result sketched above, with exact matching between the reductions of MA and the silent steps of the link-calculus (modulo some standard structural laws imposed by the MA structural congruence).

Ambients within Brackets. We just extend the syntax of MA with the possibility to enclose a process P within a pair of parentheses:

$$P ::= \cdots \mid (\!|P|\!)$$

making the presence of parentheses inessential w.r.t the behaviour of the process. To this aim, we introduce the additional structural congruence axioms:

$$(\!|(\nu n)P|\!) \equiv (\nu n)(\!|P|\!) \qquad\qquad P \equiv Q \Rightarrow (\!|P|\!) \equiv (\!|Q|\!)$$

Finally, we define the notion of *passive context* $\mathbb{C}$, to adjust the basic reduction rules to deal with the presence of an arbitrary number of balanced parentheses.

$$\mathbb{C}, \mathbb{D}, \mathbb{E} ::= \bullet \mid (\!(\mathbb{C})\!) \mid \mathbb{C}|P \mid P|\mathbb{C}$$

and write $\mathbb{C}(P)$ to denote the process obtained by replacing the hole $\bullet$ in $\mathbb{C}$ with P. Thus we add the suitable reduction rules:

$$\frac{}{\mathbb{D}(n[\,\mathbb{C}(\mathsf{in}\,m.P)\,])\mid \mathbb{E}(m[\,R]) \to \mathbb{D}(\mathbf{0})\mid \mathbb{E}(m[\,n[\,\mathbb{C}(P)\,]\mid R])} \quad \text{(In)}$$

$$\frac{}{m[\,\mathbb{D}(n[\,\mathbb{C}(\mathsf{out}\,m.P)\,])\,] \to n[\,\mathbb{C}(P)\,]\mid m[\,\mathbb{D}(\mathbf{0})\,]} \quad \text{(Out)}$$

$$\frac{}{\mathbb{C}(\mathsf{open}\,n.P)\mid \mathbb{D}(n[\,Q]) \to \mathbb{C}(P)\mid \mathbb{D}(\!(Q)\!)} \quad \text{(Open)} \qquad \frac{P \to Q}{(\!(P)\!) \to (\!(Q)\!)} \quad \text{(Brac)}$$

Structural Encoding. The encoding of a (possibly parenthesised) MA process is defined by delegating the management of an ambient $n[\cdot]$ to a suitable link-calculus process. The multi-party interaction between capabilities and ambients is regulated via communication on dedicated ports. Informally, a process P "resides" in some location, to which it addresses all its requests about the next actions to perform. The process that simulates the ambient $n[\cdot]$ also resides in some location $\tilde{p}$: it also defines an inner location $\tilde{a}$, where its content resides. A location $\tilde{a}$ denotes the 5-tuple of ports $a_{in}, a_{[in]}, a_{out}, a_{[out]}, a_{opn}$, where:

1. a_{in} is used for the interaction between the capability "in m" to enter the ambient m and the ambient $n[\cdot]$ where it is contained;
2. $a_{[in]}$ for the interaction between the ambient $n[\cdot]$ that contains the capability "in m" and the ambient $m[\cdot]$ to be entered;
3. a_{out} for the interaction between the capability "out m" to exit the ambient m and the ambient $n[\cdot]$ where it is contained;
4. $a_{[out]}$ for the interaction between the ambient $n[\cdot]$ that contains the capability "out m" and the ambient $m[\cdot]$ (that contains both) to be exited;
5. a_{opn} for the interaction between the capability "open n" to open the ambient n and the sibling ambient $n[\cdot]$ to be dissolved.

In the encoding, dissolving ambients amounts to resorting to *forwarders*, located between the dissolved locations and the "parent" ones, that redirect all the interactions that involve the processes originally inside the dissolved ambient. Forwarders need to forward requests through arbitrarily long chains of indirection. These are requests arriving from "below" (i.e., from the processes inside the dissolved ambient) to be forwarded up and requests arriving from "above" (i.e., from the processes that want to interact with the processes inside the dissolved ambient) that must be forwarded down: the former case applies to all ports, whereas the second case applies only to ports $a_{[in]}$ and a_{opn}.

The encoding $[\![\, P\,]\!]_{\tilde{a}}$ of P is parametric with respect to its location, i.e., the tuple of ports to be used to communicate with the enclosing ambient (or forwarders). Name passing is used to match the name of the ambient for which the

$$\llbracket 0 \rrbracket_{\tilde{a}} \quad\triangleq 0 \qquad\qquad \llbracket n[P] \rrbracket_{\tilde{a}} \quad\triangleq (\nu\,\tilde{b})(Amb(n,\tilde{b},\tilde{a})|\llbracket P \rrbracket_{\tilde{b}})$$

$$\llbracket P|Q \rrbracket_{\tilde{a}} \quad\triangleq \llbracket P \rrbracket_{\tilde{a}}|\llbracket Q \rrbracket_{\tilde{a}} \qquad \llbracket \text{in }m.P \rrbracket_{\tilde{a}} \quad\triangleq \,^{\tau}\backslash_{a_{in}}\langle m,\tilde{x}\rangle.\llbracket P \rrbracket_{\tilde{a}}$$

$$\llbracket (\nu\,n)P \rrbracket_{\tilde{a}} \triangleq (\nu\,n)\llbracket P \rrbracket_{\tilde{a}} \qquad \llbracket \text{out }m.P \rrbracket_{\tilde{a}} \triangleq \,^{\tau}\backslash_{a_{out}}\langle m,\tilde{x}\rangle.\llbracket P \rrbracket_{\tilde{a}}$$

$$\llbracket !P \rrbracket_{\tilde{a}} \qquad\triangleq \operatorname{rec}X.(\llbracket P \rrbracket_{\tilde{a}}|X) \qquad \llbracket \text{open }n.P \rrbracket_{\tilde{a}} \triangleq \,^{\tau}\backslash_{a_{opn}}\langle n\rangle.\llbracket P \rrbracket_{\tilde{a}}$$

$$\llbracket \langle\!\langle P\rangle\!\rangle \rrbracket_{\tilde{a}} \qquad\triangleq (\nu\,\tilde{b})(Fwd(\tilde{b},\tilde{a})|\llbracket P \rrbracket_{\tilde{b}})$$

$$Amb(n,\tilde{a},\tilde{p}) \triangleq \,^{a_{in}}\backslash_{p_{[in]}}\langle \underline{m},\tilde{z}\rangle.Amb(n,\tilde{a},\tilde{z}) + \,^{p_{[in]}}\backslash_{\tau}\langle n,\tilde{a}\rangle.Amb(n,\tilde{a},\tilde{p}) +$$
$$\,^{a_{out}}\backslash_{p_{[out]}}\langle \underline{m},\tilde{z}\rangle.Amb(n,\tilde{a},\tilde{z}) + \,^{a_{[out]}}\backslash_{\tau}\langle n,\tilde{p}\rangle.Amb(n,\tilde{a},\tilde{p}) +$$
$$\,^{p_{opn}}\backslash_{\tau}\langle n\rangle.Fwd(\tilde{a},\tilde{p})$$

$$Fwd(\tilde{a},\tilde{p}) \triangleq \,^{a_{in}}\backslash_{p_{in}}\langle \underline{n},\tilde{x}\rangle.Fwd(\tilde{a},\tilde{p}) +$$
$$\,^{a_{[in]}}\backslash_{p_{[in]}}\langle \underline{n},\tilde{x}\rangle.Fwd(\tilde{a},\tilde{p}) + \,^{p_{[in]}}\backslash_{a_{[in]}}\langle \underline{n},\tilde{x}\rangle.Fwd(\tilde{a},\tilde{p}) +$$
$$\,^{a_{out}}\backslash_{p_{out}}\langle \underline{n},\tilde{x}\rangle.Fwd(\tilde{a},\tilde{p}) + \,^{a_{[out]}}\backslash_{p_{[out]}}\langle \underline{n},\tilde{x}\rangle.Fwd(\tilde{a},\tilde{p}) +$$
$$\,^{a_{opn}}\backslash_{p_{opn}}\langle \underline{n}\rangle.Fwd(\tilde{a},\tilde{p}) + \,^{p_{opn}}\backslash_{a_{opn}}\langle \underline{n}\rangle.Fwd(\tilde{a},\tilde{p})$$

Fig. 4. Structural encoding of MA in `link`-calculus

capability is applicable with the name of the ambient where we would like to apply it. Moreover, in the case of enter/exit capabilities, name passing is necessary to inform the ambient that moves about the location where it is relocated.

We implicitly assume that all the restricted names introduced by the encoding are (globally) "fresh". The encoding is defined by straightforward structural induction in Fig. 4, where the processes $Amb(n,\tilde{a},\tilde{p})$ and $Fwd(\tilde{a},\tilde{p})$ represent, respectively, an ambient n located at $\tilde{p}$ and providing its content with location $\tilde{a}$, and a forwarder between the dissolved location $\tilde{a}$ and the "parent" location $\tilde{p}$.

As said above, applying the rule *(In)* to a process $n[\text{in }m.P\,|\,Q]\,|\,m[R]$ requires at least a three-party interaction, whose corresponding encodings are commented below. To help intuition, we can represent ambients as graphs, that reflect the hierarchical structure of nested ambients and that record the locations of ambients reside and the inner locations where ambients contents reside. We now illustrate how the encoding works in Fig. 5, where the labels to be merged are in correspondence with the processes that issued them, and the processes are arranged according to the hierarchy of processes:

(i) The process in $m.P$ with the capability to enter the ambient m is encoded by $^{\tau}\backslash_{a_{in}}\langle m,\tilde{x}\rangle.\llbracket P \rrbracket_{\tilde{a}}$, where the emphasis is on the name m (variables in $\tilde{x}$ are not important) and a_{in} is the port for the entering interaction.

(ii) The parent ambient $n[\cdot]$ to be moved is encoded by the restricted process $(\nu\,\tilde{a})(Amb(n,\tilde{a},\tilde{b})|\llbracket P \rrbracket_{\tilde{a}}|\llbracket Q \rrbracket_{\tilde{a}})$, where $Amb(n,\tilde{a},\tilde{b})$ includes the sub-process $^{a_{in}}\backslash_{b_{[in]}}\langle \underline{y},\tilde{z}\rangle.Amb(n,\tilde{a},\tilde{z})$ as a choice.

(iii) The ambient $m[\cdot]$ to be entered is encoded by $(\nu\,\tilde{c})(Amb(m,\tilde{c},\tilde{b})|\,\llbracket R \rrbracket_{\tilde{c}})$, where $Amb(m,\tilde{c},\tilde{b})$ includes the process $^{b_{[in]}}\backslash_{\tau}\langle m,\tilde{c}\rangle.Amb(m,\tilde{c},\tilde{b})$ as a choice, where m must match with the first field of the message in the first item.

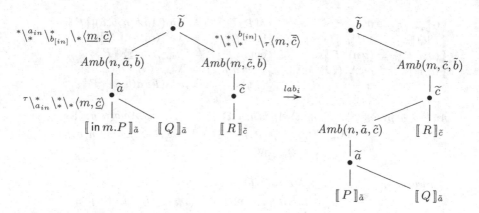

Fig. 5. Three party interaction for entering an ambient, where lab_i stands for $(\nu\,\tilde{a})(\nu\,\tilde{c})^{\tau}\backslash^{a_{in}}_{a_{in}}\backslash^{b_{[in]}}_{b_{[in]}}\backslash_{\tau} = {}^{\tau}\backslash^{b_{[in]}}_{\tau}\backslash^{b_{[in]}}_{b_{[in]}}\backslash_{\tau}$

The three process prefixes fit together: $\tilde{b}$ is the parent location of both ambients n and m, whose contents reside respectively at the sub-locations $\tilde{a}$ and $\tilde{c}$; the messages agree on the first value m (variable y is instantiated by m) and the links can be merged to form a valid link chain. Similarly, the variables $\tilde{x}$ and $\tilde{z}$ are instantiated by the location $\tilde{c}$. Intuitively, the complete interaction can be described as ${}^{\tau}\backslash^{a_{in}}_{a_{in}}\backslash^{b_{[in]}}_{b_{[in]}}\backslash_{\tau}\langle m,\tilde{c}\rangle$. However, names $\tilde{a}$ and $\tilde{c}$ are restricted, hence the link chain becomes ${}^{\tau}\backslash^{\tau}_{\tau}\backslash^{b_{[in]}}_{b_{[in]}}\backslash_{\tau}$ (by the application of the rule *(Res)* when $(\nu\,a)$ is encountered) and the tuple records the extrusion of $\tilde{c}$, i.e., it becomes $\langle m,\tilde{c}\rangle$ (by the application of the rule *(Open)* when $(\nu\,c)$ is encountered). In the end, the whole transition is just labelled by ${}^{\tau}\backslash^{\tau}_{\tau}\backslash^{b_{[in]}}_{b_{[in]}}\backslash_{\tau}$, because the link chain is solid, and thus the (ground) tuple $\langle m,\tilde{c}\rangle$ is discarded by rule *(Close)* (that also restores the restriction on the extruded location $\tilde{c}$ that was removed by the *(Open)* rule).

Similarly, as described in Fig. 6, the rule *(Out)* requires the encodings for: 1) the process $out\,m.P$ capable to leave the ambient m; 2) its parent ambient $n[\cdot]$ to be moved; 3) the "grand-parent" ambient $m[\cdot]$ (enclosing $n[\cdot]$) to be exited.

Finally (see Fig. 6), in the rule *(Open)* besides the encodings for the processes $open\,n.P$ and $n[\,Q\,]$, we should take care of the relocation of all processes included in the ambient n. We obtain this by using a forwarder, as explained below. The process with the open capability is encoded by ${}^{\tau}\backslash_{b_{opn}}\langle n\rangle.[\![\,P\,]\!]_{\tilde{b}}$, and the ambient to be dissolved by $(\nu\,\tilde{a})(Amb(n,\tilde{a},\tilde{b})|[\![\,P\,]\!]_{\tilde{a}})$, where $Amb(n,\tilde{a},\tilde{b})$ includes ${}^{b_{opn}}\backslash_{\tau}\langle n\rangle.Fwd(\tilde{a},\tilde{b})$ as a choice. The process $Fwd(\tilde{a},\tilde{b})$ presents a case for each kind of interaction, and it is used to suitably redirect all the interactions that involve the processes originally inside the dissolved ambient n. Trivially, rule *(Close)* can be applied, since ${}^{\tau}\backslash^{*}_{b_{opn}}\backslash_{*}\langle n\rangle \bullet {}^{*}\backslash^{b_{opn}}_{*}\backslash_{\tau}\langle n\rangle = {}^{\tau}\backslash^{b_{opn}}_{b_{opn}}\backslash_{\tau}\langle n\rangle$ that is a solid link chain (in this case there are no extruded names in the tuple).

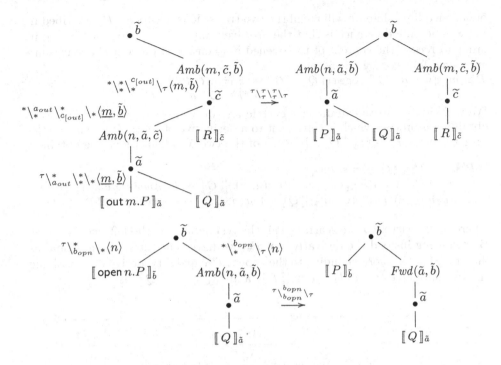

Fig. 6. (Top part) Three party interaction for exiting an ambient, (bottom part) three party interaction for opening an ambient

In the following, t stands for the topmost location in which the ambients reside.

Definition 5. *We let* $\equiv_l$ *be the structural congruence on* link-*calculus processes induced by the axioms:*

$$P \mid \mathbf{0} \equiv_l P \qquad P \mid Q \equiv_l Q \mid P \qquad (P \mid Q) \mid R \equiv_l P \mid (Q \mid R)$$
$$(\nu n)\mathbf{0} \equiv_l \mathbf{0} \qquad (\nu n)(\nu m)P \equiv_l (\nu m)(\nu n)P$$
$$(\nu n)(P \mid Q) \equiv_l P \mid (\nu n)Q, \quad if \ \ n \notin fn(P) \qquad recX.P \equiv_l P\{recX.P/P\}$$

Lemma 8. $P \equiv_l Q$ *implies* $P \sim_l Q$.

Proposition 3. *The encoding in Fig. 4 is well-defined, in the sense that if* $P \equiv Q$ *then* $[\![\,P\,]\!]_{\tilde{t}} \equiv_l [\![\,Q\,]\!]_{\tilde{t}}$.

Without loss of generality, in the following we will consider link-calculus processes as taken up to the structural congruence $\equiv_l$.

Example 9. Take again the MA process $P \triangleq m[\,s[\,\text{in}\,n.R\,]\mid T\,] \mid \text{open}\,m.Q \mid n[\,G\,]$ in Ex. 8 for putting our encoding at work. Let

$$P_\pi \triangleq [\![\,P\,]\!]_{\tilde{t}} = [\![\,m[\,s[\,\text{in}\,n.R\,]\mid T\,]\,]\!]_{\tilde{t}} \,|\, [\![\,\text{open}\,m.Q\,]\!]_{\tilde{t}} \,|\, [\![\,n[\,G\,]\,]\!]_{\tilde{t}} =$$
$$(\nu\,\tilde{a})(Amb(m,\tilde{a},\tilde{t}) | [\![\,s[\,\text{in}\,n.R\,]\mid T\,]\!]_{\tilde{a}}) \,|\, [\![\,\text{open}\,m.Q\,]\!]_{\tilde{t}} \,|\, [\![\,n[\,G\,]\,]\!]_{\tilde{t}} =$$
$$(\nu\,\tilde{a})(Amb(m,\tilde{a},\tilde{t}) | ([\![\,T\,]\!]_{\tilde{a}} | ((\nu\,\tilde{b})(Amb(s,\tilde{b},\tilde{a}) |^\tau\backslash_{b_{in}} \langle n,\tilde{\underline{x}}\rangle.[\![\,R\,]\!]_{\tilde{b}}))| $$
$$^\tau\backslash_{t_{opn}} \langle m\rangle.[\![\,Q\,]\!]_{\tilde{t}} | (\nu\,\tilde{c})(Amb(n,\tilde{c},\tilde{t}) | [\![\,G\,]\!]_{\tilde{c}})$$

be its encoding. Here we will recall the two transitions of process P, described in Ex. 8. The only difference is that the dissolved ambient is surrounded by $(\!|.|\!)$, in order to record the content of the opened ambient. The process P can execute an open n followed by a in n one:

$$P \equiv m[\,s[\,in\,n.R\,]\mid T\,]\mid open\,m.Q\mid n[\,G\,] \to (\!|s[\,in\,n.R\,]\mid T|\!)\mid Q\mid n[\,G\,]$$
$$\to (\!|T|\!)\mid Q\mid n[\,s[\,R\,]\mid G\,]\equiv P'$$

We will show that we can directly encode any derivatives on an MA process P obtaining a process which is congruent to a derivative of $[\![\,P\,]\!]_{\tilde{t}}$. If we encode the target process $P' \equiv (\!|T|\!)\mid Q\mid n[s[R]\mid G]$ of the two MA transitions, we obtain:

$$
\begin{aligned}
[\![\,P'\,]\!]_{\tilde{t}} &= [\![\,(\!|T|\!)\,]\!]_{\tilde{t}}\mid[\![\,Q\,]\!]_{\tilde{t}}\mid[\![\,n[s[R]\mid G]\,]\!]_{\tilde{t}}\\
&= (\nu\,\tilde{b})(Fwd(\tilde{b},\tilde{t})\mid[\![\,T\,]\!]_{\tilde{b}})\mid[\![\,Q\,]\!]_{\tilde{t}}(\nu\,\tilde{c})(Amb(n,\tilde{c},\tilde{t})\mid[\![\,G\,]\!]_{\tilde{c}}\mid(\nu\,\tilde{b})(Amb(s,\tilde{b},\tilde{c})\mid[\![\,R\,]\!]_{\tilde{b}}))\\
&\equiv_l (\nu\,\tilde{b})(\nu\,\tilde{c})(\nu\,\tilde{a})(Fwd(\tilde{b},\tilde{t})\mid[\![\,T\,]\!]_{\tilde{b}}\mid[\![\,Q\,]\!]_{\tilde{t}}\mid(Amb(n,\tilde{c},\tilde{t})\mid[\![\,G\,]\!]_{\tilde{c}}\mid Amb(s,\tilde{b},\tilde{c})\mid[\![\,R\,]\!]_{\tilde{b}}) \quad (1)
\end{aligned}
$$

where in the last step, we rearrange all the restrictions at the left end.

By following instead the derivatives of P_π, we have that the first transition, as illustrated below, corresponding to the "open m" capability, derived by involving the two subprocesses $Amb(m,\tilde{a},\tilde{t})\mid[\![\,T\,]\!]_{\tilde{a}}$ and $^\tau\!\backslash_{t_{opn}}\langle m\rangle.[\![\,Q\,]\!]_{\tilde{t}}$ is:

$$
\cfrac{
\cfrac{^\tau\!\backslash_{t_{opn}}\langle m\rangle.Q_\pi \xrightarrow{\ ^\tau\!\backslash_{t_{opn}}\langle m\rangle\ } Q_\pi \ (\text{Act}) \qquad Amb(m,\tilde{a},\tilde{t}) \xrightarrow{\ ^{t_{opn}}\!\backslash_\tau\langle m\rangle\ } Fwd(\tilde{a},\tilde{t}) \ (\text{Act})}
{
\cfrac{^\tau\!\backslash_{t_{opn}}\langle m\rangle.Q_\pi\mid Amb(m,\tilde{a},\tilde{t}) \xrightarrow{\ ^\tau\!\backslash^{t_{opn}}_{t_{opn}}\!\backslash_\tau\ } Q_\pi\mid Fwd(\tilde{a},\tilde{t})}
{
\cfrac{^\tau\!\backslash_{t_{opn}}\langle m\rangle.Q_\pi\mid Amb(m,\tilde{a},\tilde{t}))\mid T_\pi \xrightarrow{\ ^\tau\!\backslash^{t_{opn}}_{t_{opn}}\!\backslash_\tau\ } Q_\pi\mid Fwd(\tilde{a},\tilde{t})\mid T_\pi}
{
\cfrac{(\nu\,a)(^\tau\!\backslash_{t_{opn}}\langle m\rangle.Q_\pi\mid Amb(m,\tilde{a},\tilde{t})\mid T_\pi) \xrightarrow{\ ^\tau\!\backslash^{t_{opn}}_{t_{opn}}\!\backslash_\tau\ } (\nu\,a)(Q_\pi\mid Fwd(\tilde{a},\tilde{t})\mid T_\pi)}
{(\nu\,a)(^\tau\!\backslash_{t_{opn}}\langle m\rangle.Q_\pi\mid Amb(m,\tilde{a},\tilde{t})\mid T_\pi)\mid C_\pi \xrightarrow{\ ^\tau\!\backslash^{t_{opn}}_{t_{opn}}\!\backslash_\tau\ } (\nu\,a)(Q_\pi\mid Fwd(\tilde{a},\tilde{t})\mid T_\pi)\mid C_\pi}\ (\text{Par})}
\ (\text{Res})}
\ (\text{Par})}
\ (\text{Close})}
$$

where $C_\pi \equiv (\nu\,\tilde{b})(Amb(s,\tilde{b},\tilde{a})\mid^\tau\!\backslash_{bin}\langle n,\tilde{x}\rangle.[\![\,R\,]\!]_{\tilde{b}}\mid(\nu\,\tilde{c})(Amb(n,\tilde{c},\tilde{t})\mid[\![\,G\,]\!]_{\tilde{c}})$, $Q_\pi \equiv [\![\,Q\,]\!]_{\tilde{t}}$ and $T_\pi \equiv [\![\,T\,]\!]_{\tilde{a}}$.

The next transition (not reported for the sake of brevity) performs the corresponding "in n" capability, leading to the target process:

$$(\nu\,\tilde{a})(\nu\,\tilde{b})(\nu\,\tilde{c})([\![\,R\,]\!]_{\tilde{b}}\mid Amb(s,\tilde{b},\tilde{c})\mid Fwd(\tilde{a},\tilde{t})\mid Amb(n,\tilde{c},\tilde{t})\mid[\![\,T\,]\!]_{\tilde{a}}\mid[\![\,G\,]\!]_{\tilde{c}}\mid[\![\,Q\,]\!]_{\tilde{t}}) \quad (2)$$

In our case also the $Fwd(\tilde{a},\tilde{t})$ process is involved, as the process $Amb(s,\tilde{b},\tilde{a})$ refers to the ambient name $\tilde{a}$ that does not exist any longer. It is easy to see that (1) differs from (2) only for bound names and variables.

Operational Correspondence. We can filter out non solid transitions (representing partial interactions) of encoded processes by letting: $[\![\,P\,]\!]\triangleq(\nu\,\tilde{t})[\![\,P\,]\!]_{\tilde{t}}$.

We say that a link chain s is *silent* if it consists of τ actions only, and write $P\xrightarrow{\tau}Q$ if $P\xrightarrow{s}Q$ for some silent s. (Note that any silent link chain is solid.)

Lemma 9. *If $[\![\,P\,]\!]\xrightarrow{s}Q$, then s is silent.*

Proof. The proof is by cases on the type of capability simulated by the encoded process $[\![P]\!]$ (open $n.Q$, in $n.Q$, out $n.Q$). The proof of each case is, in turn, by induction on the length of the derivation of the transition.

Theorem 1. *Let P be a MA process, then $P \to P'$ if and only if there exists Q such that $[\![P]\!] \xrightarrow{s} Q$, and $Q \equiv [\![P']\!]$.*

The *only if* part can be proved by induction on the proof of reduction $P \to P'$. The idea underlying the proof of the *if* part goes as follows. At the extremities of the step a silent action τ should appear and the only actions able to perform τ on the left are the movement capabilities (offering the links $^{\tau}\backslash_{a_{in}}$, $^{\tau}\backslash_{a_{out}}$, $^{\tau}\backslash_{a_{opn}}$), while on the right we need an ambient offering the corresponding links $^{a'_{[in]}}\backslash_{\tau}$, $^{a'_{[out]}}\backslash_{\tau}$, $^{a'_{opn}}\backslash_{\tau}$. An ambient is needed to pass from the action a_{in} (resp. a_{out}) to the action $a'_{[in]}$ (resp. $a'_{[out]}$), while the forwarders do not change the type of actions. Participants to the action can be arranged in parallel, according to their position in s, thanks to the $\equiv_l$. By induction on the proof, we can rebuild the tree of the bracketed ambients involved in the reduction. Working modulo the structural congruence we can always build the synchronisation, starting from one of the extremities, thus rebuilding the reduction under analysis.

Linked bisimilarity induces a behavioural equivalence on MA processes via our encoding: two MA processes can be retained as equivalent if their encodings are so. We conjecture that linked bisimilarity is finer than barbed congruence, since it would distinguish e.g., $(\nu n)n[\text{in } m.\mathbf{0}]$ from $\mathbf{0}$. Moreover, link bisimilarity has a 'strong' flavour (silent moves are matched exactly), as opposed to the 'weak' flavour of barbed congruence. We can define a weak version of linked bisimilarity in the standard way, but then the weak version would not be preserved by sum and still it would distinguish e.g., $(\nu n)n[\text{in } m.\mathbf{0}]$ from $\mathbf{0}$. We think the mismatch is mostly due to the quite arbitrary choice of barbs to be observed in MA, i.e., the names of the topmost ambients: in our encoding, an ambient is just an interacting process, and its name is just a piece of information among others used to match capabilities requests.

5   Concluding Remarks and Related Works

We have presented the link-calculus as a lightweight enrichment of traditional dyadic process calculi able to deal with open multiparty interactions. We consider the link-calculus as a basis to investigate more general forms of interaction. An important field of application of our calculus is that of Systems Biology, where biological interactions are often multi-party. We would like to generalise link prefixes to *link-chain* prefixes, to encode some simple pattern of interaction directly, and also to express non linear communication patterns in the prefixes.

Among the recently presented network-aware extensions of classical calculi such as [14] (to handle explicit distribution, remote operations and process mobility), and [11] (to deal with permanent nodes crashing and links breaking), the closest proposal to ours is in [23], an extension of π-calculus, where links are named and are distinct from usual I/O actions, and there is one sender and one receiver (the output includes the final receiver name). In our calculus, links can carry message tuples, and each participant can play both the sender and the receiver rôle. Our semantics recalls their concurrent semantics, where transmissions can be observed in the form of a multi-set of routing paths. In our case the collected links are organised in a link chain. In [6] the authors present a general framework to extend synchronisation algebras [27] with name mobility, that could be easily adapted to many other high-level kinds of synchronisation, like ours, but with a more complex machinery. More sophisticated forms of synchronisations, with a fixed number of processes, are introduced in π-calculus in [24] (joint input) and in [7] (polyadic synchronisation). The focus of [19] is instead on the expressiveness of an asynchronous CCS equipped with joint inputs allowing the interactions of n processes, proving that there is no truly distributed implementation of operators synchronising more than three processes. As in the join-calculus [13], and differently from our approach, participants can act either as senders or as receivers. In [17], a conservative extension of CCS, with multiparty synchronisation is introduced. The mechanism is realised as a sequence of dyadic synchronisations and, furthermore, puts some constraints that make the parallel operator non associative. Our approach also recalls the asynchronous semantics of CCS-like process calculi (see e.g., [26,10]). In both cases, the idea is that one process decides which interaction to try, and the other processes have to match. We introduce the capability of creating chains of links, useful to model communication patterns and information routing. Finally, [2] introduces a distributed version of the π-calculus for names to be exported, where names are equipped with the information needed to point back to its local environment, thus keeping track of the origin of mobile agents in a multi-hop travel.

Several approaches, among which we recall [12,20,25,3,4], are specifically devoted to provide an LTS semantics to MA, as a basis for bisimulation congruence. A contextual equivalence for MA is instead presented in [16]. While, in all these works *ad hoc* semantics are introduced, our proposed encoding is just an illustrative example of application of our network-aware calculus. Our link labels naturally accommodate the encoding. Furthermore, their LTSs are higher-order, since they can move processes (e.g., in [20], a transition can lead from a process to a context, while in [3] contexts are used as labels). A different approach is in [18], where a coalgebraic denotational semantics for the MA is presented.

References

1. Bodei, C., Brodo, L., Degano, P., Gao, H.: Detecting and preventing type flaws at static time. Journal of Computer Security 18(2), 229–264 (2010)
2. Bodei, C., Degano, P., Priami, C.: Names of the π-calculus agents handled locally. Theor. Comput. Sci. 253(2), 155–184 (2001)
3. Bonchi, F., Gadducci, F., Monreale, G.V.: Labelled transitions for mobile ambients (as synthesized via a graphical encoding). ENTCS 242(1), 73–98 (2009)
4. Bonchi, F., Gadducci, F., Monreale, G.V.: Reactive systems, barbed semantics, and the mobile ambients. In: de Alfaro, L. (ed.) FOSSACS 2009. LNCS, vol. 5504, pp. 272–287. Springer, Heidelberg (2009)

5. Brodo, L.: On the expressiveness of the π-calculus and the mobile ambients. In: Johnson, M., Pavlovic, D. (eds.) AMAST 2010. LNCS, vol. 6486, pp. 44–59. Springer, Heidelberg (2011)

6. Bruni, R., Lanese, I.: Parametric synchronizations in mobile nominal calculi. Theor. Comput. Sci. 402(2-3), 102–119 (2008)

7. Carbone, M., Maffeis, S.: On the expressive power of polyadic synchronisation in pi-calculus. Nordic Journal of Computing 10(2), 70–98 (2003)

8. Cardelli, L.: Brane calculi. In: Danos, V., Schachter, V. (eds.) CMSB 2004. LNCS (LNBI), vol. 3082, pp. 257–278. Springer, Heidelberg (2005)

9. Cardelli, L., Gordon, A.D.: Mobile ambients. Theor. Comput. Sci. 240(1), 177–213 (2000)

10. de Boer, F.S., Palamidessi, C.: On the asynchronous nature of communication in concurrent logic languages: A fully abstract model based on sequences. In: Baeten, J.C.M., Klop, J.W. (eds.) CONCUR 1990. LNCS, vol. 458, pp. 99–114. Springer, Heidelberg (1990)

11. De Nicola, R., Gorla, D., Pugliese, R.: Basic observables for a calculus for global computing. Information and Computation 205(10), 1491–1525 (2007)

12. Ferrari, G.-L., Montanari, U., Tuosto, E.: A LTS semantics of ambients via graph synchronization with mobility. In: Restivo, A., Ronchi Della Rocca, S., Roversi, L. (eds.) ICTCS 2001. LNCS, vol. 2202, pp. 1–16. Springer, Heidelberg (2001)

13. Fournet, C., Gonthier, G.: The reflexive CHAM and the join-calculus. In: Proc. of POPL 1996, pp. 372–385. ACM Press (1996)

14. Francalanza, A., Hennessy, M.: A theory of system behaviour in the presence of node and link failure. Information and Computation 206(6), 711–759 (2008)

15. Gardner, P., Laneve, C., Wischik, L.: Linear forwarders. Inf. Comput. 205(10), 1526–1550 (2007)

16. Gordon, A., Cardelli, L.: Equational properties of mobile ambients. Math. Struct. in Comp. Sci. 13(3), 371–408 (2003)

17. Gorrieri, R., Versari, C.: An Operational Petri Net Semantics for A^2CCS. Fundam. Inform. 109(2), 135–160 (2011)

18. Hausmann, D., Mossakowski, T., Schröder, L.: A coalgebraic approach to the semantics of the ambient calculus. Theor. Comput. Sci. 366(1-2), 121–143 (2006)

19. Laneve, C., Vitale, A.: The expressive power of synchronizations. In: Proc. of LICS 2010, pp. 382–391. IEEE Computer Society (2010)

20. Merro, M., Nardelli, F.Z.: Behavioral theory for mobile ambients. Journal of the ACM 52(6), 961–1023 (2005)

21. Milner, R.: A Calculus of Communication Systems. LNCS, vol. 92. Springer, Heidelberg (1980)

22. Milner, R.: Communicating and mobile systems: the π-calculus. CUP (1999)

23. Montanari, U., Sammartino, M.: Network conscious pi-calculus: a concurrent semantics. ENTCS 286, 291–306 (2012)

24. Nestmann, U.: On the expressive power of joint input. ENTCS 16(2) (1998)

25. Rathke, J., Sobociński, P.: Deriving structural labelled transitions for mobile ambients. Information and Computation 208(10), 1221–1242 (2010)

26. Vaandrager, F.W.: On the relationship between process algebra and input/output automata. In: Proc. of LICS 1991, pp. 387–398. IEEE Computer Society (1991)

27. Winskel, G.: Synchronization trees. Theor. Comput. Sci. 34(1-2), 33–82 (1984)

28. Wischik, L., Gardner, P.: Explicit fusions. Theor. Comput. Sci. 340(3), 606–630 (2005)

Behaviour Protection
in Modular Rule-Based System Specifications

Francisco Durán[1], Fernando Orejas[2], and Steffen Zschaler[3]

[1] Departamento de Lenguajes y Ciencias de la Computación
Universidad de Málaga
duran@lcc.uma.es
[2] Departament de Llenguatges i Sistemes Informàtics
Universitat Politècnica de Catalunya
orejas@lsi.upc.edu
[3] Department of Informatics
King's College London
szschaler@acm.org

Abstract. Model-driven engineering (MDE) and, in particular, the notion of do-main-specific modelling languages (DSMLs) is an increasingly popular approach to systems development. DSMLs are particularly interesting because they allow encoding domain-knowledge into a modelling language and enable full code generation and analysis based on high-level models. However, as a result of the domain-specificity of DSMLs, there is a need for many such languages. This means that their use only becomes economically viable if the development of new DSMLs can be made efficient. One way to achieve this is by reusing functionality across DSMLs. On this background, we are working on techniques for modularising DSMLs into reusable units. Specifically, we focus on DSMLs whose semantics are defined through in-place model transformations. In this paper, we present a formal framework of morphisms between graph-transformation systems (GTSs) that allow us to define a novel technique for conservative extensions of such DSMLs. In particular, we define different behaviour-aware GTS morphisms and prove that they can be used to define conservative extensions of a GTS.

1 Introduction

Model-Driven Engineering (MDE) [41] has raised the level of abstraction at which systems are developed, moving development focus from the programming-language level to the development of software models. Models and specifications of systems have been around the software industry from its very beginning, but MDE articulates them so that the development of information systems can be at least partially automated. Thus models are being used not only to specify systems, but also to simulate, analyze, modify and generate code of such systems. A particularly useful concept in MDE are domain-specific modelling languages (DSMLs) [7]. These languages offer concepts specifically targeted at a particular domain. On the one hand this makes it easier for domain experts to express their problems and requirements. On the other hand, the higher amount of knowledge embedded in each concept allows for much more complete generation of

N. Martí-Oliet and M. Palomino (Eds.): WADT 2012, LNCS 7841, pp. 24–49, 2013.

executable solution code from a DSML model [27] as compared to a model expressed in a general-purpose modelling language.

DSMLs can only be as effective as they are specific for a particular domain. This implies that there is a need for a large number of such languages to be developed. However, development of a DSML takes additional effort in a software-development project. DSMLs are only viable if their development can be made efficient. One way of achieving this is by allowing them to be built largely from reusable components. Consequently, there has been substantial research on how to modularise language specifications. DSMLs are often defined by specifying their syntax (often separated into concrete and abstract syntax) and their semantics. While we have reasonably good knowledge of how to modularise DSML syntax, the modularisation of language semantics is an as yet unsolved issue.

DSML semantics can be represented in a range of different ways—for example using UML behavioural models [17,20], abstract state machines [8,2], Kermeta [34], or in-place model transformations [33,37]. In the context of MDE it seems natural to describe the semantics by means of models, so that they may be integrated with the rest of the MDE environment and tools. We focus on the use of in-place model transformations.

Graph transformation systems (GTSs) were proposed as a formal specification technique for the rule-based specification of the dynamic behaviour of systems [10]. Different approaches exist for modularisation in the context of the graph-grammar formalism [5,40,12]. All of them have followed the tradition of modules inspired by the notion of algebraic specification module [15]. A module is thus typically considered as given by an export and an import interface, and an implementation body that realises what is offered in the export interface, using the specification to be imported from other modules via the import interface. For example, Große-Rhode, Parisi-Presicce, and Simeoni introduce in [22] a notion of *module* for typed graph transformation systems, with interfaces and implementation bodies; they propose operations for union, composition, and refinement of modules. Other approaches to modularisation of graph transformation systems include PROGRES Packages [42], GRACE Graph Transformation Units and Modules [31], and DIEGO Modules [43]. See [26] for a discussion on these proposals.

For the kind of systems we deal with, the type of module we need is much simpler. For us, a module is just the specification of a system, a GTS, without import and export interfaces. Then, we build on GTS morphisms to compose these modules, and specifically we define parametrised GTSs. The instantiation of such parameterized GTS is then provided by an amalgamation construction. We present formal results about graph-transformation systems and morphisms between them. Specifically, we provide definitions for behaviour-reflecting and -protecting GTS morphisms and show that they can be used to infer semantic properties of these morphisms. We give a construction for the amalgamation of GTSs, as a base for the composition of GTSs, and we prove it to protect behaviour under certain circumstances. Although we motivate and illustrate our approach using the *e-Motions* language [38,39], our proposal is language-independent, and all the results are presented for GTSs and adhesive HLR systems [32,14].

Different forms of GTS morphisms have been used in the literature, taking one form or another depending on their concrete application. Thus, we find proposals centered on refinements (see., e.g., [25,21,22]), views (see, e.g., [19]), and substitutability (see [18]). See [18] for a first attempt to a systematic comparison of the different proposals and notations. None of these notions fit our needs, and none of them coincide with our behaviour-aware GTS morphisms.

Moreover, as far as we know, parameterised GTSs and GTS morphisms, as we discuss them, have not been studied before. Heckel and Cherchago introduce parameterised GTSs in [24], but their notion has little to do with our parameterised GTSs. In their case, the parameter is a signature, intended to match service descriptions. They however use a double-pullback semantics, and have a notion of substitution morphism which is related to our behaviour preserving morphism.

Our work is originally motivated by the specification of non-functional properties (NFPs), such as performance or throughput, in DSMLs. We have been looking for ways in which to encapsulate the ability to specify non-functional properties into reusable DSML modules. Troya et al. used the concept of observers in [45,46] to model non-functional properties of systems described by GTSs in a way that could be analysed by simulation. In [9], we have built on this work and ideas from [48] to allow the modular encapsulation of such observer definitions in a way that can be reused in different DSML specifications. In this paper, we present a full formal framework of such language extensions. Nevertheless, this framework is independent of the specific example of non-functional property specifications, but instead applies to any conservative extension of a base GTS.

The way in which we think about composition of reusable DSML modules has been inspired by work in aspect-oriented modelling (AOM). In particular, our ideas for expressing parametrised metamodels are based on the proposals in [3,29]. Most AOM approaches use syntactic notions to automate the establishment of mappings between different models to be composed, often focusing primarily on the structural parts of a model. While our mapping specifications are syntactic in nature, we focus on composition of behaviours and provide semantic guarantees. In this sense, our work is perhaps most closely related to the work on MATA [47] or semantic-based weaving of scenarios [30].

The rest of the paper begins with a presentation of a motivating example expressed in the *e-Motions* language in Section 2. Section 3 introduces a brief summary of graph transformation and adhesive HLR categories. Section 4 introduces behaviour-reflecting GTS morphisms, the construction of amalgamations in the category of GTSs and GTS morphisms, and several results on these amalgamations, including the one stating that the morphisms induced by these amalgamations protect behaviour, given appropriate conditions. The paper finishes with some conclusions and future work in Section 5.

2 NFP Specification with *e-Motions*

In this section, we use *e-Motions* [38,39] to provide a motivating example, adapted from [45], as well as intuitions for the formal framework developed. However, as stated in the previous section, the framework itself is independent of such a language.

e-Motions is a Domain Specific Modeling Language (DSML) and graphical framework developed for Eclipse that supports the specification, simulation, and formal

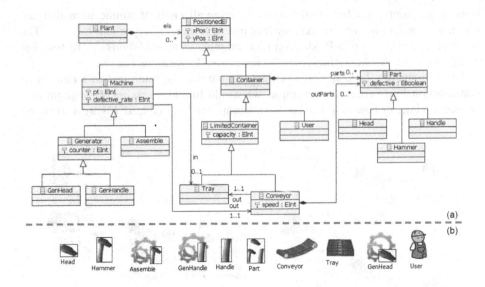

Fig. 1. Production Line (a) metamodel and (b) concrete syntax (from [45])

analysis of DSMLs. Given a MOF metamodel (abstract syntax) and a GCS model (a graphical concrete syntax) for it, the behaviour of a DSML is defined by in-place graph transformation rules. Although we briefly introduce the language here, we omit all those details not relevant to this paper. We refer the interested reader to [36,39] or http://atenea.lcc.uma.es/e-Motions for additional details.

Figure 1(a) shows the metamodel of a DSML for specifying Production Line systems for producing hammers out of hammer heads and handles, which are generated in respective machines, transported along the production line via conveyors, and temporarily stored in trays. As usual in MDE-based DSMLs, this metamodel defines all the concepts of the language and their interconnections; in short, it provides the language's *abstract* syntax. In addition, a *concrete* syntax is provided. In the case of our example, this is sufficiently well defined by providing icons for each concept (see Figure 1(b)); connections between concepts are indicated through arrows connecting the corresponding icons. Figure 2 shows a model conforming to the metamodel in Figure 1(a) using the graphical notation introduced in the GCS model in Figure 1(b).

The behavioural semantics of the DSML is then given by providing transformation rules specifying how models can evolve. Figure 3 shows an example of such a rule. The rule consists of a left-hand side matching a situation before the execution of the rule and a right-hand side showing the result of applying the rule. Specifically, this rule shows how a new hammer is assembled: a hammer generator a has an incoming tray of parts and is connected to an outgoing conveyor belt. Whenever there is a handle and a head available, and there is space in the conveyor for at least one part, the hammer generator can assemble them into a hammer. The new hammer is added to the parts set of the outgoing conveyor belt in time T, with T some value in the range [a.pt - 3, a.pt + 3], and where pt is an attribute representing the production time of a machine. The complete semantics of our production-line DSML is constructed from a number of such rules covering all kinds of atomic steps that can occur, e.g., generating new pieces, moving pieces from a conveyor to a tray, etc. The complete specification of a Production Line example using *e-Motions* can be found at `http://atenea.lcc.uma.es/E-motions/PLSExample`.

For a Production Line system like this one, we may be interested in a number of non-functional properties. For example, we would like to assess the throughput of a production line, or how long it takes for a hammer to be produced. Figure 4(a) shows

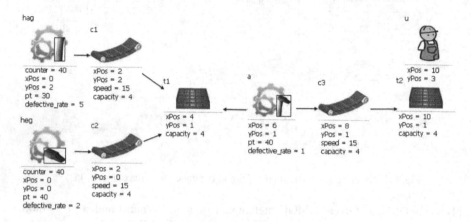

Fig. 2. Example of production line configuration

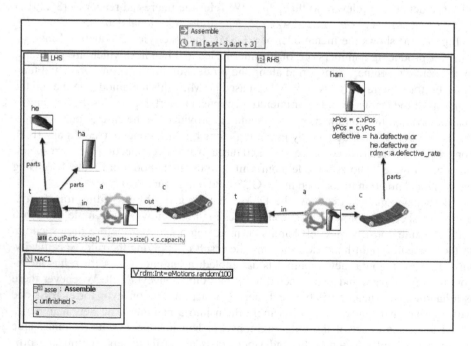

Fig. 3. Assemble rule indicating how a new hammer is assembled (from [45])

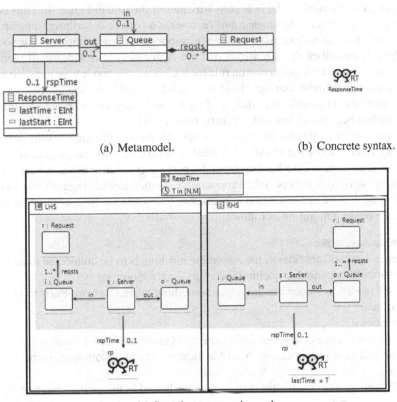

(a) Metamodel.   (b) Concrete syntax.

(c) Sample response time rule.

Fig. 4. Generic model of response time observer

the metamodel for a DSML for specifying production time. It is defined as a parametric model (i.e., a model template), defined independently of the Production Line system. It uses the notion of response time, which can be applied to different systems with different meanings. The concepts of Server, Queue, and Request and their interconnections are parameters of the metamodel, and they are shaded in grey for illustration purposes. Figure 4(b) shows the concrete syntax for the response time observer object. Whenever that observer appears in a behavioural rule, it will be represented by that graphical symbol.

Figure 4(c) shows one of the transformation rules defining the semantics of the response time observer. It states that if there is a server with an in queue and an out queue and there initially are some requests (at least one) in the in queue, and the out queue contains some requests after rule execution, the last response time should be recorded to have been equal to the time it took the rule to execute. Similar rules need to be written to capture other situations in which response time needs to be measured, for example, where a request stays at a server for some time, or where a server does not have an explicit in or out queue.

Note that, as in the metamodel in Figure 4(a), part of the rule in Figure 4(c) has been shaded in grey. Intuitively, the shaded part represents a pattern describing transformation rules that need to be extended to include response-time accounting.[1] The lower part of the rule describes the extensions that are required. So, in addition to reading Figure 4(c) as a 'normal' transformation rule (as we have done above), we can also read it as a *rule transformation*, stating: "Find all rules that match the shaded pattern and add ResponseTime objects to their left- and right-hand sides as described." In effect, observer models become higher-order transformations [44].

To use our response-time language to allow specification of production time of hammers in our Production Line DSML, we need to weave the two languages together. For this, we need to provide a binding from the parameters of the response-time metamodel (Figure 4(a)) to concepts in the Production Line metamodel (Figure 1(a)). In this case, assuming that we are interested in measuring the response time of the Assemble machine, the binding might be as follows:

- Server to Assemble;
- Queue to LimitedContainer as the Assemble machine is to be connected to an arbitrary LimitedContainer for queuing incoming and outgoing parts;
- Request to Part as Assemble only does something when there are Parts to be processed; and
- Associations:
    • The in and out associations from Server to Queue are bound to the corresponding in and out associations from Machine to Tray and Conveyor, respectively; and
    • The association from Queue to Request is bound to the association from Container to Part.

As we will see in Section 4, given DSMLs defined by a metamodel plus a behaviour, the weaving of DSMLs will correspond to amalgamation in the category of DSMLs and DSML morphisms. Figure 5 shows the amalgamation of an inclusion morphism between the model of an observer DSML, M_{Obs}, and its parameter sub-model M_{Par}, and the binding morphism from M_{Par} to the DSML of the system at hand, M_{DSML}, the Production Line DSML in our example. The amalgamation object $M_{\overline{DSML}}$ is obtained by the construction of the amalgamation of the corresponding metamodel morphisms and the amalgamation of the rules describing the behaviour of the different DSMLs.

In our example, the amalgamation of the metamodel corresponding morphisms is shown in Figure 6 (note that the binding is only partially depicted). The weaving process has added the ResponseTime concept to the metamodel. Notice that the weaving process also ensures that only sensible woven metamodels can be produced: for a given binding of parameters, there needs to be a match between the constraints expressed in the observer metamodel and the DSML metamodel. We will discuss this issue in more formal detail in Section 4.

The binding also enables us to execute the rule transformations specified in the observer language. For example, the rule in Figure 3 matches the pattern in Figure 4(c),

---

[1]   Please, notice the use of the cardinality constraint 1..∗ in the rule in Figure 4(c). It is out of the scope of this paper to discuss the syntactical facilities of the *e-Motions* system.

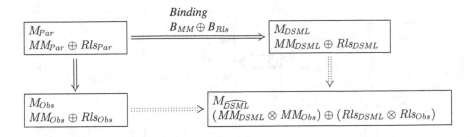

Fig. 5. Amalgamation in the category of DSMLs and DSML morphisms

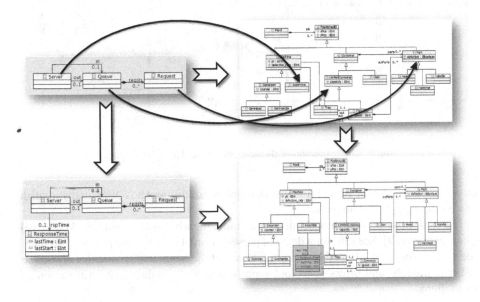

Fig. 6. Weaving of metamodels (highlighting added for illustration purposes)

given this binding: In the left-hand side, there is a Server (Assemble) with an in-Queue (Tray) that holds two Requests (Handle and Head) and an out-Queue (Conveyor). In the right-hand side, there is a Server (Assemble) with an in-Queue (Tray) and an out-Queue (Conveyor) that holds one Request (Hammer). Consequently, we can apply the rule transformation from the rule in Figure 4(c). As we will explain in Section 4, the semantics of this rule transformation is provided by the rule amalgamation illustrated in Figure 7, where we can see how the obtained amalgamated rule is similar to the Assemble rule but with the observers in the RespTime rule appropriately introduced.

Clearly, such a separation of concerns between a specification of the base DSML and specifications of languages for non-functional properties is desirable. We have used the response-time property as an example here. Other properties can be defined easily in a similar vein as shown in [45] and at http://atenea.lcc.uma.es/index. php/Main_Page/Resources/E-motions/PLSObExample. In the following

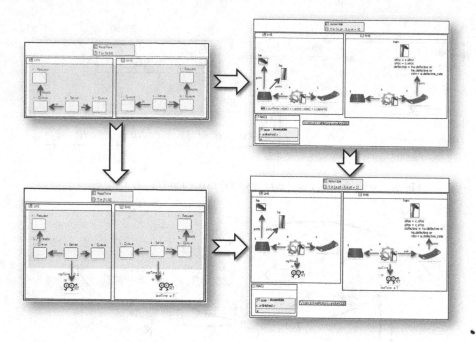

Fig. 7. Amalgamation of the Assemble and RespTime rules

sections, we discuss the formal framework required for this and how we can distinguish safe bindings from unsafe ones.

The *e-Motions* models thus obtained are automatically transformed into Maude [4] specifications [39]. See [36] for a detailed presentation of how Maude provides an accurate way of specifying both the abstract syntax and the behavioral semantics of models and metamodels, and offers good tool support both for simulating and for reasoning about them.

3   Graph Transformation and Adhesive HLR Categories

Graph transformation [40] is a formal, graphical and natural way of expressing graph manipulation based on rewriting rules. In graph-based modelling (and meta-modelling), graphs are used to define the static structures, such as class and object ones, which represent visual alphabets and sentences over them. We formalise our approach using the typed graph transformation approach, specifically the Double Pushout (DPO) algebraic approach, with positive and negative (nested) application conditions [11,23]. We however carry on our formalisation for weak adhesive high-level replacement (HLR) categories [12]. Some of the proofs in this paper assume that the category of graphs at hand is adhesive HLR. Thus, in the rest of the paper, when we talk about graphs or typed graphs, keep in mind that we actually mean some type of graph whose corresponding category is adhesive HLR. Specifically, the category of typed attributed graphs, the one of interest to us, was proved to be adhesive HLR in [16].

3.1 Generic Notions

The concepts of adhesive and (weak) adhesive HLR categories abstract the foundations of a general class of models, and come together with a collection of general semantic techniques [32,14]. Thus, e.g., given proofs for adhesive HLR categories of general results such as the Local Church-Rosser, or the Parallelism and Concurrency Theorem, they are automatically valid for any category which is proved an adhesive HLR category. This framework has been a break-through for the DPO approach of algebraic graph transformation, for which most main results can be proved in these categorical frameworks, and then instantiated to any HLR system.

Definition 1. *(**Van Kampen square**) Pushout (1) is a van Kampen square if, for any commutative cube with (1) in the bottom and where the back faces are pullbacks, we have that the top face is a pushout if and only if the front faces are pullbacks.*

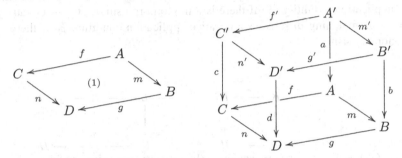

Definition 2. *(**Adhesive HLR category**) A category C with a morphism class M is called adhesive HLR category if*

- *M is a class of monomorphisms closed under isomorphisms and closed under composition and decomposition,*
- *C has pushouts and pullbacks along M-morphisms, i.e., if one of the given morphisms is in M, then also the opposite one is in M, and M-morphisms are closed under pushouts and pullbacks, and*
- *pushouts in C along M-morphisms are van Kampen squares.*

In the DPO approach to graph transformation, a rule p is given by a span $(L \xleftarrow{l} K \xrightarrow{r} R)$ with graphs L, K, and R, called, respectively, left-hand side, interface, and right-hand side, and some kind of monomorphisms (typically, inclusions) l and r. A graph transformation system (GTS) is a pair (P, π) where P is a set of rule names and π is a function mapping each rule name p into a rule $L \xleftarrow{l} K \xrightarrow{r} R$.

An application of a rule $p : L \xleftarrow{l} K \xrightarrow{r} R$ to a graph G via a match $m : L \to G$ is constructed as two gluings (1) and (2), which are pushouts in the corresponding graph category, leading to a direct transformation $G \xRightarrow{p,m} H$.

$$p \quad : \quad L \xleftarrow{l} K \xrightarrow{r} R$$
$$m \downarrow \quad (1) \quad \downarrow \quad (2) \quad \downarrow$$
$$G \longleftarrow D \longrightarrow H$$

We only consider injective matches, that is, monomorphisms. If the matching m is understood, a DPO transformation step $G \xRightarrow{p,m} H$ will be simply written $G \xRightarrow{p} H$. A transformation sequence $\rho = \rho_1 \dots \rho_n : G \Rightarrow^* H$ via rules $p_1, \dots, p_n$ is a sequence of transformation steps $\rho_i = (G_i \xRightarrow{p_i,m_i} H_i)$ such that $G_1 = G$, $H_n = H$, and consecutive steps are composable, that is, $G_{i+1} = H_i$ for all $1 \le i < n$. The category of transformation sequences over an adhesive category $\mathbf{C}$, denoted by $\mathbf{Trf(C)}$, has all graphs in $|\mathbf{C}|$ as objects and all transformation sequences as arrows.

Transformation rules may have application conditions. We consider rules of the form $(L \xleftarrow{l} K \xrightarrow{r} R, ac)$, where $(L \xleftarrow{l} K \xrightarrow{r} R)$ is a normal rule and ac is a *(nested) application condition* on L. Application conditions may be positive or negative (see Figure 8). Positive application conditions have the form $\exists a$, for a monomorphism $a : L \to C$, and demand a certain structure in addition to L. Negative application conditions of the form $\nexists a$ forbid such a structure. A match $m : L \to G$ satisfies a positive application condition $\exists a$ if there is a monomorphism $q : C \to G$ satisfying $q \circ a = m$. A matching m satisfies a negative application condition $\nexists a$ if there is no such monomorphism.

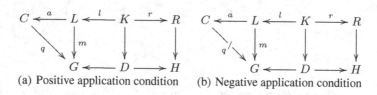

(a) Positive application condition   (b) Negative application condition

Fig. 8. Positive and negative application conditions

Given an application condition $\exists a$ or $\nexists a$, for a monomorphism $a : L \to C$, another application condition ac can be established on C, giving place to nested application conditions [23]. For a basic application condition $\exists(a, ac_C)$ on L with an application condition ac_C on C, in addition to the existence of q it is required that q satisfies ac_C. We write $m \models ac$ if m satisfies ac. $ac_C \cong ac'_C$ denotes the semantical equivalence of ac_C and ac'_C on C.

To improve readability, we assume projection functions ac, *lhs* and *rhs*, returning, respectively, the application condition, the left-hand side and the right-hand side of a rule. Thus, given a rule $r = (L \xleftarrow{l} K \xrightarrow{r} R, ac)$, $ac(r) = ac$, $lhs(r) = L$, and $rhs(r) = R$.

Given an application condition ac_L on L and a monomorphism $t : L \to L'$, then there is an application condition $\mathsf{Shift}(t, ac_L)$ on L' such that for all $m' : L' \to G$, $m' \models \mathsf{Shift}(t, ac_L) \leftrightarrow m = m' \circ t \models ac_L$.

$$ac_L \quad \triangleright \quad L \xrightarrow{\quad t \quad} L' \quad \triangleleft \quad \mathsf{Shift}(t, ac_L)$$

$$m \searrow \qquad \swarrow m'$$

$$G$$

Parisi-Presicce proposed in [35] a notion of rule morphism very similar to the one below, although we consider rules with application conditions, and require the commuting squares to be pullbacks.

Definition 3. *(**Rule morphism**) Given transformation rules* $p_i = (L_i \xleftarrow{l_i} K_i \xrightarrow{r_i} R_i, ac_i)$, *for* $i = 0, 1$, *a rule morphism* $f : p_0 \to p_1$ *is a tuple* $f = (f_L, f_K, f_R)$ *of graph monomorphisms* $f_L : L_0 \to L_1$, $f_K : K_0 \to K_1$, *and* $f_R : R_0 \to R_1$ *such that the squares with the span morphisms* l_0, l_1, r_0, *and* r_1 *are pullbacks, as in the diagram below, and such that* $ac_1 \Rightarrow \mathsf{Shift}(f_L, ac_0)$.

$$
\begin{array}{ccccccc}
p_0 & : & ac_0 & \rhd & L_0 & \xleftarrow{l_0} K_0 \xrightarrow{r_0} & R_0 \\
f \downarrow & & & & f_L \downarrow \quad pb \quad f_K \downarrow \quad pb & & \downarrow f_R \\
p_1 & : & ac_1 & \rhd & L_1 & \xleftarrow[l_1]{} K_1 \xrightarrow[r_1]{} & R_1
\end{array}
$$

The requirement that the commuting squares are pullbacks is quite natural from an intuitive point of view: the intuition of morphisms is that they should preserve the "structure" of objects. If we think of rules not as a span of monomorphisms, but in terms of their intuitive semantics (i.e., L\K is what should be deleted from a given graph, R\K is what should be added to a given graph and K is what should be preserved), then asking that the two squares are pullbacks means, precisely, to preserve that structure. I.e., we preserve what should be deleted, what should be added and what must remain invariant. Of course, pushouts also preserve the created and deleted parts. But they reflect this structure as well, which we do not want in general.

Fact 1. *With componentwise identities and composition, rule morphisms define the category* **Rule**.

Proof Sketch. Follows trivially from the fact that $ac \cong \mathsf{Shift}(id_L, ac)$, pullback composition, and that given morphisms $f' \circ f$ such that

$$
\begin{array}{ccccccc}
p_0 & : & ac_0 & \rhd & L_0 & \xleftarrow{l_0} K_0 \xrightarrow{r_0} & R_0 \\
f \downarrow & & & & f_L \downarrow \quad pb \quad \downarrow \quad pb & & \downarrow \\
p_1 & : & ac_1 & \rhd & L_1 & \xleftarrow[l_1]{} K_1 \xrightarrow[r_1]{} & R_1 \\
f' \downarrow & & & & f'_L \downarrow \quad pb \quad \downarrow \quad pb & & \downarrow \\
p_2 & : & ac_2 & \rhd & L_2 & \xleftarrow[l_2]{} K_2 \xrightarrow[r_2]{} & R_2
\end{array}
$$

then we have $\mathsf{Shift}(f'_L, \mathsf{Shift}(f_L, ac_0)) \cong \mathsf{Shift}(f'_L \circ f_L, ac_0)$. $\square$

A key concept in the constructions in the following section is that of *rule amalgamation* [1,12]. The amalgamation of two rules p_1 and p_2 glues them together into a single rule $\tilde{p}$ to obtain the effect of the original rules. I.e., the simultaneous application of p_1 and p_2 yields the same successor graph as the application of the amalgamated rule $\tilde{p}$.

The possible overlapping of rules p_1 and p_2 is captured by a rule p_0 and rule morphisms $f : p_0 \to p_1$ and $g : p_0 \to p_2$.

Definition 4. *(Rule   amalgamation)   Given   transformation   rules p_i :*
$(L_i \overset{l_i}{\leftarrow} K_i \overset{r_i}{\to} R_i, ac_i)$, *for $i = 0, 1, 2$, and rule morphisms $f : p_0 \to p_1$ and $g : p_0 \to p_2$, the amalgamated production $p_1 +_{p_0} p_2$ is the production $(L \overset{l}{\leftarrow} K \overset{r}{\to} R, ac)$ in the diagram below, where subdiagrams (1), (2) and (3) are pushouts, l and r are induced by the universal property of (2) so that all subdiagrams commute, and $ac = \mathsf{Shift}(\widehat{f}_L, ac_2) \wedge \mathsf{Shift}(\widehat{g}_L, ac_1)$.*

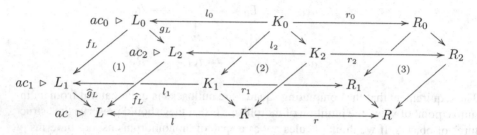

Notice that in the above diagram all squares are either pushouts or pullbacks (by the van Kampen property) which means that all their arrows are monomorphisms (by being an adhesive HLR category).

We end this section by introducing the notion of rule identity.

Definition 5. *(Rule-identity morphism) Given graph transformation rules*
$p_i = (L_i \overset{l_i}{\leftarrow} K_i \overset{r_i}{\to} R_i, ac_i)$, *for $i = 0, 1$, and a rule morphism $f : p_0 \to p_1$, with $f = (f_L, f_K, f_R)$, p_0 and p_1 are said to be identical, denoted $p_0 \equiv p_1$, if f_L, f_K, and f_R are identity morphisms and $ac_0 \cong ac_1$.*

3.2   Typed Graph Transformation Systems

A (directed unlabeled) *graph* $G = (V, E, s, t)$ is given by a set of nodes (or vertices) V, a set of edges E, and source and target functions $s, t : E \to V$. Given graphs $G_i = (V_i, E_i, s_i, t_i)$, with $i = 1, 2$, a graph homomorphism $f : G_1 \to G_2$ is a pair of functions $(f_V : V_1 \to V_2, f_E : E_1 \to E_2)$ such that $f_V \circ s_1 = s_2 \circ f_E$ and $f_V \circ t_1 = t_2 \circ f_E$. With componentwise identities and composition this defines the category **Graph**.

Given a distinguished graph TG, called *type graph*, a TG-*typed graph* (G, g_G), or simply *typed graph* if TG is known, consists of a graph G and a typing homomorphism $g_G : G \to TG$ associating with each vertex and edge of G its type in TG. However, to enhance readability, we will use simply g_G to denote a typed graph (G, g_G), and when the typing morphism g_G can be considered implicit, we will often refer to it just as G. A TG-typed graph morphism between TG-typed graphs $(G_i, g_i : G_i \to TG)$, with $i = 1, 2$, denoted $f : (G_1, g_1) \to (G_2, g_2)$ (or simply $f : g_1 \to g_2$), is a graph morphism $f : G_1 \to G_2$ which preserves types, i.e., $g_2 \circ f = g_1$. **Graph**$_{TG}$ is the category of

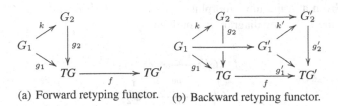

(a) Forward retyping functor.   (b) Backward retyping functor.

Fig. 9. Forward and backward retyping functors

TG-typed graphs and TG-typed graph morphisms, which is the comma category **Graph** over TG.

If the underlying graph category is adhesive (resp., adhesive HLR, weakly adhesive) then so are the associated typed categories [12], and therefore all definitions in Section 3.1 apply to them. A TG-typed graph transformation rule is a span $p = L \xleftarrow{l} K \xrightarrow{r} R$ of injective TG-typed graph morphisms and a (nested) application condition on L. Given TG-typed graph transformation rules $p_i = (L_i \xleftarrow{l_i} K_i \xrightarrow{r_i} R_i, ac_i)$, with $i = 1, 2$, a typed rule morphism $f : p_1 \to p_2$ is a tuple (f_L, f_K, f_R) of TG-typed graph monomorphisms such that the squares with the span monomorphisms l_i and r_i, for $i = 1, 2$, are pullbacks, and such that $ac_2 \Rightarrow \mathsf{Shift}(f_L, ac_1)$. TG-typed graph transformation rules and typed rule morphisms define the category **Rule**$_{TG}$, which is the comma category **Rule** over TG.

Following [5,22], we use forward and backward retyping functors to deal with graphs over different type graphs. A graph morphism $f : TG \to TG'$ induces a forward retyping functor $f^> : \mathbf{Graph}_{TG} \to \mathbf{Graph}_{TG'}$, with $f^>(g_1) = f \circ g_1$ and $f^>(k : g_1 \to g_2) = k$ by composition, as shown in the diagram in Figure 9(a). Similarly, such a morphism f induces a backward retyping functor $f^< : \mathbf{Graph}_{TG'} \to \mathbf{Graph}_{TG}$, with $f^<(g_1') = g_1$ and $f^<(k' : g_1' \to g_2') = k : g_1 \to g_2$ by pullbacks and mediating morphisms as shown in the diagram in Figure 9(b). Retyping functors also extends to application conditions and rules, so we will write things like $f^>(ac)$ or $f^<(p)$ for some application condition ac and production p. Notice, for example, that given a graph morphism $f : TG \to TG'$, the forward retyping of a production $p = (L \xleftarrow{l} K \xrightarrow{r} R, ac)$ over TG is a production $f_{TG}^>(p) = (f_{TG}^>(L) \xleftarrow{f_{TG}^>(l)} f_{TG}^>(K) \xrightarrow{f_{TG}^>(r)} f_{TG}^>(R), f_{TG}^>(ac))$ over TG', defining an induced morphism $f^p : p \to f_{TG}^>(p)$ in **Rule**. Since f^p is a morphism between rules in $|\mathbf{Rule}_{TG}|$ and $|\mathbf{Rule}_{TG'}|$, it is defined in **Rule**, forgetting the typing. Notice also that $f_{TG}^>(ac) \cong \mathsf{Shift}(f_L^p, ac)$.

As said above, to improve readability, if $G \to TG$ is a TG-typed graph, we sometimes refer to it just by its typed graph G, leaving TG implicit. As a consequence, if $f : TG \to TG'$ is a morphism, we may refer to the TG'-typed graph $f^>(G)$, even if this may be considered an abuse of notation.

The following results will be used in the proofs in the following section.

Proposition 1. *(From [22]) (Adjunction) Forward and backward retyping functors are left and right adjoints; i.e., for each $f : TG \to TG'$ we have $f^> \dashv f^< : TG \to TG'$.*

Remark 1. Given a graph monomorphism $f : TG \to TG'$, for all $k' : G'_1 \to G'_2$ in **Graph$_{\mathbf{TG'}}$**, the following diagram is a pullback:

$$
\begin{array}{ccc}
f^<(G'_1) & \xrightarrow{f^<(k')} & f^<(G'_2) \\
\downarrow & pb & \downarrow \\
G'_1 & \xrightarrow{k'} & G'_2
\end{array}
$$

This is true just by pullback decomposition.

Remark 2. Given a graph monomorphism $f : TG \to TG'$, and given monomorphisms $k' : G'_0 \to G'_1$ and $h' : G'_0 \to G'_2$ in **Graph$_{\mathbf{TG'}}$**, if the following diagram on the left is a pushout then the diagram on the right is also a pushout:

$$
\begin{array}{ccc}
G'_0 & \xrightarrow{k'} & G'_1 \\
h' \downarrow & po & \downarrow \widehat{h'} \\
G'_2 & \xrightarrow{\widehat{k'}} & \widehat{G}
\end{array}
\qquad
\begin{array}{ccc}
f^<(G'_0) & \xrightarrow{f^<(k')} & f^<(G'_1) \\
f^<(h') \downarrow & po & \downarrow f^<(\widehat{h'}) \\
f^<(G'_2) & \xrightarrow{f^<(\widehat{k'})} & f^<(\widehat{G})
\end{array}
$$

Notice that since in an adhesive HLR category all pushouts along M-morphisms are van Kampen squares, the commutative square created by the pullbacks and induced morphisms by the backward retyping functor imply the second pushout.

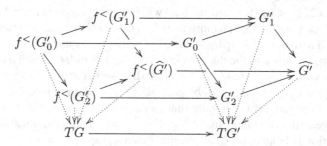

Remark 3. Given a graph monomorphism $f : TG \to TG'$, and given monomorphisms $k : G_0 \to G_1$ and $h : G_0 \to G_2$ in **Graph$_{\mathbf{TG}}$**, if the diagram on the left is a pushout (resp., a pullback) then the diagram on the right is also a pushout (resp., a pullback):

$$
\begin{array}{ccc}
G_0 & \xrightarrow{k} & G_1 \\
h \downarrow & & \downarrow \widehat{h} \\
G_2 & \xrightarrow{\widehat{k}} & \widehat{G}
\end{array}
\qquad
\begin{array}{ccc}
f^>(G_0) & \xrightarrow{f^>(k)} & f^>(G_1) \\
f^>(h) \downarrow & & \downarrow f^>(\widehat{h}) \\
f^>(G_2) & \xrightarrow{f^>(\widehat{k})} & f^>(\widehat{G})
\end{array}
$$

Remark 4. Given a graph monomorphism $f : TG \to TG'$, and a TG'-typed graph transformation rule $p = (L \xleftarrow{l} K \xrightarrow{r} R, ac)$, if a matching $m : L \to C$ satisfies ac, that is, $m \models ac$, then, $f^<(m) \models f^<(ac)$.

4 GTS Morphisms and Preservation of Behaviour

A typed graph transformation system over a type graph TG, is a graph transformation system where the given graph transformation rules are defined over the category of TG-typed graphs. Since in this paper we deal with GTSs over different type graphs, we will make explicit the given type graph. This means that, from now on, a typed GTS is a triple (TG, P, π) where TG is a type graph, P is a set of rule names and π is a function mapping each rule name p into a rule $(L \xleftarrow{l} K \xrightarrow{r} R, ac)$ typed over TG.

The set of transformation rules of each GTS specifies a behaviour in terms of the derivations obtained via such rules. A GTS morphism defines then a relation between its source and target GTSs by providing an association between their type graphs and rules.

Definition 6. *(GTS morphism) Given typed graph transformation systems $GTS_i = (TG_i, P_i, \pi_i)$, for $i = 0, 1$, a GTS morphism $f : GTS_0 \to GTS_1$, with $f = (f_{TG}, f_P, f_r)$, is given by a morphism $f_{TG} : TG_0 \to TG_1$, a surjective mapping $f_P : P_1 \to P_0$ between the sets of rule names, and a family of rule morphisms $f_r = \{f^p : f_{TG}^>(\pi_0(f_P(p))) \to \pi_1(p)\}_{p \in P_1}$.*

Given a GTS morphism $f : GTS_0 \to GTS_1$, each rule in GTS_1 extends a rule in GTS_0. However if there are internal computation rules in GTS_1 that do not extend any rule in GTS_0, we can always consider that the empty rule is included in GTS_0, and assume that those rules extend the empty rule.

Please note that rule morphisms are defined on rules over the same type graph (see Definition 3). To deal with rules over different type graphs we retype one of the rules to make them be defined over the same type graph.

Typed GTSs and GTS morphisms define the category **GTS**. The GTS amalgamation construction provides a very convenient way of composing GTSs.

Definition 7. *(GTS Amalgamation). Given transformation systems $GTS_i = (TG_i, P_i, \pi_i)$, for $i = 0, 1, 2$, and GTS morphisms $f : GTS_0 \to GTS_1$ and $g : GTS_0 \to GTS_2$, the amalgamated GTS $\widehat{GTS} = GTS_1 +_{GTS_0} GTS_2$ is the GTS $(\widehat{TG}, \widehat{P}, \widehat{\pi})$ constructed as follows. We first construct the pushout of typing graph morphisms $f_{TG} : TG_0 \to TG_1$ and $g_{TG} : TG_0 \to TG_2$, obtaining morphisms $\widehat{f}_{TG} : TG_2 \to \widehat{TG}$ and $\widehat{g}_{TG} : TG_1 \to \widehat{TG}$. The pullback of set morphisms $f_P : P_1 \to P_0$ and $g_P : P_2 \to P_0$ defines morphisms $\widehat{f}_P : \widehat{P} \to P_2$ and $\widehat{g}_P : \widehat{P} \to P_1$. Then, for each rule p in $\widehat{P}$, the rule $\widehat{\pi}(p)$ is defined as the amalgamation of rules $\widehat{f}_{TG}^>(\pi_2(\widehat{f}_P(p)))$ and $\widehat{g}_{TG}^>(\pi_1(\widehat{g}_P(p)))$ with respect to the kernel rule $\widehat{f}_{TG}^>(g_{TG}^>(\pi_0(g_P(\widehat{f}_P(p)))))$.*

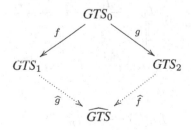

Among the different types of GTS morphisms, let us now focus on those that reflect behaviour. Given a GTS morphism $f : GTS_0 \to GTS_1$, we say that it reflects behaviour if for any derivation that may happen in GTS_1 there exists a corresponding derivation in GTS_0.

Definition 8. *(Behaviour-reflecting GTS morphism) Given graph transformation systems* $GTS_i = (TG_i, P_i, \pi_i)$, *for* $i = 0, 1$, *a GTS morphism* $f : GTS_0 \to GTS_1$ *is behaviour-reflecting if for all graphs* G, H *in* $|\mathbf{Graph}_{TG_1}|$, *all rules* p *in* P_1, *and all matches* $m : lhs(\pi_1(p)) \to G$ *such that* $G \overset{p,m}{\Longrightarrow} H$, *then* $f_{TG}^<(G) \xrightarrow{f_P(p), f_{TG}^<(m)} f_{TG}^<(H)$ *in* GTS_0.

Morphisms between GTSs that only add to the transformation rules elements not in their source type graph are behaviour-reflecting. We call them *extension morphisms*.

Definition 9. *(Extension GTS morphism) Given graph transformation systems* $GTS_i = (TG_i, P_i, \pi_i)$, *for* $i = 0, 1$, *a GTS morphism* $f : GTS_0 \to GTS_1$, *with* $f = (f_{TG}, f_P, f_r)$, *is an extension morphism if* f_{TG} *is a monomorphism and for each* $p \in P_1$, $\pi_0(f_P(p)) \equiv f_{TG}^<(\pi_1(p))$.

That an extension morphism is indeed a behaviour-reflecting morphism is shown by the following lemma.

Lemma 1. *All extension GTS morphisms are behaviour-reflecting.*

Proof Sketch. Given graph transformation systems $GTS_i = (TG_i, P_i, \pi_i)$, for $i = 0, 1$, let a GTS morphism $f : GTS_0 \to GTS_1$ be an extension morphism. Then, we have to prove that for all graphs G, H in $|\mathbf{Graph}_{TG_1}|$, all rules p in P_1, and all matches $m : lhs(\pi_1(p)) \to G$, if $G \overset{p,m}{\Longrightarrow} H$ then $f_{TG}^<(G) \xrightarrow{f_P(p), f_{TG}^<(m)} f_{TG}^<(H)$.

Assuming transformation rules $\pi_1(p) = (L_1 \xleftarrow{l_1} K_1 \xrightarrow{r_1} R_1, ac_1)$ and $\pi_0(f_P(p)) = (L_0 \xleftarrow{l_0} K_0 \xrightarrow{r_0} R_0, ac_0)$, and given the derivation

$$ac_1 \quad \triangleright \quad L_1 \xleftarrow{l_1} K_1 \xrightarrow{r_1} R_1$$
$$ m\downarrow \quad po \quad \downarrow \quad po \quad \downarrow$$
$$ G \longleftarrow D \longrightarrow H$$

since f is an extension morphism, and therefore f_{TG} is a monomorphism, and l_1 and m are also monomorphisms, by Remark 2 and Definition 8, we have the diagram

$$ac_0 \quad \triangleright \quad L_0 \xleftarrow{l_1} K_0 \xrightarrow{r_1} R_0$$
$$\cong \qquad \qquad \| \qquad \qquad \| \qquad \qquad \|$$
$$f_{TG}^<(ac_1) \quad \triangleright \quad f_{TG}^<(L_1) \xleftarrow{f_{TG}^<(l_1)} f_{TG}^<(K_1) \xrightarrow{f_{TG}^<(r_1)} f_{TG}^<(R_1)$$
$$\phantom{f_{TG}^<(ac_1) \quad \triangleright \quad} f_{TG}^<(m)\downarrow \quad po \quad \downarrow \quad po \quad \downarrow$$
$$\phantom{f_{TG}^<(ac_1) \quad \triangleright \quad} f_{TG}^<(G) \longleftarrow f_{TG}^<(D) \longrightarrow f_{TG}^<(H)$$

Then, given the pushouts in the above diagram and Remark 4, we have the derivation
$$f_{TG}^{<}(G) \xrightarrow{f_P(p), f_{TG}^{<}(m)} f_{TG}^{<}(H).$$
□

Notice that Definition 9 provides specific checks on individual rules. In the concrete case we presented in Section 2, the inclusion morphism between the model of an observer DSML, M_{Obs}, and its parameter sub-model M_{Par}, may be very easily checked to be an extension, by making sure that the features "added" in the rules will be removed by the backward retyping functor. In this case the check is particularly simple because of the subgraph relation between the type graphs, but for a morphism as the binding morphism between M_{Par} and the DSML of the system at hand, M_{DSML}, the check would also be relatively simple. Basically, the backward retyping of each rule in M_{DSML}, i.e., the rule resulting from removing all elements not target of the binding map, must coincide with the corresponding rule, and the application conditions must be equivalent.

Since the amalgamation of GTSs is the basic construction for combining them, it is very important to know whether the reflection of behaviour remains invariant under amalgamations.

Proposition 2. *Given transformation systems $GTS_i = (TG_i, P_i, \pi_i)$, for $i = 0, 1, 2$, and the amalgamation $\widehat{GTS} = GTS_1 +_{GTS_0} GTS_2$ of GTS morphisms $f : GTS_0 \to GTS_1$ and $g : GTS_0 \to GTS_2$, if f_{TG} is a monomorphism and g is an extension morphism, then $\widehat{g}$ is also an extension morphism.*

$$
\begin{array}{ccc}
GTS_0 & \xrightarrow{\quad f \quad} & GTS_1 \\
{\scriptstyle g}\big\downarrow & & \big\downarrow{\scriptstyle \widehat{g}} \\
GTS_2 & \dashrightarrow[\widehat{f}]{} & \widehat{GTS}
\end{array}
$$

Proof Sketch. Let it be $\widehat{GTS} = (\widehat{TG}, \widehat{P}, \widehat{\pi})$. We have to prove that for each $p \in \widehat{P}$, $\pi_1(\widehat{g}_P(p)) \equiv \widehat{g}_{TG}^{<}(\widehat{\pi}(p))$. By construction, rule $\widehat{\pi}(p)$ is obtained from the amalgamation of rules $\widehat{g}_{TG}^{>}(\pi_1(\widehat{g}_P(\widehat{p})))$ and $\widehat{f}_{TG}^{>}(\pi_2(\widehat{f}_P(\widehat{p})))$. More specifically, without considering application conditions by now, the amalgamation of such rules is accomplished by constructing the pushouts of the morphisms for the left-hand sides, for the kernel graphs, and for the right-hand sides.

By Remark 2, we know that if the diagram

$$
\begin{array}{ccc}
\widehat{f}_{TG}^{>}(\widehat{g}_{TG}^{>}(L_0)) \cong \widehat{g}_{TG}^{>}(\widehat{f}_{TG}^{>}(L_0)) & \xrightarrow{\widehat{g}_{TG}^{>}(f_L^{\widehat{g}_P(p)})} & \widehat{g}_{TG}^{>}(L_1) \\
{\scriptstyle \widehat{f}_{TG}^{>}(g_L^{\widehat{f}_P(p)})}\big\downarrow & & \big\downarrow{\scriptstyle \widehat{g}_L^p} \\
\widehat{f}_{TG}^{>}(L_2) & \xrightarrow{\quad \widehat{f}_L^p \quad} & \widehat{L}
\end{array}
$$

is a pushout, then if we apply the backward retyping functor $\widehat{g}^<_{TG}$ to all its components (graphs) and morphisms, the resulting diagram is also a pushout.

$$
\begin{array}{ccc}
\widehat{g}^<_{TG}(\widehat{g}^>_{TG}(f^>_{TG}(L_0))) & \xrightarrow{\widehat{g}^<_{TG}(\widehat{g}^>_{TG}(f^{\widehat{g}_P(p)}_L))} & \widehat{g}^<_{TG}(\widehat{g}^>_{TG}(L_1)) \\
{\scriptstyle \widehat{g}^<_{TG}(\widehat{f}^>_{TG}(g^{\widehat{f}_P(p)}_L))} \Big\downarrow & & \Big\downarrow {\scriptstyle \widehat{g}^<_{TG}(\widehat{g}^p_L)} \\
\widehat{g}^<_{TG}(\widehat{f}^>_{TG}(L_2)) & \xrightarrow[\widehat{g}^<_{TG}(\widehat{f}^p_L)]{} & \widehat{g}^<_{TG}(\widehat{L})
\end{array}
$$

Because, by Proposition 1, for every $f : TG \to TG'$ and every TG-type graph G and morphism g, since f_{TG} is assumed to be a monomorphism, $f^<(f^>(G)) = G$ and $f^<(f^>(g)) = g$, we have $\widehat{g}^<_{TG}(\widehat{g}^>_{TG}(f^>_{TG}(L_0))) = f^>_{TG}(L_0)$, $\widehat{g}^<_{TG}(\widehat{g}^>_{TG}(f^{\widehat{g}_P(p)}_L)) = f^{\widehat{g}_P(p)}_L$, and $\widehat{g}^<_{TG}(\widehat{g}^>_{TG}(L_1)) = L_1$. By pullback decomposition in the corresponding retyping diagram, $\widehat{g}^<_{TG}(\widehat{f}^>_{TG}(L_2)) = f^>_{TG}(g^<_{TG}(L_2))$.

Thus, we are left with this other pushout:

$$
\begin{array}{ccc}
f^>_{TG}(L_0) & \xrightarrow{f^{\widehat{g}_P(p)}_L} & L_1 \\
{\scriptstyle f^>_{TG}(g^<_{TG}(g^{\widehat{f}_P(p)}_L))} \Big\downarrow & & \Big\downarrow {\scriptstyle \widehat{g}^<_{TG}(\widehat{g}^p_L)} \\
f^>_{TG}(g^<_{TG}(L_2)) & \xrightarrow[\widehat{g}^<_{TG}(\widehat{f}^p_L)]{} & \widehat{g}^<_{TG}(\widehat{L})
\end{array}
$$

Since g is an extension, $L_0 \cong g^<_{TG}(L_2)$, which, because f_{TG} is a monomorphism, implies $f^>_{TG}(L_0) \cong f^>_{TG}(g^<_{TG}(L_2))$. This implies that $\widehat{g}^<_{TG}(\widehat{L}) \cong L_1$.

Similar diagrams for kernel objects and right-hand sides lead to similar identity morphisms for them. It only remains to see that $ac(\pi_1(\widehat{g}_P(p))) \cong ac(\widehat{g}^<_{TG}(\widehat{\pi}(p)))$.

By the rule amalgamation construction, $\widehat{ac} = \widehat{f}^>_{TG}(ac_2) \wedge \widehat{g}^>_{TG}(ac_1)$. Since g is an extension morphism, $ac_2 \cong g^>_{TG}(ac_0)$. Then, $\widehat{ac} \cong \widehat{f}^>_{TG}(\widehat{g}^>_{TG}(ac_0)) \wedge \widehat{g}^>_{TG}(ac_1)$. For f, as for any other rule morphism, we have $ac_1 \Rightarrow f^>_{TG}(ac_0)$. By the Shift construction, for any match $m_1 : L_1 \to C_1$, $m_1 \models ac_1$ iff $\widehat{g}^>_{TG}(m_1) \models \widehat{g}^>_{TG}(ac_1)$ and, similarly, for any match $m_0 : L_0 \to C_0$, $m_0 \models ac_0$ iff $f^>_{TG}(m_0) \models f^>_{TG}(ac_0)$. Then, $ac_1 \Rightarrow f^>_{TG}(ac_0) \cong \widehat{g}^>_{TG}(ac_1) \Rightarrow \widehat{g}^>_{TG}(f^>_{TG}(ac_0)) \cong \widehat{g}^>_{TG}(ac_1) \Rightarrow \widehat{f}^>_{TG}(\widehat{g}^>_{TG}(ac_0))$. And therefore, since $\widehat{ac} = \widehat{f}^>(\widehat{g}^>_{TG}(ac_0)) \wedge \widehat{g}^>_{TG}(ac_1)$ and $\widehat{g}^>_{TG}(ac_1) \Rightarrow \widehat{f}^>_{TG}(\widehat{g}^>_{TG}(ac_0))$, we conclude $\widehat{ac} \cong \widehat{g}^>_{TG}(ac_1)$. □

When a DSL is extended with observers and other alien elements whose goal is to measure some property, or to verify certain invariant property, we need to guarantee that such an extension does not change the semantics of the original DSL. Specifically, we need to guarantee that the behaviour of the resulting system is exactly the same, that is, that any derivation in the source system also happens in the target one (behaviour preservation), and any derivation in the target system was also possible in the source one (behaviour reflection). The following definition of behaviour-protecting GTS morphism captures the intuition of a morphism that both reflects and preserves behaviour, that is, that establishes a bidirectional correspondence between derivations in the source and target GTSs.

Definition 10. *(Behaviour-protecting GTS morphism) Given typed graph transformation systems $GTS_i = (TG_i, P_i, \pi_i)$, for $i = 0, 1$, a GTS morphism $f : GTS_0 \to GTS_1$ is behaviour-protecting if for all graphs G and H in $|\mathbf{Graph}_{TG_1}|$, all rules p in P_1, and all matches $m : lhs(\pi_1(p)) \to G$,*

$$g_{TG}^{<}(G) \xrightarrow{g_P(p), g_{TG}^{<}(m)} g_{TG}^{<}(H) \iff G \xRightarrow{p,m} H$$

We find in the literature definitions of behaviour-preserving morphisms as morphisms in which the rules in the source GTS are included in the set of rules of the target GTS. Although these morphisms trivially preserve behaviour, they are not useful for our purposes. Works like [25] or [22], mainly dealing with refinements of GTSs, only consider cases in which GTSs are extended by adding new transformation rules. In our case, in addition to adding new rules, we are enriching the rules themselves.

The main result in this paper is related to the protection of behaviour, and more precisely on the behaviour-related guarantees on the induced morphisms.

Theorem 1. *Given typed transformation systems $GTS_i = (TG_i, P_i, \pi_i)$, for $i = 0, 1, 2$, and the amalgamation $\widehat{GTS} = GTS_1 +_{GTS_0} GTS_2$ of GTS morphisms $f : GTS_0 \to GTS_1$ and $g : GTS_0 \to GTS_2$, if f is a behaviour-reflecting GTS morphism, f_{TG} is a monomorphism, and g is an extension and behaviour-protecting morphism, then $\widehat{g}$ is behaviour-protecting as well.*

$$
\begin{array}{ccc}
GTS_0 & \xrightarrow{\ f\ } & GTS_1 \\[4pt]
\downarrow{\scriptstyle g} & & \dashdownarrow{\scriptstyle \widehat{g}} \\[4pt]
GTS_2 & \dashrightarrow[\ \widehat{f}\ ]{} & \widehat{GTS}
\end{array}
$$

Proof Sketch. Since g is an extension morphism and f_{TG} is a monomorphism, by Proposition 2, $\widehat{g}$ is also an extension morphism, and therefore, by Lemma 1, also behaviour-reflecting. We are then left with the proof of behaviour preservation.

Given a derivation $G_1 \xRightarrow{p_1, m_1} H_1$ in GTS_1, with $\pi_1(p_1) = (L_1 \xleftarrow{l_1} K_1 \xrightarrow{r_1} R_1, ac_1)$, since $f : GTS_0 \to GTS_1$ is a behaviour-reflecting morphism, there is a corresponding derivation in GTS_0. Specifically, the rule $f_P(p_1)$ can be applied on $f_{TG}^{<}(G_1)$ with match $f_{TG}^{<}(m_1)$ satisfying the application condition of production $\pi_0(f_P(p_1))$, and resulting in a graph $f_{TG}^{<}(H_1)$.

$$f_{TG}^{<}(G_1) \xrightarrow{f_P(p_1), f_{TG}^{<}(m_1)} f_{TG}^{<}(H_1)$$

Moreover, since g is a behaviour-protecting morphism, this derivation implies a corresponding derivation in GTS_2.

By the amalgamation construction in Definition 7, the set of rules of $\widehat{GTS}$ includes, for each p in $\widehat{P}$, the amalgamation of (the forward retyping of) the rules $\pi_1(\widehat{g}_P(p)) = (L_1 \xleftarrow{l_1} K_1 \xrightarrow{r_1} R_1, ac_1)$ and $\pi_2(\widehat{f}_P(p)) = (L_2 \xleftarrow{l_2} K_2 \xrightarrow{r_2} R_2, ac_2)$, with kernel rule $\pi_0(f_P(\widehat{g}_P(p))) = \pi_0(g_P(\widehat{f}_P(p))) = (L_0 \xleftarrow{l_0} K_0 \xrightarrow{r_0} R_0, ac_0)$.

First, notice that for any $\widehat{TG}$ graph G, G is the pushout of the graphs $\widehat{g}_{TG}^{<}(G)$, $\widehat{f}_{TG}^{<}(G)$ and $f_{TG}^{<}(\widehat{g}_{TG}^{<}(G))$ (with the obvious morphisms). This can be proved using a van Kampen square, where in the bottom we have the pushout of the type graphs, the vertical faces are the pullbacks defining the backward retyping functors and on top we have that pushout.

Thus, for each graph G in $\widehat{GTS}$, if a transformation rule in GTS_1 can be applied on $\widehat{g_{TG}^<}(G)$, the corresponding transformation rule should be applicable on G in $\widehat{GTS}$. The following diagram focus on the lefthand sides of the involved rules.

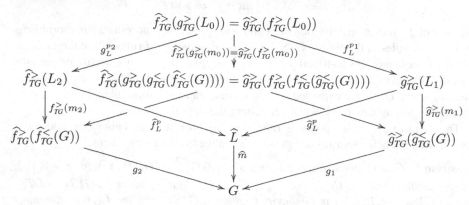

As we have seen above, rules $\widehat{g}_P(p)$, $\widehat{f}_P(p)$, and $\widehat{f}_P(g_P(p)) = \widehat{g}_P(f_P(p))$ are applicable on their respective graphs using the matchings depicted in the above diagram. Since, by the amalgamation construction, the top square is a pushout, and $g_1 \circ \widehat{g_{TG}^>}(m_1) \circ f_L^{p_1} = g_2 \circ \widehat{f_{TG}^>}(m_2) \circ g_L^{p_2}$, then there is a unique morphism $\widehat{m} : \widehat{L} \to G$ making $g_1 \circ \widehat{g_{TG}^>}(m_1) = \widehat{m} \circ \widehat{g}_L^p$ and $g_2 \circ \widehat{f_{TG}^>}(m_2) = \widehat{m} \circ \widehat{f}_L^p$. This $\widehat{m}$ will be used as matching morphism in the derivation we seek.

By construction, the application condition $\widehat{ac}$ of the amalgamated rule p is the conjunction of the shiftings of the application conditions of $\widehat{g}_P(p)$ and $\widehat{f}_P(p)$. Then, since

$$m_1 \models ac_1 \iff \widehat{m} \models \mathsf{Shift}(\widehat{g}_L^p, ac_1)$$

and

$$m_2 \models ac_2 \iff \widehat{m} \models \mathsf{Shift}(\widehat{f}_L^p, ac_2),$$

and therefore

$$m_1 \models ac_1 \wedge m_2 \models ac_2 \iff \widehat{m} \models \widehat{ac}.$$

We can then conclude that rule p is applicable on graph G with match $\widehat{m}$ satisfying its application condition $\widehat{ac}$. Indeed, given the rule $\pi(p) = (\widehat{L} \xleftarrow{\widehat{l}} \widehat{K} \xrightarrow{\widehat{r}} \widehat{R}, \widehat{ac})$ we have the following derivation:

$$\widehat{ac} \quad \triangleright \quad \widehat{L} \xleftarrow{l_1} \widehat{K} \xrightarrow{r_1} \widehat{R}$$
$$\widehat{m} \downarrow \quad po \quad \downarrow \quad po \quad \downarrow$$
$$G \xleftarrow{\quad} D \xrightarrow{\quad} H$$

Let us finally check then that D and H are as expected. To improve readability, in the following diagrams we eliminate the retyping functors. For instance, for the rest of the theorem L_0 denotes $\widehat{f_{TG}^>}(\widehat{g_{TG}^>}(L_0)) = \widehat{g_{TG}^>}(\widehat{f_{TG}^>}(L_0))$, L_1 denotes $\widehat{g_{TG}^>}(L_1)$, etc.

First, let us focus on the pushout complement of $\widehat{l} : \widehat{K} \to \widehat{L}$ and $\widehat{m} : \widehat{L} \to G$. Given rules $\widehat{g}_P(p)$, $\widehat{f}_P(p)$, and $\widehat{f}_P(g_P(p)) = \widehat{g}_P(f_P(p))$ and rule morphisms between them as above, the following diagram shows both the construction by amalgamation of the morphism $\widehat{l} : \widehat{K} \to \widehat{L}$, and the construction of the pushout complements for morphisms l_i and m_i, for $i = 0 \ldots 2$.

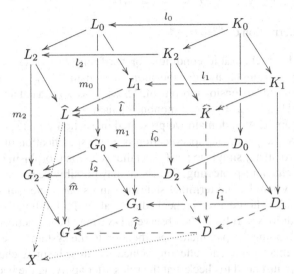

By the pushout of $D_0 \to D_1$ and $D_0 \to D_2$, and given the commuting subdiagram

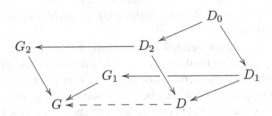

there exists a unique morphism $D \to G$ making the diagram commute. This D is indeed the object of the pushout complement we were looking for. By the pushout of $K_0 \to K_1$ and $K_0 \to K_2$, there is a unique morphism from $\widehat{K}$ to D making the diagram commute. We claim that these morphisms $\widehat{K} \to D$ and $D \to G$ are the pushout complement of $\widehat{K} \to \widehat{L}$ and $\widehat{L} \to G$. Suppose that the pushout of $\widehat{K} \to \widehat{L}$ and $\widehat{K} \to D$ were $\widehat{L} \to X$ and $D \to X$ for some graph X different from G. By the pushout of $K_1 \to D_1$ and $K_1 \to L_1$ there is a unique morphism $G_1 \to X$ making the diagram commute. By the pushout of $K_2 \to D_2$ and $K_2 \to L_2$ there is a unique morphism $G_2 \to X$ making the diagram commute. By the pushout of $G_0 \to G_1$ and $G_0 \to G_2$, there is a unique morphism $G \to X$. But since $\widehat{L} \to X$ and $D \to X$ are the pushout of $\widehat{K} \to \widehat{L}$ and $\widehat{K} \to D$, there is a unique morphism $X \to G$ making the diagram commute. Therefore, we can conclude that X and G are isomorphic.

By a similar construction for the righthand sides we get the pushout

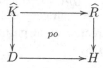

and therefore the derivation $\widehat{G} \overset{\widehat{p},\widehat{m}}{\Longrightarrow} \widehat{H}$.   □

Theorem 1 provides a checkable condition for verifying the conservative nature of an extension in our example, namely the monomorphism $M_{Par} \rightarrow M_{Obs}$ being a behaviour-protecting and extension morphism, $M_{Par} \rightarrow M_{DSML}$ a behaviour-reflecting morphism, and $MM_{Par} \rightarrow MM_{DSML}$ a monomorphism.

In the concrete application domain we presented in Section 2 this result is very important. Notice that the parameter specification is a sub-specification of the observers DSL, making it particularly simple to verify that the inclusion morphism is an extension and also that it is behaviour-protecting. The check may possibly be reduced to checking that the extended system has no terminal states not in its parameter sub-specification. Application conditions should also be checked equivalent. Forbidding the specification of application conditions in rules in the observers DSL may be a practical shortcut.

The morphism binding the parameter specification to the system to be analysed can very easily be verified behaviour-reflecting. Once the morphism is checked to be a monomorphism, we just need to check that the rules after applying the backward retyping morphism exactly coincide with the rules in the source GTS. Checking the equivalence of the application conditions may require human intervention. Notice that with appropriate tools and restrictions, most of these restrictions, if not all, can be automatically verified. We may even be able to restrict editing capabilities so that only correct bindings can be specified.

Once the observers DSL are defined and checked, they can be used as many times as wished. Once they are to be used, we just need to provide the morphism binding the parameter DSL and the target system. As depicted in Figures 6 for the metamodels the binding is just a set of pairs, which may be easily supported by appropriate graphical tools. The binding must be completed by similar correspondences for each of the rules. Notice that once the binding is defined for the metamodels, most of the rule bindings can be inferred automatically.

Finally, given the appropriate morphisms, the specifications may be merged in accordance to the amalgamation construction in Definition 7. The resulting system is guaranteed to both reflect and preserve the original behaviour by Theorem 1.

5   Conclusions and Future Work

In this paper, we have presented formal notions of morphisms between graph transformation systems (GTSs) and a construction of amalgamations in the category of GTSs and GTS morphisms. We have shown that, given certain conditions on the morphisms involved, such amalgamations reflect and protect behaviour across the GTSs. This result is useful because it can be applied to define a notion of conservative extensions of GTSs, which allow adding spectative behaviour (cf. [28]) without affecting the core transformation behaviour expressed in a GTS.

There are of course a number of further research steps to be taken—both in applying the formal framework to particular domains and in further development of the framework itself. In terms of application, we need to provide methods to check the preconditions of Theorem 1, and if possible automatically checkable conditions that imply these, so that behaviour protection of an extension can be checked effectively. This will enable the development of tooling to support the validation of language or transformation compositions. On the part of the formal framework, we need to study relaxations of our definitions so as to allow cases where there is a less than perfect match between the base DSML and the DSML to be woven in. Inspired by [28], we are also planning to study different categories of extensions, which do not necessarily need to be spectative (conservative), and whether syntactic characterisations exist for them, too.

Acknowledgments. We are thankful to Andrea Corradini for his very helpful comments. We would also like to thank Javier Troya and Antonio Vallecillo for fruitful discussions and previous and on-going collaborations this work relies on. This work has been partially supported by CICYT projects TIN2011-23795 and TIN2007-66523, and by the AGAUR grant to the research group ALBCOM (ref. 00516).

References

1. Boehm, P., Fonio, H.-R., Habel, A.: Amalgamation of graph transformations with applications to synchronization. In: Ehrig, H., Floyd, C., Nivat, M., Thatcher, J. (eds.) CAAP 1985 and TAPSOFT 1985. LNCS, vol. 185, pp. 267–283. Springer, Heidelberg (1985)
2. Chen, K., Sztipanovits, J., Abdelwalhed, S., Jackson, E.: Semantic anchoring with model transformations. In: Hartman, A., Kreische, D. (eds.) ECMDA-FA 2005. LNCS, vol. 3748, pp. 115–129. Springer, Heidelberg (2005)
3. Clarke, S., Walker, R.J.: Generic aspect-oriented design with Theme/UML. In: Aspect-Oriented Software Development, pp. 425–458. Addison-Wesley (2005)
4. Clavel, M., Durán, F., Eker, S., Lincoln, P., Martí-Oliet, N., Meseguer, J., Talcott, C.: All About Maude. LNCS, vol. 4350. Springer, Heidelberg (2007)
5. Corradini, A., Ehrig, H., Löwe, M., Montanari, U., Padberg, J.: The category of typed graph grammars and its adjunctions with categories of derivations. In: Cuny, et al. (eds.) [6], pp. 56–74
6. Cuny, J., Ehrig, H., Engels, G., Rozenberg, G.: Graph Grammars 1994. LNCS, vol. 1073. Springer, Heidelberg (1996)
7. van Deursen, A., Klint, P., Visser, J.: Domain-specific languages: An annotated bibliography. SIGPLAN Not. 35(6), 26–36 (2000)
8. Di Ruscio, D., Jouault, F., Kurtev, I., Bézivin, J., Pierantonio, A.: Extending AMMA for supporting dynamic semantics specifications of DSLs. Tech. Rep. 06.02, Laboratoire d'Informatique de Nantes-Atlantique (LINA) (April 2006)
9. Durán, F., Zschaler, S., Troya, J.: On the reusable specification of non-functional properties in DSLs. In: Czarnecki, K., Hedin, G. (eds.) SLE 2012. LNCS, vol. 7745, pp. 332–351. Springer, Heidelberg (2013)
10. Ehrig, H.: Introduction to the algebraic theory of graph grammars (a survey). In: Claus, V., Ehrig, H., Rozenberg, G. (eds.) Graph Grammars 1978. LNCS, vol. 73, pp. 1–69. Springer, Heidelberg (1979)

11. Ehrig, H., Ehrig, K., Habel, A., Pennemann, K.H.: Theory of constraints and application conditions: From graphs to high-level structures. Fundamenta Informaticae 74(1), 135–166 (2006)
12. Ehrig, H., Ehrig, K., Prange, U., Taentzer, G.: Fundamentals of Algebraic Graph Transformation. Springer (2005)
13. Ehrig, H., Engels, G., Kreowski, H.J., Rozenberg, G. (eds.): Handbook of Graph Grammars and Computing by Graph Transformation. Applications, Languages and Tools, vol. II. World Scientific (1999)
14. Ehrig, H., Habel, A., Padberg, J., Prange, U.: Adhesive high-level replacement categories and systems. In: Ehrig, H., Engels, G., Parisi-Presicce, F., Rozenberg, G. (eds.) ICGT 2004. LNCS, vol. 3256, pp. 144–160. Springer, Heidelberg (2004)
15. Ehrig, H., Mahr, B.: Fundamentals of Algebraic Specification 2. Module Specifications and Constraints. Springer (1990)
16. Ehrig, H., Prange, U., Taentzer, G.: Fundamental theory for typed attributed graph transformation. In: Ehrig, H., Engels, G., Parisi-Presicce, F., Rozenberg, G. (eds.) ICGT 2004. LNCS, vol. 3256, pp. 161–177. Springer, Heidelberg (2004)
17. Engels, G., Hausmann, J.H., Heckel, R., Sauer, S.: Dynamic meta modeling: A graphical approach to the operational semantics of behavioral diagrams in UML. In: Evans, A., Caskurlu, B., Selic, B. (eds.) UML 2000. LNCS, vol. 1939, pp. 323–337. Springer, Heidelberg (2000)
18. Engels, G., Heckel, R., Cherchago, A.: Flexible interconnection of graph transformation modules. In: Kreowski, H.-J., Montanari, U., Orejas, F., Rozenberg, G., Taentzer, G. (eds.) Formal Methods in Software and Systems Modeling. LNCS, vol. 3393, pp. 38–63. Springer, Heidelberg (2005)
19. Engels, G., Heckel, R., Taentzer, G., Ehrig, H.: A combined reference model- and view-based approach to system specification. International Journal of Software Engineering and Knowledge Engineering 7(4), 457–477 (1997)
20. Fischer, T., Niere, J., Torunski, L., Zündorf, A.: Story diagrams: A new graph rewrite language based on the unified modeling language and java. In: Ehrig, H., Engels, G., Kreowski, H.-J., Rozenberg, G. (eds.) TAGT 1998. LNCS, vol. 1764, pp. 296–309. Springer, Heidelberg (2000)
21. Große-Rhode, M., Parisi-Presicce, F., Simeoni, M.: Spatial and temporal refinement of typed graph transformation systems. In: Brim, L., Gruska, J., Zlatuška, J. (eds.) MFCS 1998. LNCS, vol. 1450, pp. 553–561. Springer, Heidelberg (1998)
22. Große-Rhode, M., Parisi-Presicce, F., Simeoni, M.: Formal software specification with refinements and modules of typed graph transformation systems. Journal of Computer and System Sciences 64(2), 171–218 (2002)
23. Habel, A., Pennemann, K.H.: Correctness of high-level transformation systems relative to nested conditions. Mathematical Structures in Computer Science 19(2), 245–296 (2009)
24. Heckel, R., Cherchago, A.: Structural and behavioural compatibility of graphical service specifications. Journal of Logic and Algebraic Programming 70(1), 15–33 (2007)
25. Heckel, R., Corradini, A., Ehrig, H., Löwe, M.: Horizontal and vertical structuring of typed graph transformation systems. Mathematical Structures in Computer Science 6(6), 613–648 (1996)
26. Heckel, R., Engels, G., Ehrig, H., Taentzer, G.: Classification and comparison of modularity concepts for graph transformation systems. In: Ehrig, et al. (eds.) [13], ch. 17, pp. 669–690
27. Hemel, Z., Kats, L.C.L., Groenewegen, D.M., Visser, E.: Code generation by model transformation: A case study in transformation modularity. Software and Systems Modelling 9(3), 375–402 (2010)
28. Katz, S.: Aspect categories and classes of temporal properties. In: Rashid, A., Aksit, M. (eds.) Transactions on AOSD I. LNCS, vol. 3880, pp. 106–134. Springer, Heidelberg (2006)

29. Klein, J., Kienzle, J.: Reusable aspect models. In: Aspect-Oriented Modeling Workshop at MODELS 2007 (2007)
30. Klein, J., Hélouët, L., Jézéquel, J.M.: Semantic-based weaving of scenarios. In: Proc. 5th Int'l Conf. Aspect-Oriented Software Development (AOSD 2006). ACM (2006)
31. Kreowski, H., Kuske, S.: Graph transformation units and modules. In: Ehrig, et al. (eds.) [13], ch. 15, pp. 607–638
32. Lack, S., Sobociński, P.: Adhesive categories. In: Walukiewicz, I. (ed.) FOSSACS 2004. LNCS, vol. 2987, pp. 273–288. Springer, Heidelberg (2004)
33. de Lara, J., Vangheluwe, H.: Automating the transformation-based analysis of visual languages. Formal Aspects of Computing 22(3-4), 297–326 (2010)
34. Muller, P.-A., Fleurey, F., Jézéquel, J.-M.: Weaving executability into object-oriented metalanguages. In: Briand, L.C., Williams, C. (eds.) MoDELS 2005. LNCS, vol. 3713, pp. 264–278. Springer, Heidelberg (2005)
35. Parisi-Presicce, F.: Transformations of graph grammars. In: Cuny, et al. (eds.) [6], pp. 428–442
36. Rivera, J.E., Durán, F., Vallecillo, A.: Formal specification and analysis of domain specific models using Maude. Simulation 85(11-12), 778–792 (2009)
37. Rivera, J.E., Guerra, E., de Lara, J., Vallecillo, A.: Analyzing rule-based behavioral semantics of visual modeling languages with Maude. In: Gašević, D., Lämmel, R., Van Wyk, E. (eds.) SLE 2008. LNCS, vol. 5452, pp. 54–73. Springer, Heidelberg (2009)
38. Rivera, J.E., Durán, F., Vallecillo, A.: A graphical approach for modeling time-dependent behavior of DSLs. In: Proceedings of the IEEE Symposium on Visual Languages and Human-Centric Computing, VL/HCC 2009, pp. 51–55. IEEE (2009)
39. Rivera, J.E., Durán, F., Vallecillo, A.: On the behavioral semantics of real-time domain specific visual languages. In: Ölveczky, P.C. (ed.) WRLA 2010. LNCS, vol. 6381, pp. 174–190. Springer, Heidelberg (2010)
40. Rozenberg, G. (ed.): Handbook of Graph Grammars and Computing by Graph Transformations. Foundations, vol. I. World Scientific (1997)
41. Schmidt, D.C.: Model-driven engineering. IEEE Computer 39(2), 25–31 (2006)
42. Schürr, A., Winter, A., Zündorf, A.: The PROGRES-approach: Language and environment. In: Ehrig, et al. (eds.) [13], ch. 13, pp. 487–550
43. Taentzer, G., Schürr, A.: DIEGO, another step towards a module concept for graph transformation systems. Electronic Notes on Theoretical Computer Science 2, 277–285 (1995)
44. Tisi, M., Jouault, F., Fraternali, P., Ceri, S., Bézivin, J.: On the use of higher-order model transformations. In: Paige, R.F., Hartman, A., Rensink, A. (eds.) ECMDA-FA 2009. LNCS, vol. 5562, pp. 18–33. Springer, Heidelberg (2009)
45. Troya, J., Rivera, J.E., Vallecillo, A.: Simulating domain specific visual models by observation. In: Proc. 2010 Spring Simulation Multiconference (SpringSim 2010), pp. 128:1–128:8. ACM (2010)
46. Troya, J., Vallecillo, A., Durán, F., Zschaler, S.: Model-driven performance analysis of rule-based domain specific visual models. Information and Software Technology 55(1), 88–110 (2013)
47. Whittle, J., Jayaraman, P., Elkhodary, A., Moreira, A., Araújo, J.: MATA: A unified approach for composing UML aspect models based on graph transformation. In: Katz, S., Ossher, H., France, R., Jézéquel, J.-M. (eds.) Transactions on AOSD VI. LNCS, vol. 5560, pp. 191–237. Springer, Heidelberg (2009)
48. Zschaler, S.: Formal specification of non-functional properties of component-based software systems: A semantic framework and some applications thereof. Software and Systems Modelling (SoSyM) 9, 161–201 (2009)

Quantitative Modal Transition Systems

(Invited Extended Abstract)

Kim G. Larsen[1] and Axel Legay[2]

[1] Aalborg University, Denmark
kgl@cs.aau.dk
[2] INRIA/IRISA, France
axel.legay@inria.fr

Abstract. This extended abstract offers a brief survey presentation of the specification formalism of modal transition systems and its recent extensions to the quantitative setting of timed as well as stochastic systems. Some applications will also be briefly mentioned.

1 Modal Transition Systems: The Origines

Modal transition systems [46] provids a behavioural compositional specification formalism for reactive systems. They grew out of the notion of relativized bisimulation, which allows for simple specifications of components by allowing the notion of bisimulation to take the restricted use that a given context may have in its.

A modal transition system is essentially a (labelled) transition system, but with two types of transitions: so-called *may* transitions, that any implementation may (or may not) have, and *must* transitions, that any implementation must have. In fact, ordinary labelled transition systems (or implementations) are modal transition systems where the set of may- and must-transitions coincide. Modal transition systems come equipped with a bisimulation-like notion of (modal) refinement, reflecting that the more must-transitions and the fewer may-transitions a modal specification has the more refined and closer to a final implementation it is.

Example 1. Consider the modal transition system shown in Figure 1 which models the requirements of a simple email system in which emails are first received and then delivered – must and may transitions are represented by solid and dashed arrows, respectively. Before delivering the email, the system may check or process the email, e.g. for en- or decryption, filtering of spam emails, or generating automatic answers using as an auto-reply feature. Any implementation of this email system specification must be able to receive and deliver email, and it may also be able to check arriving email before delivering it. No other behavior is allowed. Such an implementation is given in Figure 2.

Modal transition systems play a major role in various areas. However, the model is best known by its application in compositional reasoning, which has been recognized in the ARTIST Network of Excellence and several other related European projects. In fact, modal transition systems have all the ingredients of a complete compositinal

N. Martí-Oliet and M. Palomino (Eds.): WADT 2012, LNCS 7841, pp. 50–58, 2013.

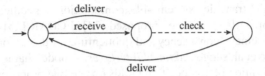

Fig. 1. Modal transition system modeling a simple email system, with an optional behavior: Once an email is received it may e.g. be scanned for containing viruses, or automatically decrypted, before it is delivered to the receiver

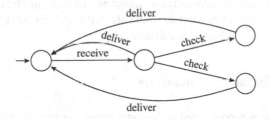

Fig. 2. An implementation of the simple email system in Figure 1 in which we explicitly model two distinct types of email pre-processing

specification theory allowing for logical compositions (e.g. conjunction) [42], structural compositions (e.g. parallel) [37] as well as quotienting permitting specifications of composite systems to be transformed into necessary and sufficient specification of components [35]. Thus, modal transition systems have all the benefits of both logical and behavioural specification formalisms [18]. Though modal refinement – like bisimulation – is polynomial-time decidable for finite-state modal transition systems, it only provides a sound but not complete test for the true semantic refinement between modal specification, in terms of set inclusion between their implementation-sets (so-called thorough refinement). For several years, the complexity of thorough refinement – as well as the consistency – between modal specifications was an open problem, which after a series of attempts [44,3] [2] was shown to be EXPTIME-complete [14].

   In the rest of this overview, we will briefly introduce several quantitative extensions of modal transition systems that have been recently proposed in the literature. Sections 2 to 5 mainly focus on modal transition systems as a specification theory, while Sections 6 and 7 outline some other extensions.

2   Timed Modal Specifications

It is well acknowledged that real-time can be a crucial parameter in practice, for example in embedded systems. This motivates the study of extended modal transition systems to introduce real-time features.

   Timed extensions of modal transitions were introduced early on [23] as timed extension of the process algebra CCS. Unfortunately the supporting tool EPSILON was entirely relying on the so-called region-abstraction, making scalability extremely poor. Most recently, taking advantage of the powerfull game-theoretical engine of UPPAAL

Tiga [10,21] a "modal-transition system"-like compositional specification theory based on Timed I/O Automata [41] has been proposed [24]. ECDAR [25] gives an efficient tool support for refinement, consistency and quotienting for this theory.

In [27], de Alfaro et al. suggested *timed interfaces*, a model that is similar to the one of TIOTSs. Our definition of composition builds on the one proposed there. However, the work in [27] is incomplete. Indeed, there was no notion of implementation and refinement. Moreover, conjunction and quotient were not studied. Finally, the theory has only been implemented in a prototype tool which does not handle continuous time, while our contribution takes advantage of the powerful game engine of UPPAAL Tiga.

The work of [16] suggests an alternative timed extension of modal transition systems (though still relying on regions for refinement algorithms). This work is less elaborated and implementation was not considered there.

3 Weighted Modal Specifications

The previous section was concerned with modal transition systems whose implementations are timed automata. There are various other extensions of automata, among which one finds weighted automata, that are classical automata whose transitions are equipped with integer weights. In [40], modal transition systems were extended with integer intervals in order to capture and abstract weighted automata. The works also proposes structural and logical composition as well as refinement for the extended model. Latter, in [7], the work was generalized to weighted modal transition systems, whose transition can be equipped with any type of quantity.

Albeit the extensions mentioned above allow for a quantitative treatment of automata behaviors, the operations on weighted modal transition systems remain qualitative. Especially, the refinement relation of modal transition systems is qualitative in nature, i.e. an implementation does, or does not, refine a specification. Such a view may be fragile in the sense that the inevitable approximation of systems by models, combined with the fundamental unpredictability of hardware platforms, make difficult to transfer conclusions about the behavior to the actual system. Hence, this approach is arguably unsuited for modern software systems. In [5], the first quantitative extension of modal automata was proposed. This model allows to capture quantitative aspects during the refinement and implementation process, thus leveraging the problems of the qualitative setting.

In [5], satisfaction and refinement are lifted from the well-known qualitative setting to the quantitative setting, by introducing a notion of distance between weighted modal transition systems. It is also shown that quantitative versions of parallel composition, as well as quotient (the dual operator to parallel composition), inherit the properties from the Boolean setting.

Example 2 (taken from [5]). Consider again the modal transition system of Figure 1, but this time with quantities, see Figure 3: Every transition label is extended by an integer intervals modeling upper and lower bounds on the time required for performing the corresponding actions. For instance, the reception of a new email (action *receive*) must take between one and three time units, the checking of the email (action *check*) is allowed to take up to five time units.

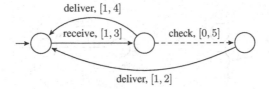

Fig. 3. Specification of a simple email system, similar to Figure 1, but extended by integer intervals modeling time units for performing the corresponding actions

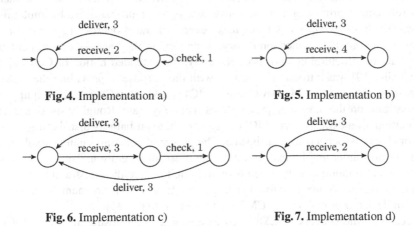

Fig. 4. Implementation a)   **Fig. 5.** Implementation b)

Fig. 6. Implementation c)   **Fig. 7.** Implementation d)

Four implementations of the simple email system in Figure 3

In this quantitative setting, there is a problem with using a *Boolean* notion of refinement: If one only can decide *whether or not* an implementation refines a specification, then the quantitative aspects get lost in the refinement process. As an example, consider the email system implementations in Figures 4 to 7. Implementation (a) does not refine the specification, as there is an error in the discrete structure of actions: after receiving an email, the system can check it indefinitely without ever delivering it. Also, implementations (b) and (c) do not refine the specification: (b) takes too long to receive any email, while (c) does not deliver the email fast. Implementation (d), on the other hand, is a perfect refinement of the specification.

The work in [5] uses an accumulating distance, but the contribution was latter generalized to any type of distance in [6]. The extended model allowed, as an example, to define various notions of robustness for specification theories. Complexity of refinement and efficient implementations remain open problems.

4 Probabilistic Modal Specifications

In [39], modal transitions systems were extended into a specification formalism for Markov Chains by the introduction of so-called probabilistic specifications (now known

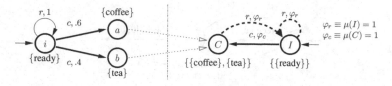

Fig. 8. Implementation PA and specification APA of a coffee machine

as Interval Markov Chains), where concrete probabilities are replaced with intervals, and with refinement providing a conservative extension or probabilistic bisimulation [45]. However, Interval Markov Chains lack several of the properties required for a complete compositional specification theory; in particular, they are not closed neither under logical nor structural composition. Recently, the extended notion of Constraint Markov Chains [19] was introduced precisely with the purpose of providing these closure properties. A Constraint Markov Chain (CMC) is a Markov Chain (MC) equipped with a constraint on the next-state probabilities from any state. Roughly speaking, an implementation for a CMC is thus a MC, whose next-state probability distribution satisfies the constraint associated with each state. The power of constrains can be exploited to obtain closure under any logical/structural composition operation. The complexity of the refinement relation largely depends on the one to solve the constraints – it is at least quadratic (resp. exponential) for syntactic (resp. thorough) refinement. The reader interested in decision probblems for CMCs is redirected to [32,31]

More recently, the concept of CMC was extended to offer abstractions for Probabilistic Automata (PA), i.e., structures that mix both stochastic and non-deterministic aspects. The work in [28] proposes Abstract Probabilistic Automata, that are a combination of modal transition systems and CMCs, modalities being used to capture the non-determinism in PAs. The model was implemented in the APAC toolset [30] and various decision problems, including stuttering and abstraction, were studied in [29,54].

Example 3 (taken from [29]). Consider the implementation (left) and specification (right) of a coffee machine given in Figure 8. The specification indicates that there are two possible transitions from initial state I: a may transition labeled with action r (reset) and a must transition labeled with action c (coin). May transitions are represented with dashed arrows and must transitions are represented with plain arrows. The probability distributions associated with these actions are specified by the constraints φ_r and φ_c, respectively.

5   Beyond Modalities

In a seminal paper [26], de Alfaro and Henzinger promoted another specification theory known under the name of *interface automata*. An interface is represented by an input/output automaton [47], i.e., an automaton whose transitions are labeled with *input* or *output* actions. The semantics of such an automaton is given by a two-players game: an *Input* player represents the environment, and an *Output* player represents the component itself. Interface automata do not encompass any notion of model, because there is no way to distinguish between an interface and its implementations.

Refinement between interface automata corresponds to the alternating refinement relation between games [1], i.e., an interface refines another if its environment is more permissive, whereas its component is more restrictive. Contrary to most interfaces theories, the game-based interpretation offers an *optimistic* treatment of composition: two interfaces can be composed if there exists at least one environment (i.e., one strategy for the Input player) in which they can interact together in a safe way (i.e., whatever the strategy of the Output player is). This is referred to as the compatibility of interfaces. A quotient, which is the adjoint of the game-based composition, has been proposed in [17] for the deterministic case.

In [43,52], modal transition systems and interface automata were combined to give rise to modal interfaces, a model that offers both the power of modalities and the optimistic composition approach of interface automata through labeling of may and must modalities with input/output features. Implementations of modal interfaces can be find in the Mika toolset [20] and the MIOs workbench [9].

It is worth mentioning that the methodology used in [52] is rather generic and can be applied to other extensions of modal automata. As an example, the APA model was also extended to a modal APA model in [29].

6   Contract Theory

We already observed that modal transition systems act as a good model for a complete specification theory. Of course, there are other similar approaches. In fact, some of them advocate that specification theories should be equipped with additional structure that makes more explicit their possible connections. This is particularly the case of the contract theory approach, which is based on a assume-guarantee (AG) reasoning.

Concretely, contract theories differ from classical specification theories in that they strictly follow the principle of separation of concerns. They separate the specification of assumptions from the specification of guarantees, a choice largely inspired by early ideas on manual proof methods by Misra, Chandy [49] and Jones [38], along with the wide acceptance of the pre-/post-condition specification in programming [48,55], and more in general semantical rules independent from language representation [22].

Recently, Benveniste et al [15] proposed a contract theory where assumptions and guarantees are represented by trace structures. While this work is of clear interest, it suffers from the absence of an effective representation for the embedded interface theory. Some extensions, such as the one proposed in [50,35], leverage this problem but in a specific manner, i.e., just for the case of a single theory.

In [4], it was shown how a theory of contracts can be built on top of a given abstract specification theory. Contracts are just pairs (A, G) of an assumption and a guarantee specification. Particularly, it was shown how the contract theory can be instantiated by using modal transition systems.

7   Others Modal Extensions and Applications

There are many other extensions of modal transition systems, which include those that encompass unbounded data or costs and parameters [8,12,13], and those that offer a more elaborated treatment of modalities [51,11].

In addition to their contribution to specification theories, modal transition systems have also played a major role in abstraction-based model checking [33,34], software differences [53], and in the design of efficient approaches for software product lines verification [36].

References

1. Alur, R., Henzinger, T.A., Kupferman, O., Vardi, M.Y.: Alternating refinement relations. In: Sangiorgi, D., de Simone, R. (eds.) CONCUR 1998. LNCS, vol. 1466, pp. 163–178. Springer, Heidelberg (1998)
2. Antonik, A., Huth, M., Larsen, K.G., Nyman, U., Wasowski, A.: 20 years of modal and mixed specifications. Bulletin of the EATCS 95, 94–129 (2008)
3. Antonik, A., Huth, M., Larsen, K.G., Nyman, U., Wasowski, A.: Modal and mixed specifications: key decision problems and their complexities. Mathematical Structures in Computer Science 20(1), 75–103 (2010)
4. Bauer, S.S., David, A., Hennicker, R., Larsen, K.G., Legay, A., Nyman, U., Wąsowski, A.: Moving from specifications to contracts in component-based design. In: de Lara, J., Zisman, A. (eds.) FASE 2012. LNCS, vol. 7212, pp. 43–58. Springer, Heidelberg (2012)
5. Bauer, S.S., Fahrenberg, U., Juhl, L., Larsen, K.G., Legay, A., Thrane, C.: Quantitative refinement for weighted modal transition systems. In: Murlak, F., Sankowski, P. (eds.) MFCS 2011. LNCS, vol. 6907, pp. 60–71. Springer, Heidelberg (2011)
6. Bauer, S.S., Fahrenberg, U., Legay, A., Thrane, C.: General quantitative specification theories with modalities. In: Hirsch, E.A., Karhumäki, J., Lepistö, A., Prilutskii, M. (eds.) CSR 2012. LNCS, vol. 7353, pp. 18–30. Springer, Heidelberg (2012)
7. Bauer, S.S., Juhl, L., Larsen, K.G., Legay, A., Srba, J.: Extending modal transition systems with structured labels. Mathematical Structures in Computer Science 22(4), 581–617 (2012)
8. Bauer, S.S., Larsen, K.G., Legay, A., Nyman, U., Wąsowski, A.: A modal specification theory for components with data. In: Arbab, F., Ölveczky, P.C. (eds.) FACS 2011. LNCS, vol. 7253, pp. 61–78. Springer, Heidelberg (2012)
9. Bauer, S.S., Mayer, P., Legay, A.: MIO workbench: A tool for compositional design with modal input/output interfaces. In: Bultan, T., Hsiung, P.-A. (eds.) ATVA 2011. LNCS, vol. 6996, pp. 418–421. Springer, Heidelberg (2011)
10. Behrmann, G., Cougnard, A., David, A., Fleury, E., Larsen, K.G., Lime, D.: UPPAAL-tiga: Time for playing games! In: Damm, W., Hermanns, H. (eds.) CAV 2007. LNCS, vol. 4590, pp. 121–125. Springer, Heidelberg (2007)
11. Beneš, N., Křetínský, J.: Modal process rewrite systems. In: Roychoudhury, A., D'Souza, M. (eds.) ICTAC 2012. LNCS, vol. 7521, pp. 120–135. Springer, Heidelberg (2012)
12. Beneš, N., Křetínský, J., Larsen, K.G., Møller, M.H., Srba, J.: Parametric modal transition systems. In: Bultan, T., Hsiung, P.-A. (eds.) ATVA 2011. LNCS, vol. 6996, pp. 275–289. Springer, Heidelberg (2011)
13. Beneš, N., Křetínský, J., Larsen, K.G., Møller, M.H., Srba, J.: Dual-priced modal transition systems with time durations. In: Bjørner, N., Voronkov, A. (eds.) LPAR-18. LNCS, vol. 7180, pp. 122–137. Springer, Heidelberg (2012)

14. Beneš, N., Křetínský, J., Larsen, K.G., Srba, J.: Checking thorough refinement on modal transition systems is EXPTIME-complete. In: Leucker, M., Morgan, C. (eds.) ICTAC 2009. LNCS, vol. 5684, pp. 112–126. Springer, Heidelberg (2009)
15. Benveniste, A., Caillaud, B., Ferrari, A., Mangeruca, L., Passerone, R., Sofronis, C.: Multiple viewpoint contract-based specification and design. In: de Boer, F.S., Bonsangue, M.M., Graf, S., de Roever, W.-P. (eds.) FMCO 2007. LNCS, vol. 5382, pp. 200–225. Springer, Heidelberg (2008)
16. Bertrand, N., Legay, A., Pinchinat, S., Raclet, J.-B.: A compositional approach on modal specifications for timed systems. In: Breitman, K., Cavalcanti, A. (eds.) ICFEM 2009. LNCS, vol. 5885, pp. 679–697. Springer, Heidelberg (2009)
17. Bhaduri, P.: Synthesis of interface automata. In: Peled, D.A., Tsay, Y.-K. (eds.) ATVA 2005. LNCS, vol. 3707, pp. 338–353. Springer, Heidelberg (2005)
18. Boudol, G., Larsen, K.G.: Graphical versus logical specifications. In: Arnold, A. (ed.) CAAP 1990. LNCS, vol. 431, pp. 57–71. Springer, Heidelberg (1990)
19. Caillaud, B., Delahaye, B., Larsen, K.G., Legay, A., Pedersen, M.L., Wasowski, A.: Constraint markov chains. Theor. Comput. Sci. 412(34), 4373–4404 (2011)
20. Caillaud, B., Raclet, J.-B.: Ensuring reachability by design. In: Roychoudhury, A., D'Souza, M. (eds.) ICTAC 2012. LNCS, vol. 7521, pp. 213–227. Springer, Heidelberg (2012)
21. Cassez, F., David, A., Fleury, E., Larsen, K.G., Lime, D.: Efficient on-the-fly algorithms for the analysis of timed games. In: Abadi, M., de Alfaro, L. (eds.) CONCUR 2005. LNCS, vol. 3653, pp. 66–80. Springer, Heidelberg (2005)
22. Cau, A., Collette, P.: Parallel composition of assumption-commitment specifications: A unifying approach for shared variable and distributed message passing concurrency. Acta Inf. 33(2), 153–176 (1996)
23. Čerāns, K., Godskesen, J.C., Larsen, K.G.: Timed modal specification - theory and tools. In: Courcoubetis, C. (ed.) CAV 1993. LNCS, vol. 697, pp. 253–267. Springer, Heidelberg (1993)
24. David, A., Larsen, K.G., Legay, A., Nyman, U., Wasowski, A.: Timed i/o automata: a complete specification theory for real-time systems. In: Johansson, K.H., Yi, W. (eds.) HSCC, pp. 91–100. ACM (2010)
25. David, A., Larsen, K.G., Legay, A., Nyman, U., Wąsowski, A.: ECDAR: An environment for compositional design and analysis of real time systems. In: Bouajjani, A., Chin, W.-N. (eds.) ATVA 2010. LNCS, vol. 6252, pp. 365–370. Springer, Heidelberg (2010)
26. de Alfaro, L., Henzinger, T.A.: Interface theories for component-based design. In: Henzinger, T.A., Kirsch, C.M. (eds.) EMSOFT 2001. LNCS, vol. 2211, pp. 148–165. Springer, Heidelberg (2001)
27. de Alfaro, L., Henzinger, T.A., Stoelinga, M.: Timed interfaces. In: Sangiovanni-Vincentelli, A.L., Sifakis, J. (eds.) EMSOFT 2002. LNCS, vol. 2491, pp. 108–122. Springer, Heidelberg (2002)
28. Delahaye, B., Katoen, J.-P., Larsen, K.G., Legay, A., Pedersen, M.L., Sher, F., Wąsowski, A.: Abstract probabilistic automata. In: Jhala, R., Schmidt, D. (eds.) VMCAI 2011. LNCS, vol. 6538, pp. 324–339. Springer, Heidelberg (2011)
29. Delahaye, B., Katoen, J.-P., Larsen, K.G., Legay, A., Pedersen, M.L., Sher, F., Wasowski, A.: New results on abstract probabilistic automata. In: ACSD, pp. 118–127. IEEE (2011)
30. Delahaye, B., Larsen, K.G., Legay, A., Pedersen, M.L., Wasowski, A.: Apac: A tool for reasoning about abstract probabilistic automata. In: QEST, pp. 151–152. IEEE Computer Society (2011)
31. Delahaye, B., Larsen, K.G., Legay, A., Pedersen, M.L., Wąsowski, A.: Decision problems for interval markov chains. In: Dediu, A.-H., Inenaga, S., Martín-Vide, C. (eds.) LATA 2011. LNCS, vol. 6638, pp. 274–285. Springer, Heidelberg (2011)
32. Delahaye, B., Larsen, K.G., Legay, A., Pedersen, M.L., Wasowski, A.: Consistency and refinement for interval markov chains. J. Log. Algebr. Program. 81(3), 209–226 (2012)

33. Godefroid, P.: Reasoning about abstract open systems with generalized module checking. In: Alur, R., Lee, I. (eds.) EMSOFT 2003. LNCS, vol. 2855, pp. 223–240. Springer, Heidelberg (2003)
34. Godefroid, P.: Generalized model checking. In: TIME, p. 3. IEEE Computer Society (2005)
35. Goessler, G., Raclet, J.-B.: Modal contracts for component-based design. In: SEFM, pp. 295–303. IEEE Computer Society (2009)
36. Gruler, A., Leucker, M., Scheidemann, K.: Modeling and model checking software product lines. In: Barthe, G., de Boer, F.S. (eds.) FMOODS 2008. LNCS, vol. 5051, pp. 113–131. Springer, Heidelberg (2008)
37. Hüttel, H., Larsen, K.G.: The use of static constructs in a modal process logic. In: Meyer, A.R., Taitslin, M.A. (eds.) Logic at Botik 1989. LNCS, vol. 363, pp. 163–180. Springer, Heidelberg (1989)
38. Jones, C.B.: Development methods for computer programs including a notion of interference. PhD thesis, Oxford University Computing Laboratory (1981)
39. Jonsson, B., Larsen, K.G.: Specification and refinement of probabilistic processes. In: LICS, pp. 266–277 (1991)
40. Juhl, L., Larsen, K.G., Srba, J.: Modal transition systems with weight intervals. J. Log. Algebr. Program. 81(4), 408–421 (2012)
41. Kaynar, D.K., Lynch, N.A., Segala, R., Vaandrager, F.W.: The Theory of Timed I/O Automata, 2nd edn. Synthesis Lectures on Distributed Computing Theory. Morgan & Claypool Publishers (2010)
42. Larsen, K.G.: Modal specifications. In: Sifakis, J. (ed.) CAV 1989. LNCS, vol. 407, pp. 232–246. Springer, Heidelberg (1990)
43. Larsen, K.G., Nyman, U., Wąsowski, A.: Modal I/O automata for interface and product line theories. In: De Nicola, R. (ed.) ESOP 2007. LNCS, vol. 4421, pp. 64–79. Springer, Heidelberg (2007)
44. Larsen, K.G., Nyman, U., Wąsowski, A.: On modal refinement and consistency. In: Caires, L., Vasconcelos, V.T. (eds.) CONCUR 2007. LNCS, vol. 4703, pp. 105–119. Springer, Heidelberg (2007)
45. Larsen, K.G., Skou, A.: Bisimulation through probabilistic testing. Inf. Comput. 94(1), 1–28 (1991)
46. Larsen, K.G., Thomsen, B.: A modal process logic. In: LICS, pp. 203–210 (1988)
47. Lynch, N., Tuttle, M.R.: An introduction to Input/Output automata. CWI-quarterly 2(3) (1989)
48. Meyer, B.: Applying "design by contract". IEEE Computer 25(10), 40–51 (1992)
49. Misra, J., Mani Chandy, K.: Proofs of networks of processes. IEEE Trans. Software Eng. 7(4), 417–426 (1981)
50. Quinton, S., Graf, S.: Contract-based verification of hierarchical systems of components. In: SEFM, pp. 377–381. IEEE Computer Society (2008)
51. Raclet, J.-B.: Residual for component specifications. Electr. Notes Theor. Comput. Sci. 215, 93–110 (2008)
52. Raclet, J.-B., Badouel, E., Benveniste, A., Caillaud, B., Legay, A., Passerone, R.: A modal interface theory for component-based design. Fundam. Inform. 108(1-2), 119–149 (2011)
53. Sassolas, M., Chechik, M., Uchitel, S.: Exploring inconsistencies between modal transition systems. Software and System Modeling 10(1), 117–142 (2011)
54. Sher, F., Katoen, J.-P.: Compositional abstraction techniques for probabilistic automata. In: Baeten, J.C.M., Ball, T., de Boer, F.S. (eds.) TCS 2012. LNCS, vol. 7604, pp. 325–341. Springer, Heidelberg (2012)
55. Xu, Q., Cau, A., Collette, P.: On unifying assumption-commitment style proof rules for concurrency. In: Jonsson, B., Parrow, J. (eds.) CONCUR 1994. LNCS, vol. 836, pp. 267–282. Springer, Heidelberg (1994)

Bounded Model Checking
of Recursive Programs with Pointers in K

Irina Măriuca Asăvoae[1,*], Frank de Boer[2,3],
Marcello M. Bonsangue[3,2], Dorel Lucanu[1], and Jurriaan Rot[3,2,**]

[1] Faculty of Computer Science - Alexandru Ioan Cuza University, Romania
{mariuca.asavoae,dlucanu}@info.uaic.ro
[2] Centrum voor Wiskunde en Informatica, The Netherlands
frb@cwi.nl
[3] LIACS — Leiden University, The Netherlands
{marcello,jrot}@liacs.nl

Abstract. We present an adaptation of model-based verification, via model checking pushdown systems, to semantics-based verification. First we introduce the algebraic notion of pushdown system specifications (PSS) and adapt a model checking algorithm for this new notion. We instantiate pushdown system specifications in the K framework by means of Shylock, a relevant PSS example. We show why K is a suitable environment for the pushdown system specifications and we give a methodology for defining the PSS in K. Finally, we give a parametric K specification for model checking pushdown system specifications based on the adapted model checking algorithm for PSS.

Keywords: pushdown systems, model checking, the K framework.

1 Introduction

The study of computation from a program verification perspective is an effervescent research area with many ramifications. We take into consideration two important branches of program verification which are differentiated based on their perspective over programs, namely model-based versus semantics-based program verification.

Model-based program verification relies on modeling the program as some type of transition system which is then analyzed with specific algorithms. Pushdown systems are known as a standard model for sequential programs with recursive procedures. Intuitively, pushdown systems are transition systems with a stack of unbounded size, which makes them strictly more expressive than finite

---

* The research of this author has been partially supported by Project POSDRU/88/ 1.5/S/47646 and by Contract ANCS POS-CCE, O2.1.2, ID nr 602/12516, ctr.nr 161/15.06.2010 (DAK).

** The research of this author has been funded by the Netherlands Organisation for Scientific Research (NWO), CoRE project, dossier number: 612.063.920.

N. Martí-Oliet and M. Palomino (Eds.): WADT 2012, LNCS 7841, pp. 59–76, 2013.
© IFIP International Federation for Information Processing 2013

state systems. More importantly, there exist fundamental decidability results for pushdown systems [1] which enable program verification via model checking [17].

Semantics-based program verification relies on specification of programming language semantics and derives the program model from the semantics specification. For example, the rewriting logic semantics project [12] studies the unification of algebraic denotational semantics with operational semantics of programming languages. The main incentive of this semantics unification is the fact that the algebraic denotational semantics is executable via tools like the Maude system [10], or the $\mathbb{K}$ framework [14]. As such, a programming language (operational) semantics specification implemented with these tools becomes an interpreter for programs via execution of the semantics. The tools come with model checking options, so the semantics specification of a programming language have for-free program verification capabilities.

The current work solves the following problem in the rewriting logic semantics project: though the semantics expressivity covers a quite vast and interesting spectrum of programming languages, the offered verification capabilities via model checking are restricted to finite state systems. Meanwhile, the fundamental results from pushdown systems provide a strong incentive for approaching the verification of this class of infinite transition systems from a semantics-based perspective. As such, we introduce the notion of *pushdown system specifications* (PSS), which embodies the algebraic specification of pushdown systems. Furthermore, we adapt a state-of-the-art model checking algorithm for pushdown systems [17] to work for PSS and present an algebraic specification of this algorithm implemented in the $\mathbb{K}$ tool [15]. Our motivating example is Shylock, a programming language with recursive procedures and pointers, introduced by the authors in [16].

Related work. $\mathbb{K}$ is a rewriting logic based framework for the design, analysis, and verification of programming languages, originating in the rewriting logic semantics project. $\mathbb{K}$ specifies transition systems and is built upon a continuation-based technique and a series of notational conventions to allow for more compact and modular executable programming language definitions. Because of the continuation-based technique, $\mathbb{K}$ specifications resemble PSS where the stack is the continuation. The most complex and thorough $\mathbb{K}$ specification developed so far is the C semantics [5].

The standard approach to model checking programs, used for $\mathbb{K}$ specifications, involves the Maude LTL model checker [4] which is inherited from the Maude back-end of the $\mathbb{K}$ tool. The Maude LTL checker, by comparison with other model checkers, presents a great versatility in defining the state properties to be verified (these being given as a rewrite theory). Moreover, the actual model checking is performed on-the-fly, so that the Maude LTL checker can verify systems with states that involve data in types of infinite cardinality under the assumption of a *finite reachable state space*. However, this assumption is infringed by PSS because of the stack which is allowed to grow unboundedly, hence the Maude LTL checker cannot be used for PSS verification.

The Moped tool for model checking pushdown systems was successfully used for a subset of C programs [17] and was adapted for Java with full recursion, but with a fixed-size number of objects, in jMoped [6]. The WPDS++ tool [8] uses a weighted pushdown system model to verify x86 executable code. However, we cannot employ any of these dedicated tools for model checking pushdown systems because we work at a higher level, namely with *specifications* of pushdown system where we do not have the actual pushdown system.

Structure of the paper. In Section 2 we introduce pushdown system specifications and an associated invariant model checking algorithm. In Section 3 we introduce the $\mathbb{K}$ framework by showing how Shylock's PSS is defined in $\mathbb{K}$. In Section 4 we present the $\mathbb{K}$ specification of the invariant model checking for PSS and show how a certain type of bounded model checking can be directly achieved.

2   Model Checking Specifications of Pushdown Systems

In this section we discuss an approach to model checking pushdown system specifications by adapting an existing model checking algorithm for ordinary pushdown systems. Recall that a *pushdown system* is an input-less pushdown automaton without acceptance conditions. Basically, a pushdown system is a transition system equipped with a finite set of control locations and a stack. The stack consists of a non-a priori bounded string over some finite stack alphabet [1,17]. The difference between a pushdown system specification and an ordinary pushdown system is that the former uses production rules with open terms for the stack and control locations. This allows for a more compact representation of infinite systems and paves the way for applications of model checking to recursive programs defined by means of structural operational semantics.

We assume a countably infinite set of variables $Var = \{v_1, v_2, \ldots\}$. A signature Σ consists of a finite set of function symbols $g_1, g_2, \ldots$, each with a fixed arity $ar(g_1), ar(g_2), \ldots$. Function symbols with arity 0 are called constants. The set of terms, denoted by $T_\Sigma(Var)$ and typically ranged over by s and t, is inductively defined from the set of variables Var and the signature Σ. A substitution σ replaces variables in a term with other terms. A term s can match term t if there exists a substitution σ such that $\sigma(t) = s$. A term t is said to be closed if no variables appear in t, and we use the convention that these terms are denoted as "hatted" terms, i.e., $\hat{t}$.

A *pushdown system specification* (PSS) is a tuple $(\Sigma, \Xi, Var, \Delta)$ where Σ and Ξ are two signatures, Var is a set of variables, and Δ is a finite set of production rules (defined below). Terms in $T_\Sigma(Var)$ define *control locations* of a pushdown system, whereas terms in $T_\Xi(Var)$ define the *stack alphabet*. A production rule in Δ is defined as a formula of the form $(s, \gamma) \Rightarrow (s', \Gamma)$, where s and s' are terms in $T_\Sigma(Var)$, γ is a term in $T_\Xi(Var)$, and Γ is a finite (possibly empty) sequence of terms in $T_\Xi(Var)$. The pair (s, γ) is the *source* of the rule, and (s', Γ) is the *target*. We require for each rule that all variables appearing in the target are included in those of the source. A rule with no variables in the source is called an *axiom*. The notions of substitution and matching are lifted to sequences of terms and to formulae as expected.

Example 1. Let $Var = \{s, t, \gamma\}$, let $\Sigma = \{0, a, +\}$ with $ar(0) = ar(a) = 0$ and $ar(+) = 2$, and let $\Xi = \{L, R\}$ with $ar(L) = ar(R) = 0$. Moreover consider the following three production rules, denoted as a set by Δ:

$$(a, \gamma) \Rightarrow (0, \varepsilon) \quad (s + t, L) \Rightarrow (s, R) \quad (s + t, R) \Rightarrow (t, LR).$$

Then $(\Sigma, \Xi, Var, \Delta)$ is a pushdown system specification.

Given a pushdown system specification $\bar{\mathcal{P}} = (\Sigma, \Xi, Var, \Delta)$, a concrete configuration is a pair $\langle \hat{s}, \hat{\Gamma} \rangle$ where $\hat{s}$ is a closed term in $T_\Sigma(Var)$ denoting the *current control state*, and $\hat{\Gamma}$ is a finite sequence of closed terms in $T_\Xi(Var)$ representing the content of the *current stack*. A transition $\langle \hat{s}, \hat{\gamma} \cdot \hat{\Gamma} \rangle \longrightarrow \langle \hat{s}', \hat{\Gamma}' \cdot \hat{\Gamma} \rangle$ between concrete configurations is derivable from the pushdown system specification $\bar{\mathcal{P}}$ if and only if there is a rule $r = (s_r, \gamma_r) \Rightarrow (s_r', \Gamma_r)$ in Δ and a substitution σ such that $\sigma(s_r) = \hat{s}$, $\sigma(\gamma_r) = \hat{\gamma}$, $\sigma(s_r') = \hat{s}'$ and $\sigma(\Gamma_r) = \hat{\Gamma}'$. The above notion of pushdown system specification can be extended in the obvious way by allowing also conditional production rules and equations on terms.

Continuing on Example 1, we can derive the following sequence of transitions:

$$\langle a + (a + a), R \rangle \longrightarrow \langle a + a, LR \rangle \longrightarrow \langle a, RR \rangle \longrightarrow \langle 0, R \rangle.$$

Note that no transition is derivable from the last configuration $\langle 0, R \rangle$.

A pushdown system specification $\bar{\mathcal{P}}$ is said to be *locally finite* w.r.t. a concrete configuration $\langle \hat{s}, \hat{\Gamma} \rangle$, if the set of all closed terms appearing in the configurations reachable from $\langle \hat{s}, \hat{\Gamma} \rangle$ by transitions derivable from the rules of $\bar{\mathcal{P}}$ is finite. Note that this does not imply that the set of concrete configurations reachable from a configuration $\langle \hat{s}, \hat{\Gamma} \rangle$ is finite, as the stack is not bounded. However all reachable configurations are constructed from a *finite* set of control locations and a *finite* stack alphabet. An ordinary *finite pushdown system* is thus a pushdown system specification which is locally finite w.r.t. a concrete *initial* configuration $\hat{c}_0$, and such that all rules are axioms, i.e., all terms appearing in the source and target of the rules are closed.

For example, if we add $(s, L) \Rightarrow (s + a, L)$ to the rules of the pushdown system specification $\bar{\mathcal{P}}$ defined in Example 1, then it is not hard to see that there are infinitely many different location reachable from $\langle a, L \rangle$, meaning that $\bar{\mathcal{P}}$ is not locally finite w.r.t. the initial configuration $\langle a, L \rangle$. However, if instead we add the rule $(s, L) \Rightarrow (s, LL)$ then all reachable configurations from $\langle a, L \rangle$ will only use a or 0 as control locations and L as the only element of the stack alphabet. In this case $\bar{\mathcal{P}}$ is locally finite w.r.t. the initial configuration $\langle a, L \rangle$.

2.1 A Model Checking Algorithm for PSS

Next we describe a model checking algorithm for (locally finite) pushdown system specifications. We adapt the algorithm for checking LTL formulae against pushdown systems, as presented in [17], which, in turn, exploits the result from [1], where it is proved that for any finite pushdown system the set $R(\hat{c}_0)$ of all configurations reachable from the initial configuration $\hat{c}_0$ is regular. The LTL model checking algorithm in [17] starts by constructing a finite automaton which recognizes this set $R(\hat{c}_0)$. This automaton has the property that $\langle \hat{s}, \hat{\Gamma} \rangle \in R(\hat{c}_0)$ if the string $\hat{\Gamma}$ is accepted in the automaton, starting from $\hat{s}$.

According to [17], the automaton associated to $R(\hat{c}_0)$, denoted by A_{post*}, can be constructed in a forward manner starting with $\hat{c}_0$, as described in Fig. 1. We use the notation $\hat{x} \in T_\Sigma(Var)$ for closed terms representing control states in $\bar{\mathcal{P}}$, $\hat{\gamma}, \hat{\gamma}_1, \hat{\gamma}_2 \in T_\Xi(Var)$ for closed terms representing stack letters, $\hat{y}_{\hat{x}, \hat{\gamma}}$ for the new states of the A_{post*} automaton, f for the final states in A_{post*}, while $\hat{y}, \hat{z}, \hat{u}$ stand for any state in A_{post*}. The transitions in A_{post*} are denoted by $\hat{y} \overset{\hat{\gamma}}{\leadsto} \hat{z}$ or $\hat{y} \overset{\varepsilon}{\leadsto} \hat{z}$. The notation $\hat{y} \overset{\hat{\Gamma}}{\leadsto} \hat{z}$, where $\hat{\Gamma} = \hat{\gamma}_1..\hat{\gamma}_n$, stands for $\hat{y} \overset{\hat{\gamma}_1}{\leadsto} .. \overset{\hat{\gamma}_n}{\leadsto} \hat{z}$.

In Fig. 1 we present how the reachability algorithm in [17] for generating A_{post*} can be adjusted to invariant model checking pushdown system specifications. We emphasize that the transformation is minimal and consists in:

(a) The modification in the lines containing the code:
    "**for all** $\hat{z}$ such that $\langle \hat{x}, \hat{\gamma} \rangle \hookrightarrow \langle \hat{z}, \hat{\ } \rangle$ is a rule in the pushdown system **do**"
    i.e., lines 9, 12, 15 in Fig. 1, where instead of *rules in the pushdown system* we use *transitions derivable from the pushdown system specification* as follows:
    "**for all** $\hat{z}$ such that $\langle \hat{x}, \hat{\gamma} \rangle \longrightarrow \langle \hat{z}, \hat{\ } \rangle$ is derivable from $\bar{\mathcal{P}}$ **do**"
(b) The addition of lines 1, 10, 13, 16 where the state invariant ϕ is checked to hold in the newly discovered control state y.

This approach for producing the A_{post*} in a "breadth-first" manner is particularly suitable for specifications of pushdown systems as we can use the newly discovered configurations to produce transitions based on Δ, the production rules in $\bar{\mathcal{P}}$. Note that we assume, without loss of generality, that the initial stack has one symbol on it.

Note that in the algorithm Apost* of [17], the set of states of the automaton is determined statically at the beginning. This is clearly not possible starting with a PSS, because this set is not known in advance, and could be infinite if the algorithm does not terminate. Hence, the states that are generated when needed, that is, in line 9, 12 and 15, where the derivable transitions are considered.

We give next some keynotes on the algorithm in Fig. 1. The "trans" variable is a set containing the transitions to be processed. Along the execution of the algorithm **Apost***$(\phi, \bar{\mathcal{P}})$, the transitions of the A_{post*} automaton are incrementally deposited in the "rel" variable which is a set where we collect transitions in the A_{post*} automaton. The outermost **while** is executed until the end, i.e., until "trans" is empty, only if all states satisfy the control state formula ϕ. Hence, the algorithm in Fig. 1 verifies the invariant $\Box\phi$. In case ϕ is a state invariant for the pushdown system specification, the algorithm collects in "rel" the entire automaton A_{post*}. Otherwise, the algorithm stops at the first encountered state x which does not satisfy the invariant ϕ.

Note that the algorithm in Fig. 1 assumes that the pushdown system specification has only rules which push on the stack at most two stack letters. This

Algorithm Apost*$(\phi, \bar{\mathcal{P}})$

Input: a initial concrete configuration $\langle \hat{x}_0, \hat{\gamma}_0 \rangle$.

1 **if** $\hat{x}_0 \not\models \phi$ **then return** false;

2 trans $:= \{\hat{x}_0 \xrightarrow{\hat{\gamma}_0} f\}$;

3 rel $:= \emptyset$;

4 **while** trans $= \{\hat{x} \xrightarrow{\hat{\gamma}} \hat{y}\} \cup$ trans' **do**

5   trans $:=$ trans';

6   **if** $\hat{x} \xrightarrow{\hat{\gamma}} \hat{y} \notin$ rel **then**

7   rel $:=$ rel $\cup \{\hat{x} \xrightarrow{\hat{\gamma}} \hat{y}\}$;

8   **if** $\hat{\gamma} \neq \varepsilon$ **then**

9   **for all** $\hat{z}$ such that $\langle \hat{x}, \hat{\gamma} \rangle \longrightarrow \langle \hat{z}, \varepsilon \rangle$ is derivable from $\bar{\mathcal{P}}$ **do**

10   **if** $\hat{z} \not\models \phi$ **then return** false;

11   trans $:=$ trans $\cup \{\hat{z} \xrightarrow{\varepsilon} \hat{y}\}$;

12   **for all** $\hat{z}$ such that $\langle \hat{x}, \hat{\gamma} \rangle \longrightarrow \langle \hat{z}, \hat{\gamma}_1 \rangle$ is derivable from $\bar{\mathcal{P}}$ **do**

13   **if** $\hat{z} \not\models \phi$ **then return** false;

14   trans $:=$ trans $\cup \{\hat{z} \xrightarrow{\hat{\gamma}_1} \hat{y}\}$;

15   **for all** $\hat{z}$ such that $\langle \hat{x}, \hat{\gamma} \rangle \longrightarrow \langle \hat{z}, \hat{\gamma}_1 \hat{\gamma}_2 \rangle$ is derivable from $\bar{\mathcal{P}}$ **do**

16   **if** $\hat{z} \not\models \phi$ **then return** false;

17   trans $:=$ trans $\cup \{\hat{z} \xrightarrow{\hat{\gamma}_1} \hat{y}_{\hat{z},\hat{\gamma}_1}\}$;

18   rel $:=$ rel $\cup \{\hat{y}_{\hat{z},\hat{\gamma}_1} \xrightarrow{\hat{\gamma}_2} \hat{y}\}$;

19   **for all** $\hat{u} \xrightarrow{\varepsilon} \hat{y}_{\hat{z},\hat{\gamma}_1} \in$ rel **do**

20   trans $:=$ trans $\cup \{\hat{u} \xrightarrow{\hat{\gamma}_2} \hat{y}\}$;

21   **else**

22   **for all** $\hat{y} \xrightarrow{\hat{\gamma}_1} \hat{z} \in$ rel **do**

23   trans $:=$ trans $\cup \{\hat{x} \xrightarrow{\hat{\gamma}_1} \hat{z}\}$;

24 **od**;

25 **return** true

Fig. 1. The algorithm for obtaining A_{post*} adapted for pushdown system specifications

assumption is inherited from the algorithm for A_{post*} in [17] where the requirement is imposed without loss of generality. The approach in [17] is to adopt a standard construction for pushdown systems which consists in transforming the rules that push on the stack more than two stack letters into multiple rules that push at most two letters. Namely, any rule $\hat{r}$ in the pushdown system, of the form $\langle \hat{x}, \hat{\gamma} \rangle \hookrightarrow \langle \hat{x}', \hat{\gamma}_1 .. \hat{\gamma}_n \rangle$ with $n \geq 3$, is transformed into the following rules:

$$\langle \hat{x}, \hat{\gamma} \rangle \hookrightarrow \langle \hat{x}', \hat{\nu}_{\hat{r}, n-2} \hat{\gamma}_n \rangle, \langle \hat{x}', \hat{\nu}_{\hat{r}, i} \rangle \hookrightarrow \langle \hat{x}', \hat{\nu}_{\hat{r}, i-1} \hat{\gamma}_{i+1} \rangle, \langle \hat{x}', \hat{\nu}_{\hat{r}, 1} \rangle \hookrightarrow \langle \hat{x}', \hat{\gamma}_1 \hat{\gamma}_2 \rangle$$

where $2 \leq i \leq n-2$ and $\hat{\nu}_{\hat{r}, 1}, .., \hat{\nu}_{\hat{r}, n-2}$ are new stack letters. This transformation produces a new pushdown system which simulates the initial one, hence the assumption in the A_{post*} generation algorithm does not restrict the generality.

However, the aforementioned assumption makes impossible the application of the algorithm **Apost*** to pushdown system specifications $\bar{\mathcal{P}}$ for which the stack can be increased with any number of stack symbols. The reason is that $\bar{\mathcal{P}}$ defines rule schemas and we cannot identify beforehand which rule schema applies for

15   **for all** $\hat{z}$ such that $\langle \hat{x}, \hat{\gamma} \rangle \longrightarrow \langle \hat{z}, \hat{\gamma}_1..\hat{\gamma}_n \rangle$ is derivable from $\bar{\mathcal{P}}$ with $n \geq 2$ **do**

16   **if** $\hat{z} \not\models \phi$ **then return** false;

17   trans := trans $\cup \{ \hat{z} \overset{\hat{\gamma}_1}{\leadsto} \hat{y}_{\hat{z}, \hat{\gamma}_1} \};$

18   rel := rel $\cup \{ \hat{y}_{\hat{z}, \nu(\hat{r}, i)} \overset{\hat{\gamma}_{i+2}}{\leadsto} \hat{y}_{\hat{z}, \nu(\hat{r}, i+1)} \mid 0 \leq i \leq n - 2 \};$
          where $\hat{r}$ denotes $\langle \hat{x}, \hat{\gamma} \rangle \longrightarrow \langle \hat{z}, \hat{\gamma}_1..\hat{\gamma}_n \rangle$
          and $\nu(\hat{r}, i), 1 \leq i \leq n - 2$ are new symbols
          (i.e., ν is a new function symbol s.t. $ar(\nu) = 2$)
          and $\hat{y}_{\hat{z}, \nu(\hat{r}, 0)} = \hat{y}_{\hat{z}, \hat{\gamma}_1}$ and $\hat{y}_{\hat{z}, \nu(\hat{r}, n-1)} = \hat{y}$

19   **for all** $\hat{u} \overset{\varepsilon}{\leadsto} \hat{y}_{\hat{z}, \nu(\hat{r}, i)} \in$ rel, $0 \leq i \leq n - 2$ **do**

20   trans := trans $\cup \{ \hat{u} \overset{\hat{\gamma}_{i+2}}{\leadsto} \hat{y}_{\hat{z}, \nu(\hat{r}, i+1)} \mid 0 \leq i \leq n - 2 \};$

Fig. 2. The modification required by the generalization of the algorithm **Apost***

which concrete configuration, i.e., we cannot identify the $\hat{r}$ in $\nu_{\hat{r}, i}$. Our solution is to obtain a similar transformation on-the-fly, as we apply the **Apost*** algorithm and discover instances of rule schemas which increase the stack, i.e., we discover $\hat{r}$. This solution induces a localized modification of the lines 15 through 20 of the **Apost*** algorithm, as described in Fig. 2. We denote by **Apost*gen** the **Apost*** algorithm in Fig. 1 with the lines 15 through 20 replaced by the lines in Fig. 2. The correctness of the new algorithm is a rather simple generalization of the one presented in [17].

3   Specification of Pushdown Systems in $\mathbb{K}$

In this section we introduce $\mathbb{K}$ by means of an example of a PSS defined using $\mathbb{K}$, and we justify why $\mathbb{K}$ is an appropriate environment for PSS.

A $\mathbb{K}$ specification evolves around its *configuration*, a nested bag of labeled cells denoted as $\langle content \rangle_{\mathsf{label}}$, which defines the state of the specified transition system. The movement in the transition system is triggered by the $\mathbb{K}$ rules which define transformations made to the configuration. A key component in this mechanism is introduced by a special cell, labeled k, which contains a list of *computational tasks* that are used to trigger computation steps. As such, the $\mathbb{K}$ rules that specify transitions discriminate the modifications made upon the configuration based on the *current computation task*, i.e., the first element in the k-cell. This instills the stack aspect to the k-cell and induces the resemblance with a PSS. Namely, in a $\mathbb{K}$ configuration we make the conceptual separation between the k-cell, seen as the stack, and the rest of the cells which form the control location. Consequently, we promote $\mathbb{K}$ as a suitable environment for PSS.

In the remainder of this section we describe the $\mathbb{K}$ definition of Shylock by means of a PSS that is based on the operational semantics of Shylock introduced in [16]. In Section 3.1 we present the configuration of Shylock's $\mathbb{K}$ implementation with emphasis on the separation between control locations and stack elements. In Section 3.2 we introduce the $\mathbb{K}$ rules for Shylock, while in Section 3.3 we point out a methodology of defining in $\mathbb{K}$ production rules for PSS. We use this definition to present $\mathbb{K}$ notations and to further emphasize and standardize a $\mathbb{K}$ style for defining PSS.

3.1 Shylock's $\mathbb{K}$ Configuration

The PSS corresponding to Shylock's semantics is given in terms of a programming language specification. First, we give a short overview of the syntax of Shylock as in [16], then describe how this syntax is used in Shylock's $\mathbb{K}$-configuration.

A Shylock program is finite set of *procedure declarations* of the form $p_i :: B_i$, where B_i is the *body* of procedure p_i and denotes a statement defined by the grammar:

$$B ::= a.f := b \mid a := b.f \mid a := \text{new} \mid [a = b]B \mid [a \neq b]B \mid B + B \mid B;\ B \mid p$$

We use a and b for program variables ranging over $G \cup L$, where G and L are two disjoint finite sets of global and local variables, respectively. Moreover we assume a finite set F of field names, ranged over by f. G, L, F are assumed to be defined for each program, as sets of *Ids*, and we assume a distinguished initial program procedure `main`.

Hence, the body of a procedure is a sequence of statements that can be: assignments or object creation denoted by the function "_:=_" where $ar(:=) = 2$ (we distinguish the object creation by the "new" constant appearing as the second argument of ":="); conditional statements denoted by "[_]_"; nondeterministic choice given by "_+_"; and function calls. Note that $\mathbb{K}$ proposes the BNF notation for defining the language syntax as well, with the only difference that the variables are replaced by their respective sorts.

A $\mathbb{K}$ configuration is a nested bag of labeled cells where the cell content can be one of the predefined types of $\mathbb{K}$, namely $K, Map, Set, Bag, List$. The $\mathbb{K}$ configuration used for the specification of Shylock is the following:

$$\langle K \rangle_k\ \langle\langle Map \rangle_{var}\ \langle\langle Map \rangle_{fld^*}\rangle_h\rangle_{heap}\ \langle\langle Set \rangle_G\ \langle Set \rangle_L\ \langle Set \rangle_F\ \langle Map \rangle_P\rangle_{pgm}\ \langle K \rangle_{kAbs}$$

The **pgm**-cell is designated as a program container where the cells G, L, F maintain the above described finite sets of variables and fields associated to a program, while the cell P maintains the set of procedures stored as a map, i.e., a set of map items $p \mapsto B$.

The **heap**-cell contains the current heap H which is formed by the variable assignment cell **var** and the field assignment cell **h**. The **var** cell contains the mapping from local and global variables to their associated identities ranging over $\mathbb{N}_\perp = \mathbb{N} \cup \{\perp\}$, where $\perp$ stands for "not-created". The **h** cell contains a set of **fld** cells, each cell associated to a field variable from F. The mapping associated to each field contains items of type $n \mapsto m$, where n, m range over the object identities space $\mathbb{N}_\perp$. Note that any **fld**-cell always contains the item $\perp \mapsto \perp$ and $\perp$ is never mapped to another object identity.

Intuitively, the contents of the **heap**-cell form a directed graph with nodes labeled by object identities (i.e., values from $\mathbb{N}_\perp$) and arcs labeled by field names. Moreover, the contents of the **var**-cell (i.e., the variable assignment) define *entry*

nodes in the graph. We use the notion of *visible heap*, denoted as $\mathcal{R}(H)$, for the set of nodes reachable in the heap H from the entry nodes.

The k-cell maintains the current continuation of the program, i.e., a list of syntax elements that are to be executed by the program. Note that the sort K is tantamount with an associative list of items separated by the set-aside symbol "$\curvearrowright$". The kAbs-cell is introduced for handling the heap modifications required by the semantics of certain syntactic operators. In this way, we maintain in the cell k only the "pure" syntactic elements of the language, and move into kAbs any additional computational effort used by the abstract semantics for object creation, as well as for procedure call and return.

In conclusion, the k-cell stands for the stack in a PSS $\bar{\mathcal{P}}$, while all the other cells, including kAbs, form together the control location. Hence the language syntax in $\mathbb{K}$ practically gives a sub-signature of the stack signature in $\bar{\mathcal{P}}$, while the rest of the cells give a sub-signature, the control location signature in $\bar{\mathcal{P}}$.

3.2   Shylock's $\mathbb{K}$ Rules

We present here the $\mathbb{K}$ rules which implement the abstract semantics of Shylock, according to [16]. Besides introducing the $\mathbb{K}$ notation for rules, we also emphasize on the separation of concerns induced by viewing the $\mathbb{K}$ definitions as PSS.

In $\mathbb{K}$ we distinguish between computational rules that describe state transitions, and structural rules that only prepare the current state for the next transition. Rules in $\mathbb{K}$ have a bi-dimensional localized notation that stands for "what is above a line is rewritten into what is bellow that line in a particular context given by the matching with the elements surrounding the lines". Note that the solid lines encode a *computational rule* in $\mathbb{K}$ which is associated with a rewrite rule, while the dashed lines denote a *structural rule* in $\mathbb{K}$, which is compiled by the $\mathbb{K}$-tool into a Maude equation.

The production rules in PSS are encoded in $\mathbb{K}$ by computational rules which basically express changes to the configuration triggered by an atomic piece of syntax matched at the top of the stack, i.e., the k-cell. An example of such encoding is the following rule:

$$\text{RULE } \langle \underline{a.f := b} \ \cdots \rangle_\mathsf{k} \ \langle \cdots \ v(a) \mapsto \underline{\quad\_\quad} \ \cdots \rangle_{\mathsf{fld}(f)} \ \langle v \rangle_\mathsf{var} \quad \text{when } v(a) \neq_{Bool} \bot$$

which reads as: if the first element in the cell k is the assignment $a.f := b$ then this is consumed from the stack and the map associated to the field f, i.e., the content of the cell $\mathsf{fld}(f)$, is modified by replacing whatever object identity was pointed by $v(a)$ with $v(b)$, i.e., the object identity associated to the variable b by the current variable assignment v, only when a is already created, i.e., $v(a)$ is not $\bot$. Note that this rule is conditional, the condition being introduced by the keyword "when".

We emphasize the following notational elements in $\mathbb{K}$ that appear in the above rule: "_" which stands for "anything" and the ellipses "$\cdots$". The meaning of the ellipses is basically the same as "_" the difference being that the ellipses appear

always near the cell walls and are interpreted according to the contents of the respective cell. For example, given that the content of the k-cell is a list of computational tasks separated by "$\curvearrowright$", the ellipses in the k-cell from the above rule signify that the assignment $a.f := b$ is at the top of the stack of the PSS. On the other hand, because the content of a fld cell is of sort Map which is a commutative sequence of map items, the ellipses appearing by both walls of the cell fld denote that the item $v(a) \mapsto \_$ may appear "anywhere" in the fld-cell. Meanwhile, the notation for the var cell signifies that v is the entire content of this cell, i.e., the map containing the variable assignment. Finally, "$\cdot$" stands for the null element in any $\mathbb{K}$ sort, hence "$\cdot$" replacing $a.f := b$ at the top of the k-cell stands for ε from the production rules in $\bar{\mathcal{P}}$.

All the other rules for assignment, conditions, and sequence are each implemented by means of a single computational rule which considers the associated piece of syntax at the top of the k-cell. The nondeterministic choice is implemented by means of two computational rules which replace $B_1 + B_2$ at the top of a k-cell by either B_1 or B_2.

Next we present the implementation of one of the most interesting rules in Shylock namely object creation. The common semantics for an object creation is the following: if the current computation (the first element in the cell k) is "$a:=\mathbf{new}$", then whatever object was pointed by a in the var-cell is replaced with the "never used before" object "$oNew$" obtained from the cell $\langle \_ \rangle_{\mathsf{kAbs}}$. Also, the fields part of the heap, i.e., the content of h-cell, is updated by the addition of a new map item "$oNew \mapsto \perp$". However, in the semantics proposed by Shylock, the value of $oNew$ is *the minimal address not used in the current visible heap* which is calculated by the function $\min(\mathcal{R}(H)^c)$ that ends in the normal form $\mathsf{oNew}(n)$. This represents the memory reuse mechanism which is handled in our implementation by the kAbs-cell. Hence, the object creation rules are:

$$\text{RULE } \langle a := \mathbf{new} \ \cdots \rangle_{\mathsf{k}} \ \langle H \rangle_{\mathsf{heap}} \ \Big\langle \underbrace{\quad \cdot \quad}_{\min(\mathcal{R}(H)^c)} \Big\rangle_{\mathsf{kAbs}}$$

$$\text{RULE } \langle a := \mathbf{new} \ \cdots \rangle_{\mathsf{k}} \ \langle H_h \rangle_{\mathsf{h}} \ \Big\langle \mathsf{oNew}(n) \curvearrowright \underbrace{\quad \cdot \quad}_{\text{update } H_h \text{ with } n \mapsto \perp} \Big\rangle_{\mathsf{kAbs}}$$

$$\text{RULE } \langle \underbrace{a := \mathbf{new}}_{\cdot} \ \cdots \rangle_{\mathsf{k}} \ \langle \cdots \ x \mapsto \underbrace{\quad \_ \quad}_{n} \ \cdots \rangle_{\mathsf{var}} \ \Big\langle \underbrace{H_h}_{H'_h} \Big\rangle_{\mathsf{h}} \ \langle \underbrace{\mathsf{oNew}(n) \curvearrowright \mathsf{updated}(H'_h)}_{\cdot} \rangle_{\mathsf{kAbs}}$$

where "$\min(\mathcal{R}(H)^c)$" finds n, the first integer not in $\mathcal{R}(H)$, and ends in $\mathsf{oNew}(n)$, then "$\mathbf{update} \ Bag \ \mathbf{with} \ MapItem$" adds $n \mapsto \perp$ to the map in each cell fld contained in the h-cell and ends in the normal form $\mathsf{updated}(Bag)$. Note that all the operators used in the kAbs-cell are implemented equationally, by means of structural $\mathbb{K}$-rules. In this manner, we ensure that the computational rule which consumes $a := \mathbf{new}$ from the top of the k-cell is accurately updating the control location with the required modification.

The rules for procedure call/return are presented in Fig. 3. They follow the same pattern as the one proposed in the rules for object creation. The renaming scheme defined for resolving name clashes induced by the memory reuse for

RULE $\langle p \ \cdots\rangle_k \ \langle H\rangle_{\text{heap}}\langle L\rangle_L\langle G\rangle_G\langle F\rangle_F \ \langle\cdots p \mapsto B \ \cdots\rangle_P \ \langle \underline{\cdot}\rangle_{\text{kAbs}}$
$$\overline{\text{processingCall}(H,L,G,F)}$$

RULE $\left\langle \dfrac{p}{B \curvearrowright \text{restore}(H)} \ \cdots\right\rangle_k \ \left\langle \dfrac{H}{H'} \right\rangle_{\text{heap}} \langle\cdots p \mapsto B \ \cdots\rangle_P \ \langle\underline{\text{processedCall}(H')}\rangle_{\text{kAbs}}$

RULE $\langle \text{restore}(H') \ \cdots\rangle_k \ \langle H\rangle_{\text{heap}}\langle L\rangle_L\langle G\rangle_G\langle F\rangle_F \ \langle \underline{\cdot}\rangle_{\text{kAbs}}$
$$\overline{\text{processingRet}(H,H',L,G,F)}$$

RULE $\langle \underline{\text{restore}(\_)} \ \cdots\rangle_k \ \left\langle \dfrac{H}{H'} \right\rangle_{\text{heap}} \langle\underline{\text{processedRet}(H')}\rangle_{\text{kAbs}}$

Fig. 3. K-rules for the procedure's call and return in Shylock

object creation is based in Shylock on the concept of *cut points* as introduced in [13]. Cut points are objects in the heap that are referred to from both local and global variables, and as such, are subject to modifications during a procedure call. Recording cut points in extra logical variables allows for a sound return in the calling procedure, enabling a precise abstract execution w.r.t. object identities. For more details on the semantics of Shylock we refer to [16].

3.3   Shylock as PSS

The benefit of a Shylock's K specification lies in the rules for object creation, which implement the memory reuse mechanism, and for procedure call/return, which implement the renaming scheme. Each element in the memory reuse mechanism is implemented equationally, i.e., by means of structural K rules which have equational interpretation when compiled in Maude. Hence, if we interpret Shylock as an abstract model for the standard semantics, i.e., with standard object creation, the K specification for Shylock's abstract semantics renders an equational abstraction. As such, Shylock is yet another witness to the versatility of the equational abstraction methodology [11].

Under the assumption of a bounded heap, the K specification for Shylock is a locally finite PSS and compiles in Maude into a rewriting system. Obviously, in the presence of recursive procedures, the stack grows unboundedly and, even if Shylock produces a finite pushdown system, the equivalent transition system is infinite and so is the associated rewriting system. We give next a relevant example for this idea.

Example 2. The following Shylock program, denoted as pgm0, is the basic example we use for Shylock. It involves a recursive procedure p0 which creates an object g.

$$\text{gvars: g} \qquad \text{main :: p0} \qquad \text{p0 :: g:=new; p0}$$

In a standard semantics, because the recursion is infinite, so is the set of object identities used for g. However, Shylock's memory reuse guarantees to produce a finite set of object identities, namely $\perp, 0, 1$. Hence, the pushdown system associated to pgm0 Shylock program is finite and has the following (ground) rules:

$$(g{:}\perp, \texttt{main}) \hookrightarrow (g{:}\perp, \texttt{p0}; \texttt{restore}(g{:}\perp))$$

$$(g{:}\perp, \texttt{p0}) \hookrightarrow (g{:}\perp, g := \texttt{new}; \texttt{p0}; \texttt{restore}(g{:}\perp)) \quad (g{:}\perp, g := \texttt{new}) \hookrightarrow (g{:}0, \epsilon)$$

$$(g{:}0, \texttt{p0}) \hookrightarrow (g{:}0, g := \texttt{new}; \texttt{p0}; \texttt{restore}(g{:}0)) \quad (g{:}0, g := \texttt{new}) \hookrightarrow (g{:}1, \epsilon)$$

$$(g{:}1, \texttt{p0}) \hookrightarrow (g{:}1, g := \texttt{new}; \texttt{p0}; \texttt{restore}(g{:}1)) \quad (g{:}1, g := \texttt{new}) \hookrightarrow (g{:}0, \epsilon)$$

Note that we cannot obtain the pushdown system by the exhaustive execution of Shylock[pgm0] because the exhaustive execution is infinite due to recursive procedure p0. For the same reason, Shylock[pgm0] specification does not comply with Maude's LTL model checker prerequisites. Moreover, we cannot use directly the dedicated pushdown systems model checkers as these work with the pushdown system automaton, while Shylock[pgm0] is a pushdown system specification. This example creates the premises for the discussion in the next section where we present a $\mathbb{K}$-specification of a model checking procedure amenable for pushdown systems specifications.

4   Model Checking $\mathbb{K}$ Definitions

We recall that the PSS perspective over the $\mathbb{K}$ definitions enables the verification by model checking of a richer class of programs which allow (infinite) recursion. In this section we focus on describing $kA_{post*}(\phi, \bar{\mathcal{P}})$, the $\mathbb{K}$ specification of the algorithm **Apost*gen**. Note that $kA_{post*}(\phi, \bar{\mathcal{P}})$ is parametric, where the two parameters are $\bar{\mathcal{P}}$, the $\mathbb{K}$ specification of a pushdown system, and ϕ a control state invariant. We describe $kA_{post*}(\phi, \bar{\mathcal{P}})$ along justifying the behavioral equivalence with the algorithm **Apost*gen**.

The **while** loop in **Apost*gen**, in Fig. 1, is maintained in kA_{post*} by the application of rewriting, until the term reaches the normal form, i.e. no other rule can be applied. This is ensured by the fact that from the initial configuration:

$$Init \equiv \langle \cdot \rangle_{\text{traces}} \langle \cdot \rangle_{\text{traces}'} \langle \langle x_0 \overset{\gamma_0}{\rightsquigarrow} f \rangle_{\text{trans}} \langle \cdot \rangle_{\text{rel}} \langle \cdot \rangle_{\text{memento}} \langle \phi \rangle_{\text{formula}} \langle true \rangle_{\text{return}} \rangle_{\text{collect}}$$

the rules keep applying, as long as trans-cell is nonempty.

We assume that the rewrite rules are applied at-random, so we need to direct/pipeline the flow of their application via matching and conditions. The notation RULEi [*label*] in the beginning of each rule hints, via [*label*], towards which part of the **Apost*gen** algorithm that rule is handling. In the followings we discuss each rule and justify its connection with code fragments in **Apost*gen**.

The last rule, RULE$\mathcal{P}$, performs the exhaustive unfolding for a particular configuration in cell trace. We use this rule in order to have a parametric definition of the kA_{post*} specification, where one of the parameters is $\bar{\mathcal{P}}$, i.e., the $\mathbb{K}$ specification of the pushdown system. Recall that the other parameter is the specification of the language defining the control state invariant properties ϕ which are to be verified on the produced pushdown system. RULE$\mathcal{P}$ takes $\langle x \rangle_{\text{ctrl}} \langle \gamma \curvearrowright \Gamma \rangle_{\text{k}}$

$Init \equiv \langle \cdot \rangle_{\text{traces}} \langle \cdot \rangle_{\text{traces}'} \langle \langle x_0 \overset{\gamma_0}{\leadsto} f \rangle_{\text{trans}} \langle \cdot \rangle_{\text{rel}} \langle \cdot \rangle_{\text{memento}} \langle \phi \rangle_{\text{formula}} \langle true \rangle_{\text{return}} \rangle_{\text{collect}}$

RULE1 $[\textbf{if } \hat{x} \overset{\hat{\gamma}}{\leadsto} \hat{y} \notin \text{rel } \textbf{else}]$:

$\langle \cdot \rangle_{\text{traces}} \langle \cdot \rangle_{\text{traces}'} \langle \cdots \langle \cdots \underline{\quad x \overset{\gamma}{\leadsto} y \quad} \cdots \rangle_{\text{trans}} \langle \cdots \quad x \overset{\gamma}{\leadsto} y \quad \cdots \rangle_{\text{rel}} \langle \cdot \rangle_{\text{memento}} \cdots \rangle_{\text{collect}}$

RULE2 $[\textbf{if } (\hat{x} \overset{\hat{\gamma}}{\leadsto} \hat{y} \notin \text{rel } \textbf{then}...\textbf{if } \hat{\gamma} \neq \epsilon \textbf{ else}]$:

$\langle \cdot \rangle_{\text{traces}} \langle \cdot \rangle_{\text{traces}'} \langle \cdots \langle \cdots \underset{(x \leadsto Rel[y \overset{-}{\leadsto}])}{\underline{\quad x \overset{\epsilon}{\leadsto} y \quad}} \cdots \rangle_{\text{trans}} \langle \underset{x \overset{\epsilon}{\leadsto} y}{\underline{\quad \cdot \quad}} Rel \rangle_{\text{rel}} \langle \cdot \rangle_{\text{memento}} \cdots \rangle_{\text{collect}}$

when $x \overset{\epsilon}{\leadsto} y \notin Rel$

RULE3 $[\textbf{if } \hat{x} \overset{\hat{\gamma}}{\leadsto} \hat{y} \notin \text{rel } \textbf{then}...\textbf{if } \hat{\gamma} \neq \epsilon \textbf{ then}]$:

$\langle \underset{\langle \langle x \rangle_{\text{ctrl}} \langle \gamma \rangle_{\text{k}} \rangle_{\text{trace}}}{\underline{\quad \cdot \quad}} \rangle_{\text{traces}} \langle \cdot \rangle_{\text{traces}'} \langle \cdots \langle \cdots x \overset{\gamma}{\leadsto} y \cdots \rangle_{\text{trans}} \langle \underset{x \overset{\gamma}{\leadsto} y}{\underline{\quad \cdot \quad}} Rel \rangle_{\text{rel}} \langle \underset{x \overset{\gamma}{\leadsto} y}{\underline{\quad \cdot \quad}} \rangle_{\text{memento}} \cdots \rangle_{\text{collect}}$

when $x \overset{\gamma}{\leadsto} y \notin Rel \ and_{Bool} \ \gamma \neq \varepsilon$

RULE4 $[\textbf{for all } \hat{z} \text{ s.t. } \langle \hat{x}, \hat{\gamma} \rangle \longrightarrow \langle \hat{z}, \varepsilon | \hat{\gamma} | \hat{\gamma}_1..\hat{\gamma}_n \rangle \text{ is derivable from } \bar{\mathcal{P}} \textbf{ do if } \hat{z} \not\models \phi \textbf{ then}]$:

$\langle \cdot \rangle_{\text{traces}} \langle \langle \langle z \rangle_{\text{ctrl}} \cdots \rangle_{\text{trace}} \underline{\ \_\ }\rangle_{\text{traces}'} \langle \langle \underline{\ \_\ } \rangle_{\text{trans}} \langle \underline{\ \_\ } \rangle_{\text{rel}} \langle \underline{\ \_\ } \rangle_{\text{memento}} \langle \phi \rangle_{\text{formula}} \langle \underset{false}{\underline{\quad true \quad}} \rangle_{\text{return}} \rangle_{\text{collect}}$

when $z \not\models \phi$

RULE5 $[\textbf{for all } \hat{z} \text{ s.t. } \langle \hat{x}, \hat{\gamma} \rangle \longrightarrow \langle \hat{z}, \varepsilon | \hat{\gamma}_1 \rangle \text{ is derivable from } \bar{\mathcal{P}} \textbf{ do}]$:

$\langle \cdot \rangle_{\text{traces}} \langle \cdots \underset{\cdot}{\underline{\langle \langle z \rangle_{\text{ctrl}} \langle \Gamma' \rangle_{\text{k}} \rangle_{\text{trace}}}} \cdots \rangle_{\text{traces}'} \langle \cdots \langle \cdots \underset{z \overset{\Gamma'}{\leadsto} y}{\underline{\quad \cdot \quad}} \cdots \rangle_{\text{trans}} \langle x \overset{\gamma}{\leadsto} y \rangle_{\text{memento}} \langle \phi \rangle_{\text{formula}} \cdots \rangle_{\text{collect}}$

when $|\Gamma'| \leq 1 \ and_{Bool} \ z \models \phi$

RULE6 $[\textbf{for all } \hat{z} \text{ s.t. } \langle \hat{x}, \hat{\gamma} \rangle \longrightarrow \langle \hat{z}, \hat{\gamma}_1..\hat{\gamma}_n \rangle \text{ is derivable from } \bar{\mathcal{P}} \textbf{ do}]$:

$\langle \cdot \rangle_{\text{traces}} \langle \cdots \underset{\cdot}{\underline{\langle \langle z \rangle_{\text{ctrl}} \langle \gamma' \curvearrowright \Gamma' \rangle_{\text{k}} \rangle_{\text{trace}}}} \cdots \rangle_{\text{traces}'}$

$\langle \cdots \underset{z \overset{\gamma'}{\leadsto} new(z,\gamma') \ (Rel[\overset{\varepsilon}{\leadsto} new(z,\gamma'), news(x,\gamma,z,\gamma',\Gamma')] \overset{\Gamma'}{\leadsto} news(x,\gamma,z,\gamma',\Gamma'), y)}{\underline{\qquad\qquad\qquad \cdot \qquad\qquad\qquad}} \cdots \rangle_{\text{trans}}$

$\langle Rel \underset{new(z,\gamma'), news(x,\gamma,z,\gamma',\Gamma') \overset{\Gamma'}{\leadsto} news(x,\gamma,z,\gamma',\Gamma'), y}{\underline{\qquad\qquad \cdot \qquad\qquad}} \rangle_{\text{rel}} \langle x \overset{\gamma}{\leadsto} y \rangle_{\text{memento}} \langle \phi \rangle_{\text{formula}}$

when $|\Gamma'| \geq 1 \ and_{Bool} \ z \models \phi$

RULE7 $[\textbf{for all } \hat{z} \text{ s.t. } \langle \hat{x}, \hat{\gamma} \rangle \longrightarrow \langle \hat{z}, \varepsilon | \hat{\gamma} | \hat{\gamma}_1..\hat{\gamma}_n \rangle \text{ is derivable from } \bar{\mathcal{P}} \textbf{ do}]$:

$\langle \cdot \rangle_{\text{traces}} \langle \cdot \rangle_{\text{traces}'} \langle \cdots \underset{\cdot}{\underline{\langle x \overset{\gamma}{\leadsto} y \rangle_{\text{memento}}}} \cdots \rangle_{\text{collect}}$

RULE$\mathcal{P}$ $[\textbf{all } \hat{z} \text{ s.t. } \langle \hat{x}, \hat{\gamma} \rangle \longrightarrow \langle \hat{z}, \hat{\Gamma'} \rangle \text{ is derivable from } \bar{\mathcal{P}}]$:

$\langle \cdots \underset{\cdot}{\underline{\langle \langle x \rangle_{\text{ctrl}} \langle \gamma \curvearrowright \Gamma \rangle_{\text{k}} \rangle_{\text{trace}}}} \cdots \rangle_{\text{traces}} \langle \cdots \underset{\langle \langle z_0 \rangle_{\text{ctrl}} \langle \Gamma_0 \curvearrowright \Gamma \rangle_{\text{k}} \rangle_{\text{trace}} .. \langle \langle z_n \rangle_{\text{ctrl}} \langle \Gamma_n \curvearrowright \Gamma \rangle_{\text{k}} \rangle_{\text{trace}}}{\underline{\qquad\qquad\qquad \cdot \qquad\qquad\qquad}} \cdots \rangle_{\text{traces}'}$

Fig. 4. $kA_{post^*}(\phi, \bar{\mathcal{P}})$

a configuration in $\bar{\mathcal{P}}$ and gives, based on the rules in $\bar{\mathcal{P}}$, all the configurations $\langle z_i \rangle_{\mathsf{ctrl}} \langle \Gamma_i \curvearrowright \Gamma \rangle_{\mathsf{k}}, 0 \leq i \leq n$ obtained from $\langle x \rangle_{\mathsf{ctrl}} \langle \gamma \curvearrowright \Gamma \rangle_{\mathsf{k}}$ after exactly one rewrite.

The pipeline stages are the following sequence of rules' application:

$$\text{RULE3} \, \text{RULE}\mathcal{P} (\text{RULE4} + \text{RULE5} + \text{RULE6})^* \text{RULE7}$$

The cell memento is filled in the beginning of the pipeline, RULE3, and is emptied at the end of the pipeline, RULE7. We use the matching on a nonempty memento for localizing the computation in **Apost*gen** at the lines $7 - 20$. We explain next the pipeline stages.

Firstly, note that when no transition derived from $\bar{\mathcal{P}}$ is processed by kA_{post^*} we enforce cells traces, traces' to be empty (with the matching $\langle \cdot \rangle_{\mathsf{traces}} \langle \cdot \rangle_{\mathsf{traces'}}$). This happens in RULEs 1 and 2 because the respective portions in **Apost*gen** do not need new transitions derived from $\bar{\mathcal{P}}$ to update "trans" and "rel".

The other cases, namely when the transitions derived from $\bar{\mathcal{P}}$ are used for updating "trans" and "rel", are triggered in RULE3 by placing the desired configuration in the cell traces, while the cell traces' is empty. At this point, since all the other rules match on either traces empty, or traces' nonempty, only RULE$\mathcal{P}$ can be applied. This rule populates traces' with all the next configurations obtained by executing $\bar{\mathcal{P}}$.

After the application of RULE$\mathcal{P}$, only one of the RULEs $4, 5, 6$ can apply because these are the only rules in kA_{post^*} matching an empty traces and a nonempty traces'.

Among the rules 4,5,6 the differentiation is made via conditions as follows:

RULE4 handles all the cases when the new configuration has a control location z which does not verify the state invariant ϕ (i.e., lines $10, 13, 16$ in **Apost*gen**). In this case we close the pipeline and the algorithm by emptying all the cells traces, traces, trans. Note that all the rules handling the **while** loop match on at least a nonempty cell traces, traces, or trans, with a pivot in a nonempty trans.

RULEs 5 and 6 are applied disjunctively of RULE4 because both have the condition $z \models \phi$. Next we describe these two rules. RULE5 handles the case when the semantic rule in $\bar{\mathcal{P}}$ which matches the current $< \hat{x}, \hat{\gamma} >$ does not increase the size of the stack. This case is associated with the lines 9 and 11, 12 and 14 in **Apost*gen**. RULE6 handles the case when the semantic rule in $\bar{\mathcal{P}}$ which matches the current $< \hat{x}, \hat{\gamma} >$ increases the stack size and is associated with lines 15 and $17 - 20$ in **Apost*gen**.

Both rules 5 and 6 use the memento cell which is filled upon pipeline initialization, in RULE3. The most complicated rule is RULE6, because it handles a **for all** piece of code, i.e., lines $17 - 20$ in Fig. 2. This part is reproduced by matching the entire content of cell rel with Rel, and using the projection operator:

$$Rel[\overset{\gamma}{\rightsquigarrow} z_1, .., z_n] := \{u \mid (u, \gamma, z_1) \in Rel\}, .., \{u \mid (u, \gamma, z_n) \in Rel\}$$

where $z_1, .., z_n$ in the left hand-side is a list of z-symbols, while in the right hand-side we have a list of sets. Hence, the notation:

$$(Rel[\overset{\epsilon}{\rightsquigarrow} new(z, \gamma'), news(x, \gamma, z, \gamma', \Gamma')] \overset{\Gamma'}{\rightsquigarrow} news(x, \gamma, z, \gamma', \Gamma'), y)$$

in RULE6 cell trans stands for the lines 17 and $19-20$ in Fig. 2. (Note that instead of notation $\hat{r}$ for rule $< \hat{x}, \hat{\gamma} > \longrightarrow < \hat{z}, \hat{\gamma}' \hat{\Gamma}' >$ we use the equivalent unique representation $(\hat{x}, \hat{\gamma}, \hat{z}, \hat{\gamma}', \hat{\Gamma}')$ and that instead of $\hat{y}_{\hat{z}, \nu(\hat{r}, 0)}$ we use directly $\hat{y}_{\hat{z}, \hat{\gamma}'}$, i.e., $new(z, \gamma')$, while instead of $\hat{y}_{\hat{z}, \nu(\hat{r}, n-1)}$ in Fig. 2 we use directly $\hat{y}$.) Also, the notation in cell rel: "$new(z, \gamma'), news(x, \gamma, z, \gamma', \Gamma') \overset{\Gamma'}{\rightsquigarrow} news(x, \gamma, z, \gamma', \Gamma'), y$" stands for line 18 in Fig. 2.

RULES $4, 5, 6$ match on a nonempty traces'-cell and an empty traces, and no other rule matches alike. RULE7 closes the pipeline when the traces' cell becomes empty by making the memento cell empty. Note that traces' empties because rules $4, 5, 6$ keep consuming it.

Example 3. We recall that the Shylock program pgm0 from Example 2 was not amenable by semantic exhaustive execution or Maude's LTL model checker, due to the recursive procedure p0. Likewise, model checkers for pushdown systems which can handle the recursive procedure p0 cannot be used because Shylock[pgm0], the pushdown system obtained from Shylock's PSS, is not available. However, we can employ kA_{post*} for Shylock's $\mathbb{K}$-specification in order to discover the reachable state space, the A_{post*} automata, as well as the pushdown system itself. In the Fig. 5 we describe the first steps in the execution of $kA_{post*}(true, \text{Shylock}[\text{pgm0}])$ and the reachability automaton generated automatically by $kA_{post*}(true, \text{Shylock}[\text{pgm0}])$.

4.1 Bounded Model Checking for Shylock

One of the major problems in model checking programs which manipulate dynamic structures, such as linked lists, is that it is not possible to bound a priori the state space of the possible computations. This is due to the fact that programs may manipulate the heap by dynamically allocating an unbounded number of new objects and by updating reference fields. This implies that the reachable state space is potentially infinite for Shylock programs with recursive procedures. Consequently for model checking purposes we need to impose some suitable bounds on the model of the program.

A natural bound for model checking Shylock programs, without necessarily restricting their capability of allocating an unbounded number of objects, is to impose constraints on the size of the *visible* heap [2]. Such a bound still allows for storage of an unbounded number of objects onto the call-stack, using local variables. Thus termination is guaranteed with heap-bounded model checking of the form $\models_k \Box \phi$ meaning $\models \Box \phi \wedge le(k)$, where $le(k)$ verifies if the size of the visible heap is smaller than k.

To this end, we define the set of atomic propositions ($\phi \in$) *Rite* as the smallest set defined by the following grammar:

$$r ::= \varepsilon \mid x \mid \neg x \mid f \mid r.r \mid r + r \mid r^*$$

where x ranges over variable names (to be used as tests) and f over field names (to be used as actions). The atomic proposition in *Rite* are basically expressions

$\langle\cdot\rangle_{\text{traces}}\ \langle\cdot\rangle_{\text{traces}'}\ \langle\ \langle\langle\langle g\mapsto\perp\rangle_{\text{var}}\langle\cdot\rangle_{\text{h}}\rangle_{\text{heap}}\overset{\text{main}}{\leadsto}\texttt{fin}\rangle_{\text{trans}}\langle\cdot\rangle_{\text{rel}}\langle\cdot\rangle_{\text{memento}}\langle true\rangle_{\text{formula}}\langle true\rangle_{\text{return}}\ \rangle_{\text{collect}}$

RULE3
$\Rightarrow\ \langle\langle\langle\langle g\mapsto\perp\rangle_{\text{var}}\langle\cdot\rangle_{\text{h}}\rangle_{\text{heap}}\rangle_{\text{ctrl}}\langle\texttt{main}\rangle_{\text{k}}\rangle_{\text{traces}}$

$\quad\langle\cdot\rangle_{\text{trans}}\ \langle\langle\langle g\mapsto\perp\rangle_{\text{var}}\langle\cdot\rangle_{\text{h}}\rangle_{\text{heap}}\overset{\text{main}}{\leadsto}\texttt{fin}\rangle_{\text{rel}}\ \langle\langle\langle g\mapsto\perp\rangle_{\text{var}}\langle\cdot\rangle_{\text{h}}\rangle_{\text{heap}}\overset{\text{main}}{\leadsto}\texttt{fin}\rangle_{\text{memento}}$

RULE$\mathcal{P}$
$\Rightarrow\ \langle\cdot\rangle_{\text{traces}}\ \langle\langle\langle\langle g\mapsto\perp\rangle_{\text{var}}\langle\cdot\rangle_{\text{h}}\rangle_{\text{heap}}\rangle_{\text{ctrl}}\langle\texttt{p0}\curvearrowright\texttt{restore}((\langle\langle g\mapsto\perp\rangle_{\text{var}}\langle\cdot\rangle_{\text{h}}\rangle_{\text{heap}}))_{\text{k}}\rangle_{\text{traces}'}$

RULE6
$\Rightarrow\ \langle\cdot\rangle_{\text{traces}}\ \langle\cdot\rangle_{\text{traces}'}\ \langle\langle\langle g\mapsto\perp\rangle_{\text{var}}\langle\cdot\rangle_{\text{h}}\rangle_{\text{heap}}\overset{\text{main}}{\leadsto}\texttt{fin}\rangle_{\text{memento}}$

$\quad\langle\ \langle\langle g\mapsto\perp\rangle_{\text{var}}\langle\cdot\rangle_{\text{h}}\rangle_{\text{heap}}\overset{\text{p0}}{\leadsto}\texttt{new}(\langle\langle g\mapsto\perp\rangle_{\text{var}}\langle\cdot\rangle_{\text{h}}\rangle_{\text{heap}},\texttt{p0})\ \rangle_{\text{trans}}$

$\quad\langle\ \langle\langle g\mapsto\perp\rangle_{\text{var}}\langle\cdot\rangle_{\text{h}}\rangle_{\text{heap}}\overset{\text{main}}{\leadsto}\texttt{fin}\quad\texttt{new}(\langle\langle g\mapsto\perp\rangle_{\text{var}}\langle\cdot\rangle_{\text{h}}\rangle_{\text{heap}},\texttt{p0})\overset{\text{restore}(...)}{\leadsto}\texttt{fin}\ \rangle_{\text{rel}}$

RULE7
$\Rightarrow\ \langle\cdot\rangle_{\text{traces}}\ \langle\cdot\rangle_{\text{traces}'}\ \langle\cdot\rangle_{\text{memento}}$

RULE3
$\Rightarrow\ \langle\langle\langle\langle g\mapsto\perp\rangle_{\text{var}}\langle\cdot\rangle_{\text{h}}\rangle_{\text{heap}}\rangle_{\text{ctrl}}\langle\texttt{p0}\rangle_{\text{k}}\rangle_{\text{traces}}$

$\quad\langle\cdot\rangle_{\text{trans}}\ \langle\langle\langle g\mapsto\perp\rangle_{\text{var}}\langle\cdot\rangle_{\text{h}}\rangle_{\text{heap}}\overset{\text{p0}}{\leadsto}\texttt{new}(\langle\langle g\mapsto\perp\rangle_{\text{var}}\langle\cdot\rangle_{\text{h}}\rangle_{\text{heap}},\texttt{p0})\rangle_{\text{memento}}$

$\quad\langle\ \langle\langle g\mapsto\perp\rangle_{\text{var}}\langle\cdot\rangle_{\text{h}}\rangle_{\text{heap}}\overset{\text{main}}{\leadsto}\texttt{fin}\quad\texttt{new}(\langle\langle g\mapsto\perp\rangle_{\text{var}}\langle\cdot\rangle_{\text{h}}\rangle_{\text{heap}},\texttt{p0})\overset{\text{restore}(...)}{\leadsto}\texttt{fin}$

$\quad\langle\langle g\mapsto\perp\rangle_{\text{var}}\langle\cdot\rangle_{\text{h}}\rangle_{\text{heap}}\overset{\text{p0}}{\leadsto}\texttt{new}(\langle\langle g\mapsto\perp\rangle_{\text{var}}\langle\cdot\rangle_{\text{h}}\rangle_{\text{heap}},\texttt{p0})\ \rangle_{\text{rel}}$

RULE$\mathcal{P}$
$\Rightarrow\ \langle\cdot\rangle_{\text{traces}}\langle\langle\langle\langle g\mapsto\perp\rangle_{\text{var}}\langle\cdot\rangle_{\text{h}}\rangle_{\text{heap}}\rangle_{\text{ctrl}}\langle\texttt{g}:=\texttt{new}\curvearrowright\texttt{p0}\curvearrowright\texttt{restore}((\langle\langle g\mapsto\perp\rangle_{\text{var}}\langle\cdot\rangle_{\text{h}}\rangle_{\text{heap}}))_{\text{k}}\rangle_{\text{traces}'}$

RULE6
$\Rightarrow\ \langle\cdot\rangle_{\text{traces}}\ \langle\cdot\rangle_{\text{traces}'}\ \langle\langle\langle g\mapsto\perp\rangle_{\text{var}}\langle\cdot\rangle_{\text{h}}\rangle_{\text{heap}}\overset{\text{p0}}{\leadsto}\texttt{new}(\langle\langle g\mapsto\perp\rangle_{\text{var}}\langle\cdot\rangle_{\text{h}}\rangle_{\text{heap}},\texttt{p0})\rangle_{\text{memento}}$

$\quad\langle\ \langle\langle g\mapsto\perp\rangle_{\text{var}}\langle\cdot\rangle_{\text{h}}\rangle_{\text{heap}}\overset{\text{g:=new}}{\leadsto}\texttt{new}(\langle\langle g\mapsto\perp\rangle_{\text{var}}\langle\cdot\rangle_{\text{h}}\rangle_{\text{heap}},\texttt{g}:=\texttt{new})\ \rangle_{\text{trans}}$

$\quad\langle\ \langle\langle g\mapsto\perp\rangle_{\text{var}}\langle\cdot\rangle_{\text{h}}\rangle_{\text{heap}}\overset{\text{main}}{\leadsto}\texttt{fin}\quad\texttt{new}(\langle\langle g\mapsto\perp\rangle_{\text{var}}\langle\cdot\rangle_{\text{h}}\rangle_{\text{heap}},\texttt{p0})\overset{\text{restore}(...)}{\leadsto}\texttt{fin}$

$\quad\langle\langle g\mapsto\perp\rangle_{\text{var}}\langle\cdot\rangle_{\text{h}}\rangle_{\text{heap}}\overset{\text{p0}}{\leadsto}\texttt{new}(\langle\langle g\mapsto\perp\rangle_{\text{var}}\langle\cdot\rangle_{\text{h}}\rangle_{\text{heap}},\texttt{p0})$

$\quad\texttt{new}(\langle\langle g\mapsto\perp\rangle_{\text{var}}\langle\cdot\rangle_{\text{h}}\rangle_{\text{heap}},\texttt{g}:=\texttt{new})$

$\quad\overset{\text{p0}}{\leadsto}\texttt{news}(\langle\langle g\mapsto\perp\rangle_{\text{var}}\langle\cdot\rangle_{\text{h}}\rangle_{\text{heap}},\texttt{p0},\langle\langle g\mapsto\perp\rangle_{\text{var}}\langle\cdot\rangle_{\text{h}}\rangle_{\text{heap}},\texttt{g}:=\texttt{new},\texttt{p0}\curvearrowright\texttt{restore}(...),1)$

$\quad\overset{\text{restore}(...)}{\leadsto}\texttt{new}(\langle\langle g\mapsto\perp\rangle_{\text{var}}\langle\cdot\rangle_{\text{h}}\rangle_{\text{heap}},\texttt{p0})\ \rangle_{\text{rel}}$

RULE7
$\Rightarrow\ \langle\cdot\rangle_{\text{traces}}\ \langle\cdot\rangle_{\text{traces}'}\ \langle\cdot\rangle_{\text{memento}}\overset{\text{RULE3}}{\Rightarrow}\ ...$

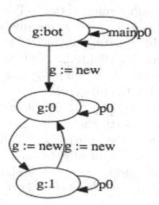

Fig. 5. The first pipeline iteration for $kA_{post*}(true,\text{Shylock}[\text{pgm0}])$ and the automatically produced reachability automaton at the end of $kA_{post*}(true,\text{Shylock}[\text{pgm0}])$. Note that for legibility reasons we omit certain cells appearing in the control state, like $\langle\langle g\rangle_G\langle\cdot\rangle_L\langle\cdot\rangle_F\langle\texttt{main}\mapsto\texttt{p0 p0}\mapsto\texttt{g}:=\texttt{new};\texttt{p0}\rangle_P\rangle_{\text{pgm}}$, which do not change along the execution. Hence, for example, the ctrl-cell is filled in RULE3 with both cells heap and pgm.

from the Kleene algebra with tests [9], where the global and local variables are used as nominals while the fields constitute the set of basic actions. The $\mathbb{K}$ specification of *Rite* is based on the circularity principle [7,3] to handle the possible cycles in the heap. We employ *Rite* with $kA_{post^*}(\phi, \bar{\mathcal{P}})$, i.e., $\phi \in Rite$, for verifying heap-shape properties for Shylock programs. For the precise definition of the interpretation of these expressions in a heap we refer to the companion paper [16]. We conclude with an example showing a simple invariant property of a Shylock program.

Example 4. The following Shylock program `pgmList` creates a potentially infinite linked list which starts in object `first` and ends with object `last`.

```
gvars: first, last    lvars: tmp    flds: next
main :: last:=new; last.next:=last; first:=last; p0
p0 :: tmp:=new; tmp.next:=first; first:=tmp; (p0 + skip)
```

This is an example of a program which induces, on some computation path, an unbounded heap. When we apply the heap-bounded model checking specification, by instantiating ϕ with the property $le(10)$, we collect all lists with a length smaller or equal than 10. We can also check the heap-shape property "($\neg$first$+$first.next*.last)". This property says that either the `first` object is not defined or the `last` object is reached from `first` via the `next` field.

5 Conclusions

In this paper we introduced pushdown system specifications (PSS) with an associated invariant model checking algorithm **Apost*gen**. We showed why the $\mathbb{K}$ framework is a suitable environment for pushdown systems specifications, but not for their verification via the for-free model checking capabilities available in $\mathbb{K}$. We gave a $\mathbb{K}$ specification of invariant model checking for pushdown system specifications, kA_{post^*}, which is behaviorally equivalent with **Apost*gen**. To the best of our knowledge, no other model checking tool has the flexibility of having structured atomic propositions and working with the generation of the state space on-the-fly.

Future work includes the study of the correctness of our translation of Shylock into the $\mathbb{K}$ framework as well as of the translation of the proposed model checking algorithm and its generalization to any LTL formula. From a more practical point of view, future applications of pushdown system specifications could be found in semantics-based transformation of real programming languages like C or Java or in benchmark-based comparisons with existing model-based approaches for program verification.

Acknowledgments. We would like to thank the anonymous reviewers for their helpful comments and suggestions.

References

1. Bouajjani, A., Esparza, J., Maler, O.: Reachability Analysis of Pushdown Automata: Application to Model Checking. In: Mazurkiewicz, A., Winkowski, J. (eds.) CONCUR 1997. LNCS, vol. 1243, pp. 135–150. Springer, Heidelberg (1997)
2. Bouajjani, A., Fratani, S., Qadeer, S.: Context-Bounded Analysis of Multithreaded Programs with Dynamic Linked Structures. In: Damm, W., Hermanns, H. (eds.) CAV 2007. LNCS, vol. 4590, pp. 207–220. Springer, Heidelberg (2007)
3. Bonsangue, M., Caltais, G., Goriac, E.-I., Lucanu, D., Rutten, J., Silva, A.: A Decision Procedure for Bisimilarity of Generalized Regular Expressions. In: Davies, J., Silva, L., da Silva Simão, A. (eds.) SBMF 2010. LNCS, vol. 6527, pp. 226–241. Springer, Heidelberg (2011)
4. Eker, S., Meseguer, J., Sridharanarayanan, A.: The Maude LTL Model Checker. Electr. Notes Theor. Comput. Sci. 71, 162–187 (2002)
5. Ellison, C., Roşu, G.: An Executable Formal Semantics of C with Applications. In: Field, J., Hicks, M. (eds.) POPL 2012, pp. 533–544. ACM (2012)
6. Esparza, J., Schwoon, S.: A BDD-Based Model Checker for Recursive Programs. In: Berry, G., Comon, H., Finkel, A. (eds.) CAV 2001. LNCS, vol. 2102, pp. 324–336. Springer, Heidelberg (2001)
7. Goguen, J., Lin, K., Roşu, G.: Circular Coinductive Rewriting. In: ASE 2000, pp. 123–132. IEEE (2000)
8. Kidd, N., Reps, T., Melski, D., Lal, A.: WPDS++: A C++ Library for Weighted Pushdown Systems (2005), http://www.cs.wisc.edu/wpis/wpds++
9. Kozen, D.: Kleene Algebra with Tests. ACM Trans. Program. Lang. Syst. 19, 427–443 (1997)
10. Maude, http://maude.cs.uiuc.edu/
11. Meseguer, J., Palomino, M., Martí-Oliet, N.: Equational Abstractions. Theor. Comput. Sci. 403(2-3), 239–264 (2008)
12. Meseguer, J., Roşu, G.: The Rewriting Logics Semantics Project. Theor. Comput. Sci. 373(3), 213–237 (2007)
13. Rinetzky, N., Bauer, J., Reps, T.W., Sagiv, S., Wilhelm, R.: A Semantics for Procedure Local Heaps and its Abstractions. In: Palsberg, J., Abadi, M. (eds.) POPL 2005, pp. 296–309. ACM (2005)
14. Roşu, G., Şerbănuţă, T.F.: An Overview of the K Semantic Framework. J. Log. Algebr. Program. 79(6), 397–434 (2010)
15. Şerbănuţă, T.F., Roşu, G.: K-Maude: A Rewriting Based Tool for Semantics of Programming Languages. In: Ölveczky, P.C. (ed.) WRLA 2010. LNCS, vol. 6381, pp. 104–122. Springer, Heidelberg (2010)
16. Rot, J., Asavoae, I.M., de Boer, F., Bonsangue, M., Lucanu, D.: Interacting via the Heap in the Presence of Recursion. In: Carbone, M., Lanese, I., Silva, A., Sokolova, A. (eds.) ICE 2012. EPTCS, vol. 104, pp. 99–113 (2012)
17. Schwoon, S.: Model-Checking Pushdown Systems. PhD thesis, Technische Universität München (2002)

A Probabilistic Strategy Language
for Probabilistic Rewrite Theories
and Its Application to Cloud Computing

Lucian Bentea[1] and Peter Csaba Ölveczky[1,2]

[1] University of Oslo
[2] University of Illinois at Urbana-Champaign

Abstract. Several formal models combine probabilistic and nondeterministic features. To allow their probabilistic simulation and statistical model checking by means of pseudo-random number sampling, all sources of nondeterminism must first be quantified. However, current tools offer limited flexibility for the user to define how the nondeterminism should be quantified. In this paper, we propose an expressive *probabilistic strategy language* that allows the user to define complex strategies for quantifying the nondeterminism in *probabilistic rewrite theories*. We have implemented PSMaude, a tool that extends Maude with a probabilistic simulator and a statistical model checker for our language. We illustrate the convenience of being able to define different probabilistic strategies on top of a system by a cloud computing example, where different load balancing policies can be specified by different probabilistic strategies. We then use PSMaude to analyze the QoS provided by different policies.

1 Introduction

Many formal analysis tools support the modeling of systems that exhibit both probabilistic and nondeterministic behaviors. To allow their probabilistic simulation and statistical model checking using pseudo-random number sampling, the nondeterminism must be quantified to obtain a fully probabilistic model. However, there is typically limited support for user-definable adversaries to quantify the nondeterminism in reasonably expressive models; such adversaries are either added by the tool or must be encoded directly into the system model.

In this paper we propose an expressive *probabilistic strategy language* for *probabilistic rewrite theories* [18,1] that allows users to define complex adversaries for a model, and therefore allows us to separate the definition of the system model from that of the adversary needed to quantify the nondeterminism in the system model.

Rewriting logic is a simple and expressive logic for concurrent systems in which the data types are defined by an algebraic equational specification and where the local transition patterns are defined by conditional labeled rewrite rules of the form $l : t \longrightarrow t'$ **if** $cond$, where l is a label and t and t' are terms representing state fragments. Maude [11] is a high-performance simulation, reachability, and

N. Martí-Oliet and M. Palomino (Eds.): WADT 2012, LNCS 7841, pp. 77–94, 2013.

LTL model checking tool for rewriting logic that has been successfully applied to many large applications (see, e.g., [25,22] for an overview).

Rewriting logic has been extended to *probabilistic rewrite theories* [18,1], where probabilities are introduced by new variables in the righthand side t' of a rewrite rule. These variables are instantiated according to a probability distribution associated with the rewrite rule. Probabilistic rewrite theories, together with the VeStA statistical model checker [27], have been applied to analyze sensor network algorithms [17] and defense mechanisms against denial of service attacks [3,13]. However, since the probabilistic rewrite theories were highly non-deterministic, adversaries had to be encoded into the model before any analysis could take place.

Probabilistic model checking suffers from state space explosion which renders it unfeasible for automated analysis of the complex concurrent systems targeted by rewriting logic. *Statistical model checking* [20,28,26] trades absolute confidence in the correctness of the model checking for computational efficiency, and essentially consists of simulating a number of different system behaviors until a certain confidence level is reached. This not only makes statistical analysis feasible, but also makes such model checking amenable to parallelization, which is exploited in the parallel version PVeStA [2] of the statistical model checker VeStA. PVeStA has recently been used to analyze an adaptive system specified as a hierarchical probabilistic rewrite theory [10].

To support the analysis of probabilistic rewrite theories where our strategy language has been used to quantify all nondeterminism, we have formalized and integrated into (Full) Maude both probabilistic simulation and statistical model checking. Our strategy language and its implementation, the PSMaude tool [4], enable a Maude-based safety/QoS modeling and analysis methodology in which:

1. A non-probabilistic rewrite theory defines all possible behaviors in a simple "uncluttered" way; this model can then be directly subjected to important safety analyses to guarantee the absence of bad behaviors.
2. Different QoS policies and/or probabilistic environments can then be defined as probabilistic strategies on top of the basic verified model for QoS reasoning by probabilistic simulation and statistical model checking.

We exemplify in Section 4 the usefulness of this methodology and of the possibility to define different complex probabilistic strategies on top of the same model with a cloud computing example, where a (non-probabilistic) rewrite theory defines all the possible ways in which requested resources can be allocated on servers in the cloud, as well as all possible environment behaviors. We can then use standard techniques to prove safety properties of this model. However, one could imagine a number of different *policies* for assigning resources to service providers and users, such as, e.g.,

- Service providers might request virtual machines uniformly across different regions (for fault-tolerance and omnipresence), or with higher probability at certain locations, or with higher probability at more stable servers.

– Service users may be assigned virtual machines either closer to their locations, on physical servers with low workload, or on reliable servers, with high probability.

Each load balancing policy can be naturally specified as a probabilistic strategy on top of the (non-probabilistic) model of the cloud infrastructure that has been proved to be "safe." We then use PSMaude to perform simulation and statistical model checking to analyze the QoS effect of the different load balancing policies.

2 Preliminaries

Rewriting Logic and Maude. A *rewrite theory* [24] is a tuple $\mathcal{R} = (\Sigma, E \cup A, L, R)$, where $(\Sigma, E \cup A)$ is a membership equational logic theory, with E a set of equations $(\forall \vec{x})\ t = t'$ **if** *cond*, and membership axioms $(\forall \vec{x})\ t : s$ **if** *cond*, where t and t' are Σ-terms, s is a sort, and *cond* is a conjunction of equalities and sort memberships, and with A a collection of *structural axioms* specifying properties of operators, like commutativity, associativity, etc., R is a set of rewrite rules $(\forall \vec{x})\ l : t \longrightarrow t'$ **if** *cond*, where $l \in L$ is a label, t and t' are terms of the same kind, *cond* is a conjunction of equalities, memberships and rewrites, and $\vec{x} = vars(t) \cup vars(t') \cup vars(cond)$. Such a rule specifies a transition from an instance of the term t to the corresponding instance of t', provided that *cond* is satisfied. $vars(t)$ denotes the set of variables in a term t; if $vars(t) = \emptyset$, then t is a *ground term*. If E is terminating, confluent and sort-decreasing modulo A, then $Can_{\Sigma,E/A}$ denotes the algebra of fully simplified ground terms, and we denote by $[t]_A$ the A-equivalence class of a term t. An *E/A-canonical ground substitution* for a set of variables $\vec{x}$ is a function $[\theta]_A : \vec{x} \to Can_{\Sigma,E/A}$; we denote by $CanGSubst_{E/A}(\vec{x})$ the set of all such functions. We also denote by $[\theta]_A$ the homomorphic extension of $[\theta]_A$ to Σ-terms. A *context* is a Σ-term with a single *hole* variable $\odot$; two contexts $\mathbb{C}$ and $\mathbb{C}'$ are A-equivalent if $A \vdash (\forall \odot)\ \mathbb{C}(\odot) = \mathbb{C}'(\odot)$. Given $[u]_A \in Can_{\Sigma,E/A}$, its *R/A-matches* are triples $([\mathbb{C}]_A, r, [\theta]_A)$ where $\mathbb{C}$ is a context, $r \in R$ is a rewrite rule, $[\theta]_A \in CanGSubst_{E/A}(\vec{x})$ is such that $E \cup A \vdash \theta(cond)$, and $[u]_A = [\mathbb{C}(\odot \leftarrow \theta(t))]_A$. We denote by $\mathcal{M}([u]_A)$ the set of all such triples, and define the set of rules that are *enabled* for a term $[u]_A$, the set of *valid contexts* for $[u]_A$ and a rule r, and the set of *valid substitutions* for $[u]_A$, a rule r, and a context $[\mathbb{C}]_A$, in the expected way:

$$\mathtt{enabled}([u]_A) = \{r \in R \mid \exists [\mathbb{C}]_A, \exists [\theta]_A : ([\mathbb{C}]_A, r, [\theta]_A) \in \mathcal{M}([u]_A)\}$$

$$C([u]_A, r) = \{[\mathbb{C}]_A \in Can_{\Sigma,E/A}(\odot) \mid \exists [\theta]_A : ([\mathbb{C}]_A, r, [\theta]_A) \in \mathcal{M}([u]_A)\}$$

$$S([u]_A, r, [\mathbb{C}]_A) = \{[\theta]_A \in CanGSubst_{E/A}(\vec{x}) \mid ([\mathbb{C}]_A, r, [\theta]_A) \in \mathcal{M}([u]_A)\}$$

Maude [11] is a high-performance simulation, reachability analysis, and LTL model checking tool for rewrite theories. We use Maude syntax, so that conditional rules are written `crl [l]:` t `=>` t' `if` *cond*. In object-oriented Maude specifications [11], the system state is a term of sort `Configuration` denoting a multiset of objects and messages, with multiset union denoted by juxtaposition. A class declaration `class C | att`$_1$ `: s`$_1$`, ..., att`$_n$ `: s`$_n$ declares a class C

with attributes $att_1, \ldots, att_n$ of sorts $s_1, \ldots, s_n$, respectively. A *subclass* inherits the attributes and rules of its superclass(es). Objects are represented as terms `< o : C | att`$_1$ `: val`$_1$`, ..., att`$_n$ `: val`$_n$ `>`, where o is the object's identifier of sort `Oid`, C is the object's class, and where $val_1, \ldots, val_n$ are the values of the object's attributes $att_1, \ldots, att_n$. For example, the rule

```
rl [l]: m(0, w)  < 0 : C | a1 : x, a2 : 0', a3 : z >   =>
                 < 0 : C | a1 : x + w, a2 : 0', a3 : z >  m'(0', x) .
```

defines a family of transitions in which a message `m`, with parameters `0` and `w`, is read and consumed by an object `0` of class `C`. The transitions change the attribute `a1` of `0` and send a new message `m'(0', x)`. "Irrelevant" attributes (such as `a3` and the righthand side occurrence of `a2`) need not be mentioned.

Markov Chains. Given $\Omega \neq \emptyset$, a σ-*algebra* over Ω is a collection $\mathcal{F} \subseteq \mathcal{P}(\Omega)$ such that $\Omega \setminus F \in \mathcal{F}$ for all $F \in \mathcal{F}$, and $\bigcup_{i \in I} F_i \in \mathcal{F}$ for all collections $\{F_i\}_{i \in I} \subseteq \mathcal{F}$ indexed by a countable set I. Given a σ-algebra $\mathcal{F}$ over Ω, a function $\mathbb{P} : \mathcal{F} \to [0, 1]$ is a *probability measure* if $\mathbb{P}(\Omega) = 1$ and $\mathbb{P}(\cup_{i \in I} F_i) = \sum_{i \in I} \mathbb{P}(F_i)$, for all collections $\{F_i\}_{i \in I} \subseteq \mathcal{F}$ of pairwise disjoint sets. We denote by $PMeas(\Omega, \mathcal{F})$ the set of all probability measures on $\mathcal{F}$ over Ω. A *probability mass function* (pmf) is a function $p : \Omega \to [0, 1]$ with $\sum_{\omega \in \Omega} p(\omega) = 1$. A family of pmf's can be used to define the behavior of a (memoryless) probabilistic system. In particular, a *discrete time Markov chain* (DTMC) is given by a countable set of *states* S, and a *transition probability* matrix $T : S \times S \to [0, 1]$, where $T(s) : S \to [0, 1]$ is a pmf for all states $s \in S$, i.e., $T(s, s')$ is the probability for the DTMC to make a transition from state s to state s'.

Probabilistic Rewrite Theories. In *probabilistic rewrite theories* (PRTs) [18] the righthand side t' of a rule $l : t \longrightarrow t'$ **if** *cond* may contain variables $\vec{y}$ that do not occur in t, and that are instantiated according to a probability measure taken from a *family* of probability measures—one for each instance of the variables in t—associated with the rule. Formally, a PRT $\mathcal{R}_\pi$ is a pair $(\mathcal{R}, \pi)$, where $\mathcal{R}$ is a rewrite theory, and π maps each rule r of $\mathcal{R}$, with $vars(t) = \vec{x}$ and $vars(t') \setminus vars(t) = \vec{y}$, to a mapping $\pi_r : [\![cond(\vec{x})]\!] \to PMeas\left(CanGSubst_{E/A}(\vec{y}), \mathcal{F}_r\right)$, where $[\![cond(\vec{x})]\!] = \{ [\theta]_A \in CanGSubst_{E/A}(\vec{x}) \mid E \cup A \vdash \theta(cond) \}$, and $\mathcal{F}_r$ is a σ-algebra over $CanGSubst_{E/A}(\vec{y})$. That is, for each substitution $[\theta]_A$ of the variables in t that satisfies *cond*, we get a probability measure $\pi_r([\theta]_A)$ for instantiating the variables $\vec{y}$. The rule r together with π_r is called a *probabilistic rewrite rule*, and is written $l : t \longrightarrow t'$ **if** *cond* **with probability** π_r. We refer to the specification of the "blackboard game" in Section 3 for an example of the syntax used to specify probabilistic rewrite rules and the probability measure π_r. An E/A-*canonical one-step rewrite* of $\mathcal{R}_\pi$ [18,1] is a labeled transition $[u]_A \xrightarrow{([\mathbb{C}]_A, r, [\theta]_A, [\rho]_A)} [v]_A$ with $m \triangleq ([\mathbb{C}]_A, r, [\theta]_A)$ a R/A-match for $[u]_A$, $[\rho]_A \in CanGSubst_{E/AS}(\vec{y})$, and $[v]_A = [\mathbb{C}(\odot \leftarrow (\theta \cup \rho)(t'))]_A$. To quantify the nondeterminism in the choice of m, the notion of *adversary* is introduced in [18,1] that samples m from a pmf that depends on the computation history.

A *memoryless* adversary[1] is a family of pmf's $\{\sigma_{[u]_A} : \mathcal{M}([u]_A) \to [0,1]\}_{[u]_A}$, where $\sigma_{[u]_A}(m)$ is the probability of picking the R/A-match m. A consequence of a result in [18] is that executing $\mathcal{R}_\pi$ under $\{\sigma_{[u]_A}\}_{[u]_A}$ is described by a DTMC.

PCTL. The *probabilistic computation tree logic* (PCTL) [16] extends CTL with an operator $\mathcal{P}$ to express properties of DTMCs. We use a subset of PCTL, without time-bounded and steady-state operators. If AP is a set of atomic propositions, ϕ is a *state* formula, and ψ is a *path* formula, PCTL formulas over AP are defined by:

$$\phi ::= true \mid a \mid \neg\phi \mid \phi \wedge \phi \mid \mathcal{P}_{\bowtie p}(\psi) \qquad \psi ::= \phi \,\mathcal{U}\, \phi \mid \mathbf{X}\phi$$

where $a \in AP$, $\bowtie \in \{<, \leq, >, \geq\}$, and $p \in [0,1]$. PCTL satisfaction is defined over DTMCs, e.g., the meaning of $\mathcal{M}, s \models \mathcal{P}_{<0.05}(true \,\mathcal{U}\, \phi)$ is that ϕ eventually becomes true in less than 5% of all the runs of the DTMC $\mathcal{M}$ from state s.

Statistical Model Checking and VESTA. Traditional model checking suffers from state space explosion problem, whereas *statistical model checking* [20,28,26] trades complete confidence for efficiency, allowing the analysis of large-scale probabilistic systems. This technique is based on simulating the model, and on performing statistical hypothesis testing to control the generation of execution traces. The simulation is stopped when a given *level of confidence* is reached for answering the model checking problem. The VESTA tool [27] supports statistical model checking and quantitative analysis of executable specifications in which all nondeterminism is quantified probabilistically. In VESTA, system properties are given in PCTL (or its continuous-time extension CSL), while quantitative analysis queries are given in the QUATEX logic [1] and ask for the average values of quantities associated with the model—VESTA simulates the model and provides estimates for these averages.

3   A Language for Specifying Memoryless Adversaries

The source of nondeterminism in a probabilistic rewrite theory is picking an R/A-match from a set of possible ones in each state. This section introduces a probabilistic strategy language that can be used to quantify this nondeterminism, i.e., for specifying memoryless adversaries of PRTs.

The probability distribution associated with picking a certain R/A-match $([\mathbb{C}]_A, r, [\theta]_A)$ can be specified using the individual (conditional) distributions for picking the rule r, the context $[\mathbb{C}]_A$, and the substitution $[\theta]_A$, which is more convenient than specifying their joint distribution. That is, by probability theory, any memoryless adversary $\{\sigma_{[u]_A}\}_{[u]_A}$ of a PRT $\mathcal{R}_\pi$ can be decomposed as:

$\sigma_{[u]_A}([\mathbb{C}]_A, r, [\theta]_A) =$
$\mathbb{P}\{\text{pick rule } r \mid \text{state is } [u]_A\} \cdot \mathbb{P}\{\text{pick context } [\mathbb{C}]_A \mid \text{state is } [u]_A, \text{rule is } r\}$
$\qquad \cdot \mathbb{P}\{\text{pick substitution } [\theta]_A \mid \text{state is } [u]_A, \text{rule is } r, \text{context is } [\mathbb{C}]_A\}$

---

[1] This is a slightly modified version of the definition of adversaries in [18,1].

We denote the factors in this product by $\mathcal{A}_{[u]_A}(r)$, $\mathcal{A}^r_{[u]_A}([\mathbb{C}]_A)$, and $\mathcal{A}^{r,[\mathbb{C}]_A}_{[u]_A}([\theta]_A)$, respectively. They give the probabilities of picking rule r in state $[u]_A$, of using context $[\mathbb{C}]_A$ to match $[u]_A$ with the lefthand side of r, and of using substitution $[\theta]_A$ to match $[u]_A$ with the lefthand side of r in context $[\mathbb{C}]_A$, respectively. We call $\{\mathcal{A}_{[u]_A} : \mathtt{enabled}([u]_A) \to [0,1]\}_{[u]_A}$, $\{\mathcal{A}^r_{[u]_A} : C([u]_A, r) \to [0,1]\}_{[u]_A, r}$, and $\{\mathcal{A}^{r,[\mathbb{C}]_A}_{[u]_A}([\theta]_A) : S([u]_A, r, [\mathbb{C}]_A) \to [0,1]\}_{[u]_A, r, [\mathbb{C}]_A}$, resp., the underlying *rule*, *context*, and *substitution adversaries* of the memoryless adversary $\{\sigma_{[u]_A}\}_{[u]_A}$.

It is cumbersome to define *absolute* probabilities for each choice (so that they add up to 1 in each state). If we have rules r_1, r_2, and r_3, and want r_1 to be applied with 3 times as high probability as r_2 (when both are enabled), which should be twice as likely as taking rule r_3, then, for a state $[u]_A$ where all rules are enabled, the probabilities would be $\{r_1 \mapsto 6/9, r_2 \mapsto 2/9, r_3 \mapsto 1/9\}$, and for a state $[u']_A$ where r_2 is not enabled, the distribution would be $\{r_1 \mapsto 6/7, r_3 \mapsto 1/7\}$, etc. This can soon become inconvenient. In our language one therefore instead defines *relative* weights for each rule, context, and substitution. That is, for any state $[u]_A$ in our example, the "weights" of the rules r_1, r_2 and r_3 could be 6, 2, and 1, respectively. Relative weights are therefore needed since the set of possible R/A-matches in a state, whose nondeterministic choice we want to quantify, is only available during the model execution. We therefore build the concrete probability distributions *on-the-fly* in each state during execution, which we sample to obtain the next state.

Language. The language we propose allows specifying memoryless rule, context, and substitution adversaries, using *strategy expressions* of the following forms:

psdrule ⟨*Identifier*⟩ := **given state:** ⟨*StatePattern*⟩
                    is: ⟨*RuleWeightDist*⟩ [**if** ⟨*Condition*⟩] [[**owise**]]

psdcontext ⟨*Identifier*⟩ := **given state:** ⟨*StatePattern*⟩
                    rule: ⟨*RulePattern*⟩
                    is: ⟨*ContextWeightDist*⟩ [**if** ⟨*Cond*⟩] [[**owise**]]

psdsubst ⟨*Identifier*⟩ := **given state:** ⟨*StatePattern*⟩
                    rule: ⟨*RulePattern*⟩
                 context: ⟨*ContextPattern*⟩
                    is: ⟨*SubstWeightDist*⟩ [**if** ⟨*Condition*⟩] [[**owise**]]

with ⟨*StatePattern*⟩ a term $t(\vec{x})$, ⟨*RulePattern*⟩ a rule label or a variable over rule labels, and ⟨*ContextPattern*⟩ a term $c(\vec{y})$ with $\odot \in \vec{y}$ and $\vec{y}\backslash\{\odot\} \subseteq \vec{x}$, or a variable over contexts. The weight expression ⟨*RuleWeightDist*⟩ is either **uniform**, or a list of ";"-separated weight assignments of the form ⟨*RuleLabel*⟩ $\mapsto$ ⟨*Weight*⟩, where ⟨*Weight*⟩ is a term $w(\vec{x})$ of sort $\mathtt{Rat}$; ⟨*ContextWeightDist*⟩ and ⟨*SubstWeightDist*⟩ have similar forms. ⟨*Condition*⟩ (abbreviated to ⟨*Cond*⟩ in the context adversary syntax above) specifies the condition under which the strategy can be applied, and [**owise**] specifies that it should be applied if no other strategy can be applied. We refer to [5] for the detailed syntax and semantics of our language, i.e., how each strategy expression defines an adversary. Our implementation also provides an executable rewriting logic semantics of our language.

PSMaude. Our tool PSMaude [4] extends Maude by adding support for specifying probabilistic rules with fixed-size probability distributions and our strategy language, a probabilistic rewrite command and a statistical PCTL model checker that can analyze a given PRT controlled by given probabilistic strategies.

Given a probabilistic module *SYSTEM-SPEC* that specifies a PRT, probabilistic strategies can be written in modules of the form:

```
(psmod PSTRAT is protecting SYSTEM-SPEC .        --- import system specification
   state StateSort .                             --- sort for system states
   psdrule      RuleStratID    := RuleStratExpr .    --- rule strategy
   psdcontext ContextStratID := ContextStratExpr . --- context strategy
   psdsubst    SubstStratID   := SubstStratExpr .  --- substitution strategy
   psd StratID := < RuleStratID | ContextStratID | SubstStratID > . --- strategy
endpsm)
```

A strategy definition, introduced with psd, associates strategies for rules, contexts, and substitutions with a strategy identifier *StratID*, that can be used in a *probabilistic strategy rewrite* command (prew [n] s using *StratID* .), which executes n one-step (probabilistic) rewrites from the state s using the strategy *StratID*. The unbounded version (uprew s using *StratID* .) rewrites until a deadlock occurs. The strategies for rules, contexts, and substitutions are introduced with psdrule, psdcontext, and psdsubst, respectively; they can be conditional, with keywords cpsdrule, cpsdcontext, and cpsdsubst, respectively. There may be several definitions for the same strategy identifier, but they should refer to disjoint cases of the arguments. [owise]-annotated strategy expressions can be used to specify how the nondeterminism is resolved when no other strategy definition is applicable. It can thus be easily ensured that a probabilistic strategy resolves *all* nondeterminism in a given system specification.

Our statistical model checking command (smc s |= φ using *StratID* .) allows for further analysis of a specification with given strategies, where φ is a PCTL formula, and satisfaction of atomic propositions is defined in a *state predicate module*:

```
(spmod SYSTEM-PRED is protecting SYSTEM-SPEC . --- import system specification
   smcstate StateSort .                         --- sort for system states
   psp φ₁ ... φₙ : Sort1 ... SortK .            --- parametric state predicates
   var S : StateSort .
   csat S |= φ₁(s₁, ..., sₖ) if f(S, (s₁, ..., sₖ)) . --- define their semantics
   ...
endspm)
```

For instance, the parametric state predicate $\varphi_1(s_1, \ldots, s_K)$ holds in a state S if and only if the condition $f(S, s_1, \ldots, s_K)$ in the above csat declaration is true, where the operator f is defined by means of (possibly conditional) equations.

The command (set type1 error b_1 .) sets the bound on type I errors (the algorithm returns "false" when the property holds), and (set type2 error b_2 .) sets the bound on type II errors (vice versa). By lowering these bounds, a higher confidence on the model checking result is achieved, but more execution samples are generated.

Example. In each step in a probabilistic version of the *blackboard game* [23], in which some numbers are on a blackboard, two arbitrary numbers x and y are replaced by $(x^2 + y)$ quo 2 with probability 3/4, and by $(x^3 + y)$ quo 2 with probability 1/4, where quo denotes integer division. The goal of the game is to obtain the highest possible number at the end. This game is formalized in the following probabilistic module, that also defines an initial state:

```
(pmod BLACKBOARD is protecting RAT .
  sort Blackboard .  subsort Nat < Blackboard .
  op empty : -> Blackboard [ctor] .
  op __ : Blackboard Blackboard -> Blackboard [ctor assoc comm id: empty] .
  vars M N K : Nat .
  prl [play] : M N => (K + N) quo 2
                    with probability K := (M * M -> 3/4 ; M * M * M -> 1/4) .
  op initState : -> Blackboard .
  eq initState = 2 3 5 7 11 13 17 .
endpm)
```

Since the multiset union `__` is associative and commutative (declared with the keywords `assoc` and `comm`), the choice of the numbers x and y from the blackboard is nondeterministic. The choice of the substitution, $\{ \text{M} \mapsto x,\ \text{N} \mapsto y \}$ or $\{ \text{M} \mapsto y,\ \text{N} \mapsto x \}$, is then also nondeterministic. We define the following probabilistic strategy `BlackboardStrat` to quantify this nondeterminism:

```
(psmod BLACKBOARD-PROB-STRAT is protecting BLACKBOARD . state Blackboard .
  var B : Blackboard . vars X Y : Nat .
  psdrule    RuleStrat := given state: B  is: (play) -> 1 .
  psdcontext CtxStrat := given state: X Y B  rule: play
                         is: ([] B) -> (1 / (X * Y)) .
  psdsubst   SubStrat := given state: X Y B  rule: play  context: [] B
                         is: {M <- X, N <- Y} -> 9 ;
                             {M <- Y, N <- X} -> 1 if X <= Y .
  psd BlackboardStrat := < RuleStrat | CtxStrat | SubStrat > .
endpsm)
```

The rule strategy `RuleStrat` assigns weight 1 to the only rule `play`. The context strategy `CtxStrat`, selecting in which context the rule `play` applies, assigns for each pair of numbers x and y on the blackboard, the relative weight $1/(x \cdot y)$ to the context that implies that the numbers x and y are replaced by the rule `play`; i.e., it gives a higher weight to contexts corresponding to picking small numbers to replace. The substitution strategy `SubStrat` selects the rule match $\{\text{M} \mapsto x,\ \text{N} \mapsto y\}$ with 9 times as high probability as $\{\text{M} \mapsto y,\ \text{N} \mapsto x\}$ when $x \leq y$.

We now explain the strategy `BlackboardStrat` in more detail. For a state $[u]_{ACU} = [2\ 3\ 5\ 7\ 11\ 13\ 17]_{ACU}$, `CtxStrat` assigns weights to each valid context as follows, where ACU refers to the structural axioms for multiset union (Associativity, Commutativity, and Unit (identity)). It first matches the state $[u]_{ACU}$ with the state pattern X Y B (where B can be `empty`), which gives several matches $\theta_1, \ldots, \theta_N$. Then all valid contexts are generated, which in this example have the form $[\odot\ t]_{ACU}$, with $\odot$ identifying the fragment of $[u]_{ACU}$ that matches the lefthand side of rule `play`, and t is the rest of the state. Next, the weight of each valid context is computed. Each context $[\odot\ t]_{ACU}$ is unified with the context pattern [] B, giving a unique match $\{B \mapsto t\}$. Of all the matches $\theta_1, \ldots, \theta_N$ obtained above, only those with $B \mapsto t$ are kept. The weight associated to $[\odot\ t]_{ACU}$ is then computed by instantiating the weight pattern 1 / (X * Y) with either one of these last substitutions. (For well-definedness, $\theta_i(1\ /\ (X\ *\ Y))$ and $\theta_j(1\ /\ (X\ *\ Y))$ should be the same, for all θ_i and θ_j with $\theta_i(B) = \theta_j(B)$.) For example, for the context $[\mathbb{C}]_{ACU} = [\odot\ 3\ 7\ 11\ 13\ 17]_{ACU}$ two such matches θ_k with $\theta_k(B) = 3\ 7\ 11\ 13\ 17$ exist: $\theta_1 = \{X \mapsto 2,\ Y \mapsto 5,\ B \mapsto 3\ 7\ 11\ 13\ 17\}$ and $\theta_2 = \{X \mapsto 5,\ Y \mapsto 2,\ B \mapsto 3\ 7\ 11\ 13\ 17\}$. The weight of the context $[\mathbb{C}]_{ACU}$ is then computed as $\theta_1(1\ /\ (X\ *\ Y)) = 1\ /\ 10$. Similarly, the weight of the context $[\odot\ 2\ 5\ 7\ 11\ 17]_{ACU}$ is $1/(3 \cdot 13) = 1/39$. After computing the weights of all contexts, they are normalized to obtain a distribution, from which a context is picked.

The substitution strategy SubStrat solves the same matching and unification problems as above, but further refines the set of matches to those that satisfy the condition X <= Y. For the context $[\mathbb{C}]_{ACU}$ above the only such match is θ_1. The weight distribution pattern associated with SubStrat is then instantiated by θ_1 to obtain a concrete weight distribution over the matches of the lefthand side of rule play: $\{M \mapsto 2,\ N \mapsto 5\} \mapsto 9$; $\{M \mapsto 5,\ N \mapsto 2\} \mapsto 1$. By normalizing the associated weights, a probability distribution is obtained, from which a match is picked, e.g., $\eta = \{M \mapsto 2,\ N \mapsto 5\}$ with probability 9/10.

Finally, a match for the probabilistic variable K of rule play is sampled from the distribution $\pi_r([\eta]_{ACU})$.

We run two simulations using unbounded probabilistic rewriting to show possible final states under the above strategy (the outputs are shown as comments):

```
(uprew initState using BlackboardStrat .) --- 276
(uprew initState using BlackboardStrat .) --- 4457
```

We then define a state predicate sumGreaterThan(i), which is true in a state S if the sum of all numbers in S is larger than i:

```
(spmod BLACKBOARD-PRED is protecting BLACKBOARD . protecting NAT .
  smcstate Blackboard .
  psp sumGreaterThan : Nat . --- declare a parametric state predicate
  var B : Blackboard . var N : Nat .
  csat B |= sumGreaterThan(N) if sum(B) > N . --- define its semantics
  op sum : Blackboard -> Nat .
  eq sum(empty) = 0 . eq sum(N B) = N + sum(B) .
endspm)
```

We first set a bound of 0.01 on both error probabilities, and then check that the sum on the blackboard never exceeds 10000 with high probability. This returns a positive result, and the estimated probability for the property to hold:

```
(set type1 error 0.01 .)
(set type2 error 0.01 .)
(smc initState |= P>= 0.9 [G ~ sumGreaterThan(10000)] using BlackboardStrat .)

Result Bool: true          Number of samples used: 11176
Confidence: 99%            Estimated probability: 79/87
```

4   Formalizing and Analyzing Cloud Computing Policies

In this section we use cloud computing to illustrate the usefulness of defining different adversaries for the same rewrite theory. The "base" rewrite theory defines all possible behaviors of the cloud and its environment, and each probabilistic strategy corresponds to a particular *load balancing policy* (and assumptions about the environment). We prove the safety of the "base" model, and use simulation and statistical model checking to analyze the QoS of different load balancing policies.

Cloud Computing Scenario. We model a cloud computing system that delivers services to *service users* and *service providers*. A service provider runs web applications on one or more *virtual machines* (VMs) hosted on *physical servers* in the cloud. Service users then use these applications via the cloud service. An example of a user is a person who uses an email application of a provider via a cloud infrastructure.

Servers are grouped into *data centers*, which are grouped into geographical *regions*. Since running applications in regions closer to the users may prove beneficial, we have included a *region selection* filter in our cloud architecture in Fig. 1. A *load balancer* distributes traffic across the data centers in a region.

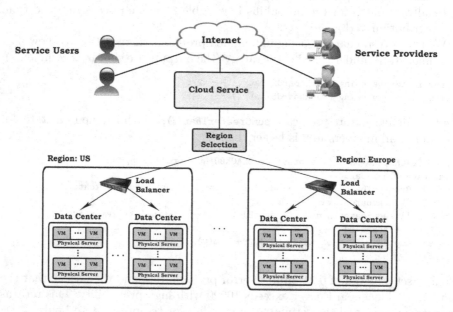

Fig. 1. A cloud computing architecture

When a *user* sends an application request, the cloud service forwards it to one of the VMs of the provider that owns the application. When a *provider* sends a request to launch a VM in a region, the cloud service forwards it to the region's load balancer, which chooses a server to host the new VM. Providers may launch a limited number of VMs, and users have a limited number of requests that the cloud can process *simultaneously*.

Formalization. We model the cloud system as a *hierarchical* object-oriented system. Service users, providers, and cloud services are declared as follows:

```
class Node | priority : Nat .
class SUser | location : Location .
class SProvider .
subclass SUser SProvider < Node .
class CService | status : CServiceDataSet, subscr : CServiceDataSet .
```

where **status** and **subscr** contain status and subscription information data about users and providers, of the forms: noReq(u, k): the number of unresolved requests of user u is k; noVM(p, l): the number of VMs of provider p is l; maxReq(u, k): the maximum number of requests that the cloud service may simultaneously process for user u is k; maxVM(p, l): the maximum number of VMs that provider p can run is l. A region has a **location**, and a **dataCenters** attribute with the data center objects in that region. The **pservers** attribute of a data center object denotes the set of physical server objects in the data center:

```
class Region | location : Location, dataCenters : Configuration .
class DCenter | pservers : Configuration .
class PServer | load : Nat, maxLoad : Nat, nextVMID : Nat, vms : Configuration .
```

with **load** the number of VMs on the server; **maxLoad** the server's capacity; **nextVMID** for generating fresh IDs for new VMs; and **vms** a set of VM objects of the class:

```
class VMachine | owner : Oid, vmReq : OidMSet, vmMaxReq : Nat, running : Bool .
```

with **owner** the ID of the provider that owns the VM; **vmReq** the object IDs of all users whose requests are running on the VM; **vmMaxReq** the maximum number of requests that can be resolved on the VM simultaneously; and **running** says whether the VM is running. We model two types of messages: *i)* user requests **req**(u,p) from user u to the application of provider p; *ii)* provider requests **launch**(p,r) from provider p to the cloud service, to launch a new VM in region r.

The following rule models how the cloud service handles a service request from user 0 to the web application of provider 0' when 0 does not exceed her limit (A < B). This request can be forwarded to *any* VM in the system (since **dataCenter**, **pservers**, and **vms** denote *sets* of objects). The user's status is updated and the selected VM updates its **vmReq** attribute by adding the user's object ID 0 to the set OIDSET[2]:

```
crl [processUserReq]:
    req(0, 0')
    < CS0 : CService | status : (noReq(0, A), AS1), subscr : (maxReq(0, B), AS2) >
```

---

[2] We do not show the declaration of variables, but follow the Maude convention that variables are in capital letters.

```
< R : Region | dataCenters : (< DCO : DCenter |
                pservers : (< PSO : PServer |
                    vms : (< VMO : VMachine | owner : O',
                            running : true, vmReq : OIDSET, vmMaxReq : D >
                        VM) > PS) > DC) >
=> < CSO : CService | status : (noReq(O, A + 1), AS1) >
   < R : Region | dataCenters : (< DCO : DCenter |
                pservers : (< PSO : PServer |
                    vms : (< VMO : VMachine | vmReq : (O OIDSET) > VM) > PS) > DC) >
if A < B /\ size(OIDSET) < D .
```

The next rule models how the cloud service handles `launch` requests from a provider O for a new VM in a region R by launching a VM on *any* server in R.

```
crl [processProviderReq]:
   launch(O, R)
   < CSO : CService | status : (noVM(O, A), AS1), subscr : (maxVM(O, B), AS2) >
   < R : Region | dataCenters : (< DCO : DCenter |
                pservers : (< PSO : PServer |
                        load : M, maxLoad : N, nextVMID : NEXTID, vms : VM >
                    PS) > DC) >
=> < CSO : CService | (status : (noVM(O, (A + 1)), AS1)) >
   < R : Region | dataCenters : (< DCO : DCenter |
                pservers : (< PSO : PServer | load : (M+1), nextVMID : (NEXTID+1),
                    vms : (< vm(PSO, NEXTID) : VMachine | owner : O, running : true,
                            vmReq : noid, vmMaxReq : MAXREQ >   VM) > PS) > DC) >
if A < B /\ M < N .
```

Model Checking Safety Properties. To ensure the "safety" of our system we use Maude's `search` command to verify that the cloud is never processing more requests from a user than allowed. For the search to terminate, we use an initial state with two users that can generate 6 and 7 requests. The following Maude command searches for a "bad" state where some user O' has more requests running (M) than allowed (N):

```
(search [1] : initState =>*
   CONFIG < CSO : CService | status : (noReq(O, M), AS1), subscr : (maxReq(O, N), AS2) >
   such that M > N .)
```

```
rewrites: 11312174 in 279134ms cpu (283262ms real) (40525 rewrites/second)
No solution.
```

Load Balancing Policies as Probabilistic Strategies. The above model describes *all possible* treatments of user and provider requests. For better system performance, one could think of different *load balancing policies*, so that, e.g., users may get with high probability a VM closer to their location, or get VMs on servers with the least workload, etc. Likewise, a provider might want VMs uniformly across the regions, or in regions with least workload, etc. Within a region, better-paying providers could be assigned VMs with high probability on stable servers, or on servers with small workload. These policies are specified by different strategies on top of our verified model.

We have defined three strategies: *i)* a strategy that does not take locations or geographical regions into account, i.e., the region to which a user request is forwarded is chosen uniformly at random; *ii)* a strategy that forwards high priority user requests to the region closest to the user with high probability, and furthermore, treats requests from high priority providers with high probability; and *iii)* a strategy that forwards high priority user requests to a region with small VM load with high probability, and processes requests from high priority providers with high probability. For each of these policies, we define two "subcases": *a)* distribute requests uniformly within the region, and *b)* with high probability, distribute requests to VMs/physical servers with small workload for the high-priority providers/users.

We first quantify the nondeterministic choices related to general/environment assumptions about the cloud system, using the following rule strategy:

```
psdrule RuleStrat := given state: CF
                    is: (resolveUserReq) -> 10 ;
                        (processUserReq)  -> 1000 ; (processProviderReq) -> 100 ;
                        (failVM)          -> 1    ; (migrateVM)          -> 1   ;
                        (newUserReq)      -> 100  ; (newProviderReq)     -> 10  .
```

Our load balancing policies are specified by different context strategies for rules `processUserReq` and `processProviderReq`. They resolve the nondeterministic choice of which request to process next, and on which VM/server the request is resolved. We only specify the policy *ii(b)*, and refer to [5] for the definitions of the other policies. The rule `processUserReq` selects *any* user request `req(u, p)`, and assigns *any* VM to handle the request, in *any* region. For policy *ii(b)*, the context strategy for rule `processUserReq` models that user requests with high priorities P are selected with probability proportional to P^2, that the probability of selecting region R is inversely proportional to the distance `distance(LOC, LOC')` between the user and R, and that requests are sent to the VMs/servers with small load `size(OIDSET)` with high probability:

```
psdcontext CtxStrat2b :=
given state: CF req(O, O') < O : SUser | location : LOC, priority : P >
              < CSO : CService | ATTRSET >
              < R : Region |  location : LOC',
                      dataCenters : (< DCO : DCenter |
                          pservers : (< PSO : PServer | ATTRSET',
                              vms : (< VMO : VMachine | vmReq : OIDSET, ATTRSET'' >
                                  VM) > PS) > DC) >
        rule: processUserReq
          is: (CF < O : SUser | location : LOC, priority : P > [])
              -> ((P * P) / ((1 + size(OIDSET)) * (1 + distance(LOC, LOC')))) .
```

The context strategy for rule `processProviderReq` models that requests from providers with high priorities P are selected with high probability proportional to P^2, and resolved by allocating VMs on servers with small load M with high probability:

```
psdcontext CtxStrat2b :=
  given state: CF < CSO : CService | ATTRSET >
              launch(O, R) < O : SProvider | priority : P >
                < R : Region |    location : LOC',
```

```
                        dataCenters : (< DC0 : DCenter |
                         pservers : (< PS0 : PServer | load : M, ATTRSET > PS)
                               > DC) >
          rule: processProviderReq
            is: (CF < O : SProvider | priority : P > []) -> ((P * P) / (1 + M)) .
```

The resulting probabilistic strategy is denoted `Strat2b`.

Simulation. We simulate our system using the strategies `Strat2b` and `Strat3b` (specifying policies *ii(b)* and *iii(b)* above). To analyze the QoS from a user U's perspective, we define a QoS measure of a request R as a function of the relative workload of the server S handling the request, and the distance between the user and S:

$$cost(U, R, S) = k \cdot \frac{\#tasksRunning(S)}{capacity(S)} + q \cdot distance(U, R)$$

The total user QoS is the sum of all such single QoS measures in the system.

We simulate the cloud system using `Strat2b` and `Strat3b`, from an initial state with 2 regions, 4 users, and 2 providers in different locations and with different priorities. Users can generate 3, 5, 8, and 3 requests, respectively, whereas providers can generate 20 requests.

(prew [200] initState using Strat2b .)   (prew [200] initState using Strat3b .)

Fig. 2 shows the total *cumulative* cost over 200 probabilistic rewrite steps (the horizontal axis), for the two strategies (using the same random seed). The smaller values correspond to `Strat2b`, which suggests that `Strat2b` is a better policy in terms of the total user QoS. We next use statistical model checking to show that, with high confidence, this is the case.

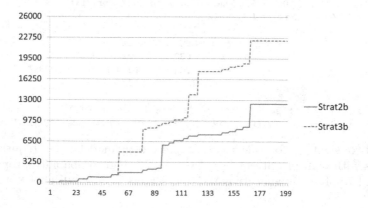

Fig. 2. Comparison of the total cloud cost with each rewrite step, for the two policies

Statistical Model Checking. We use the same initial state as for the simulation, and define a parametric state predicate effortGreaterThan as follows:

```
psp effortGreaterThan : Nat .  var CF : Configuration . var N : Nat .
csat CF |= effortGreaterThan(N) if costCloud(CF) > N .
```

such that effortGreaterThan(N) is true if the cumulative cost of the cloud exceeds N in the current state. We verify that, with confidence 0.9, under the strategy Strat2b, the effort always stays below 13000 with high probability:

```
(smc initState |= P> 0.9 [G ~ effortGreaterThan(13000)] using Strat2b .)

rewrites: 36338205 in 1024184ms cpu (1024339ms real) (35480 rewrites/second)
Result Bool: true            Number of samples used: 3520
Confidence: 90 %             Estimated probability: 1
```

and that the effort under Strat3b will exceed 15000, with high probability:

```
(smc initState |= P> 0.9 [F effortGreaterThan(15000)] using Strat3b .)

rewrites: 128202820 in 2927619ms cpu (2928049ms real) (43790 rewrites/second)
Result Bool: true            Number of samples used: 7150
Confidence: 90 %             Estimated probability: 1
```

5   Related Work

A number of tools support models that are both probabilistic and nondeterministic, including Markov automata [14], generalized stochastic Petri nets [21], or uniform labeled transition systems [6].

In probabilistic automaton-based models one can use synchronous parallel composition to quantify nondeterministic choices by composing the system with a new "scheduler" component. For example, if the model allows us to nondeterministically select action a or action b, we can quantify this nondeterminism by composing the model with a "scheduler" automaton with transitions $\{s_0 \xrightarrow{\tau\,[1/3]} do\text{-}a, \; s_0 \xrightarrow{\tau\,[2/3]} do\text{-}b, \; do\text{-}a \xrightarrow{a} s_0, \; do\text{-}b \xrightarrow{b} s_0\}$; the composed system will then do action a with probability $1/3$ and action b with probability $2/3$. Such "scheduler" components are supported by tools like MODEST [7] and PRISM [19]. Our approach contrasts with this one by: (*i*) having a more explicit separation between model and strategy, since in the above approach the strategy is just another "system" component; (*ii*) supporting a more expressive underlying specification language with unbounded data types, dynamic object/message creation and deletion, and arbitrary complex data types; and (*iii*) providing a more expressive and convenient way of specifying the strategies themselves. It is also unclear to what extent automaton-based approaches can support *hierarchical* systems.

In Uppaal-SMC [12] the nondeterminism concerning *which* component should be executed next is *implicitly* resolved by assigning a stochastic delay to each component; the one with the shortest remaining delay is then scheduled for execution. If multiple transitions are enabled for a component, then one is chosen uniformly at random. In contrast, our language allows the user to specify the probability distributions that quantify the nondeterminism in the model.

Maude itself has a non-probabilistic strategy language [15] to guide the execution of *non-probabilistic* rewrite theories; i.e., there is no support for quantifying the nondeterminism either in the model or in the strategy. VESTA [27] can analyze fully probabilistic actor PMAUDE specifications. For *flat* object-oriented systems nondeterminism is typically removed by letting each action be triggered by a message, and letting probabilistic rules add a stochastic delay to each message [1,3]. The probability of two messages being scheduled at the same "time" is then zero, and this therefore resolves nondeterminism. This method was recently extended to *hierarchical* object-oriented systems [13,10]. The differences with our work are: (i) whereas we have a clear separation between system model and adversary, the above approach encodes the adversary in the model, and hence clutters it with fictitious clocks and schedulers to obtain a fully probabilistic model; (ii) our implementation supports not only a subset of object-oriented specifications, but the entire class of PRTs with fixed-size probability distributions; and (iii) we add a simulator and statistical model checker to Maude instead of using an external tool.

ELAN [8] is a rewriting language where strategies are first class citizens that can appear in rewrite rules, so there is no separation between system model and strategies. The paper [9] adds a "probabilistic choice" operator $PC(s_1 : p_1, \ldots, s_n : p_n)$ to ELAN's strategy language, where p_i defines the probability of applying strategy s_i. This approach is different from ours in the following ways: there is no separation between "system" and "strategy;" the definition of context and substitution adversaries is not supported and therefore not all nondeterminism in a system can be quantified; and there is no support for probabilistic model checking analysis.

6   Concluding Remarks

In this paper we define what is, to the best of our knowledge, the first language for defining complex probabilistic strategies to quantify the nondeterminism in infinite-state systems with both probabilities and nondeterminism, and that includes the object-oriented systems with dynamic object creation. We propose a modular safety/QoS analysis methodology in which a simple (typically non-probabilistic) "base" model that defines all possible system behaviors can be easily verified for safety, and where different probabilistic refinements can be specified on top of the verified base model. QoS properties of the refinements can then be analyzed by statistical and exact probabilistic model checking.

We have implemented a probabilistic simulator and statistical PCTL model checker for our strategies for all probabilistic rewrite theories with discrete probability distributions. We show the usefulness of our language and methodology on a cloud computing example, where different probabilistic strategies on top of the verified base model define different load balancing policies for the cloud, and show how our tool PSMaude can be used to compare the QoS provided by different policies. This example indicates that we need to integrate timed modeling and analysis into our framework.

We also plan to extend PSMaude to allow the statistical analysis of quantitative temporal expressions specified in the QuaTEx logic [1]. This would allow, e.g., the statistical estimation of *expected values* of particular numerical quantities associated with a probabilistic rewrite theory whose nondeterminism is quantified by a given probabilistic strategy. Another direction for future work is to investigate algorithms for the *exact* (vs. statistical) probabilistic model checking of probabilistic rewrite theories for which *some*, but not necessarily all nondeterminism is quantified by a "partial" probabilistic strategy. Finally, since our proposed probabilistic strategy language only allows defining memoryless adversaries, we also aim to extend its syntax and semantics to allow defining *history-dependent* adversaries.

Acknowledgments. We thank José Meseguer and Roy Campbell for discussions on probabilistic strategies, and gratefully acknowledge partial support for this work by AFOSR Grant FA8750-11-2-0084. We also thank the anonymous reviewers for very useful comments on a previous version of the paper.

References

1. Agha, G.A., Meseguer, J., Sen, K.: PMaude: Rewrite-based specification language for probabilistic object systems. ENTCS 153(2) (2006)
2. AlTurki, M., Meseguer, J.: PVeStA: A parallel statistical model checking and quantitative analysis tool. In: Corradini, A., Klin, B., Cîrstea, C. (eds.) CALCO 2011. LNCS, vol. 6859, pp. 386–392. Springer, Heidelberg (2011)
3. AlTurki, M., Meseguer, J., Gunter, C.A.: Probabilistic modeling and analysis of DoS protection for the ASV protocol. ENTCS 234 (2009)
4. Bentea, L.: The PSMaude tool home page:
   http://folk.uio.no/lucianb/prob-strat/
5. Bentea, L., Ölveczky, P.C.: A probabilistic strategy language for probabilistic rewrite theories and its application to cloud computing. Manuscript:
   http://folk.uio.no/lucianb/publications/2012/pstrat-cloud.pdf
6. Bernardo, M., De Nicola, R., Loreti, M.: Uniform labeled transition systems for nondeterministic, probabilistic, and stochastic processes. In: Wirsing, M., Hofmann, M., Rauschmayer, A. (eds.) TGC 2010. LNCS, vol. 6084, pp. 35–56. Springer, Heidelberg (2010)
7. Bohnenkamp, H.C., D'Argenio, P.R., Hermanns, H., Katoen, J.P.: MODEST: A compositional modeling formalism for hard and softly timed systems. IEEE Trans. Software Eng. 32(10), 812–830 (2006)
8. Borovanský, P., Kirchner, C., Kirchner, H., Moreau, P.E., Ringeissen, C.: An overview of ELAN. Electronic Notes in Theoretical Computer Science 15 (1998)
9. Bournez, O., Kirchner, C.: Probabilistic rewrite strategies. Applications to ELAN. In: Tison, S. (ed.) RTA 2002. LNCS, vol. 2378, pp. 252–266. Springer, Heidelberg (2002)
10. Bruni, R., Corradini, A., Gadducci, F., Lluch Lafuente, A., Vandin, A.: Modelling and analyzing adaptive self-assembly strategies with Maude. In: Durán, F. (ed.) WRLA 2012. LNCS, vol. 7571, pp. 118–138. Springer, Heidelberg (2012)

11. Clavel, M., Durán, F., Eker, S., Lincoln, P., Martí-Oliet, N., Meseguer, J., Talcott, C.: All About Maude. LNCS, vol. 4350. Springer, Heidelberg (2007)
12. David, A., Larsen, K.G., Legay, A., Mikučionis, M., Wang, Z.: Time for statistical model checking of real-time systems. In: Gopalakrishnan, G., Qadeer, S. (eds.) CAV 2011. LNCS, vol. 6806, pp. 349–355. Springer, Heidelberg (2011)
13. Eckhardt, J., Mühlbauer, T., AlTurki, M., Meseguer, J., Wirsing, M.: Stable availability under denial of service attacks through formal patterns. In: de Lara, J., Zisman, A. (eds.) FASE 2012. LNCS, vol. 7212, pp. 78–93. Springer, Heidelberg (2012)
14. Eisentraut, C., Hermanns, H., Zhang, L.: On probabilistic automata in continuous time. In: Logic in Computer Science, pp. 342–351 (2010)
15. Eker, S., Martí-Oliet, N., Meseguer, J., Verdejo, A.: Deduction, strategies, and rewriting. Electronic Notes in Theoretical Computer Science 174(11), 3–25 (2007)
16. Hansson, H., Jonsson, B.: A logic for reasoning about time and reliability. Formal Aspects of Computing 6 (1994)
17. Katelman, M., Meseguer, J., Hou, J.: Redesign of the LMST wireless sensor protocol through formal modeling and statistical model checking. In: Barthe, G., de Boer, F.S. (eds.) FMOODS 2008. LNCS, vol. 5051, pp. 150–169. Springer, Heidelberg (2008)
18. Kumar, N., Sen, K., Meseguer, J., Agha, G.: Probabilistic rewrite theories: Unifying models, logics and tools. Technical report UIUCDCS-R-2003-2347, Department of Computer Science, University of Illinois at Urbana-Champaign (2003)
19. Kwiatkowska, M., Norman, G., Parker, D.: PRISM 4.0: Verification of probabilistic real-time systems. In: Gopalakrishnan, G., Qadeer, S. (eds.) CAV 2011. LNCS, vol. 6806, pp. 585–591. Springer, Heidelberg (2011)
20. Larsen, K.G., Skou, A.: Bisimulation through probabilistic testing. Information and Computation 94(1), 1–28 (1991)
21. Marsan, M.A., Balbo, G., Conte, G., Donatelli, S., Franceschinis, G.: Modelling with generalized stochastic Petri nets. SIGMETRICS Performance Evaluation Review 26(2), 2 (1998)
22. Martí-Oliet, N., Meseguer, J.: Rewriting logic: roadmap and bibliography. Theoretical Computer Science 285(2) (2002)
23. Martí-Oliet, N., Meseguer, J., Verdejo, A.: Towards a strategy language for Maude. Electronic Notes in Theoretical Computer Science 117, 417–441 (2005)
24. Meseguer, J.: Conditional rewriting logic as a unified model of concurrency. Theoretical Computer Science 96(1) (1992)
25. Meseguer, J.: A rewriting logic sampler. In: Van Hung, D., Wirsing, M. (eds.) ICTAC 2005. LNCS, vol. 3722, pp. 1–28. Springer, Heidelberg (2005)
26. Sen, K., Viswanathan, M., Agha, G.: On statistical model checking of stochastic systems. In: Etessami, K., Rajamani, S.K. (eds.) CAV 2005. LNCS, vol. 3576, pp. 266–280. Springer, Heidelberg (2005)
27. Sen, K., Viswanathan, M., Agha, G.A.: VeStA: A statistical model-checker and analyzer for probabilistic systems. In: QEST 2005. IEEE Computer Society (2005)
28. Younes, H.L.S., Simmons, R.G.: Probabilistic verification of discrete event systems using acceptance sampling. In: Brinksma, E., Larsen, K.G. (eds.) CAV 2002. LNCS, vol. 2404, pp. 223–235. Springer, Heidelberg (2002)

Adaptable Transition Systems[*]

Roberto Bruni[1], Andrea Corradini[1], Fabio Gadducci[1],
Alberto Lluch Lafuente[2], and Andrea Vandin[2]

[1] Dipartimento di Informatica, University of Pisa, Italy
{bruni,andrea,gadducci}@di.unipi.it
[2] IMT Institute for Advanced Studies, Lucca, Italy
{alberto.lluch,andrea.vandin}@imtlucca.it

Abstract. We present an essential model of adaptable transition systems inspired by white-box approaches to adaptation and based on foundational models of component based systems. The key feature of adaptable transition systems are *control propositions*, imposing a clear separation between ordinary, functional behaviours and adaptive ones. We instantiate our approach on interface automata yielding *adaptable interface automata*, but it may be instantiated on other foundational models of component-based systems as well. We discuss how control propositions can be exploited in the specification and analysis of adaptive systems, focusing on various notions proposed in the literature, like *adaptability*, *control loops*, and *control synthesis*.

Keywords: Adaptation, autonomic systems, control data, interface automata.

1 Introduction

Self-adaptive systems have been advocated as a convenient solution to the problem of mastering the complexity of modern software systems, networks and architectures. In particular, self-adaptation is considered a fundamental feature of *autonomic systems*, that can specialise to several other self-* properties like self-configuration, self-optimisation, self-protection and self-healing. Despite some valuable efforts (see e.g. [16,11]), there is no general agreement on the notion of adaptation, neither in general nor in software systems. There is as well no widely accepted foundational model for adaptation. Using Zadeh's words [18]: *"it is very difficult -perhaps impossible- to find a way of characterizing in concrete terms the large variety of ways in which adaptive behavior can be realized"*. Zadeh's concerns were conceived in the field of Control Theory but are valid in Computer Science as well. Zadeh' skepticism for a concrete unifying definition of adaptation is due to the attempt to subsume two aspects under the same definition: the *external* manifestations of adaptive systems (sometimes called *black-box* adaptation), and the *internal* mechanisms by which adaptation is achieved (sometimes called *white-box* adaptation).

[*] Research partially supported by the EU through the FP7-ICT Integrated Project 257414 ASCEns (Autonomic Service-Component Ensembles).

N. Martí-Oliet and M. Palomino (Eds.): WADT 2012, LNCS 7841, pp. 95–110, 2013.

The limited effort placed so far in the investigation of the foundations of adaptive software systems might be due to the fact that it is not clear what are the characterising features that distinguish adaptive systems from those that are not so. For instance, very often a software system is considered "self-adaptive" if it *"modifies its own behavior in response to changes in its operating environment"* [14], when the software system realises that *"it is not accomplishing what the software is intended to do, or better functionality or performance is possible"* [15]. But, according to this definition, almost any software system can be considered self-adaptive, since any system of a reasonable complexity can *modify its behaviour* (e.g. following one of the different branches of a conditional statement) as a *reaction to a change in its context of execution* (e.g. values of variables or parameters).

Consider the automaton of Fig. 1, which models a server providing a task execution service. Each state has the format $s\{q\}[r]$ where s can be either D (the server is down) or U (it is up), and q, r are possibly empty sequences of t symbols representing, respectively, the lists of tasks scheduled for execution and the ones received but not scheduled yet. Transitions are labelled with t? (receive a task), u! (start-up the server), s! (schedule a task), f! (notify the conclusion of a task), and d! (shut-down the server).

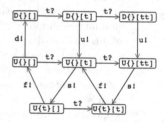

Fig. 1. Is it self-adaptive?

Annotations ? and ! denote *input* and *output* actions, respectively. Summing up, the server can receive tasks, start up, schedule tasks and notify their termination, and eventually shut down. Now, is the modelled server self-adaptive? One may argue that indeed it is, since the server schedules tasks only when it is up. Another argument can be that the server is self-adaptive since it starts up only when at least one task has to be processed, and shuts down only when no more tasks have to be processed. Or one could say that the server is not adaptive, because all transitions just implement its ordinary functional behaviour. Which is the right argument? How can we handle such diverse interpretations?

White-box adaptation. White-box perspectives on adaptation allow one to specify or inspect (part of) the internal structure of a system in order to offer a clear *separation of concerns* to distinguish changes of behaviour that are part of the application or functional logic from those which realise the adaptation logic.

In general, the behaviour of a component is governed by a program and according to the traditional, basic view, a program is made of *control* (i.e. algorithms) and *data*. The conceptual notion of adaptation we proposed in [5] requires to identify *control data* which can be changed to *adapt* the component's behaviour. *Adaptation* is, hence, the run-time modification of such control data. Therefore, a component is *adaptable* if it has a distinguished collection of control data that can be modified at run-time, *adaptive* if it is adaptable and its control data are modified at run-time, at least in some of its executions, and *self-adaptive* if it modifies its own control data at run-time.

Several programming paradigms and reference models have been proposed for adaptive systems. A notable example is the Context Oriented Programming paradigm, where the contexts of execution and code variations are first-class citizens that can be used to structure the adaptation logic in a disciplined way [17]. Nevertheless, it is not the programming language what makes a program adaptive: any computational model or programming language can be used to implement an adaptive system, just by identifying the part of the data that governs the adaptation logic, that is the control data. Consequently, the nature of control data can vary considerably, including all possible ways of encapsulating behaviour: from simple configuration parameters to a complete representation of the program in execution that can be modified at run-time, as it is typical of computational models that support meta-programming or reflective features.

The subjectivity of adaptation is captured by the fact that the collection of control data of a component can be defined in an arbitrary way, ranging from the empty set ("the system is not adaptable") to the collection of all the data of the program ("any data modification is an adaptation"). This means that white-box perspectives are as subjective as black-box ones. The fundamental difference lies in who is responsible of declaring which behaviours are part of the adaptation logic and which not: the observer (black-box) or the designer (white-box).

Consider again the system in Fig. 1 and the two possible interpretations of its adaptivity features. As elaborated in Sect. 3, in the first case control data are defined by the state of the server, while in the second case control data are defined by the two queues. If instead the system is not considered adaptive, then the set of control data is empty. This way the various interpretations are made concrete in our conceptual approach. We shall use this system as our running example.

It is worth to mention that the control data approach [5] is agnostic with respect to the form of interaction with the environment, the level of context-awareness, the use of reflection for self-awareness. It applies equally well to most of the existing approaches for designing adaptive systems and provides a satisfactory answer to the question "what is adaptation *conceptually*?". But "what is adaptation *formally*?" and "how can we reason about adaptation, *formally*?".

Contribution. This paper provides an answer to the questions we raised above. Building on our informal discussion, on a foundational model of component based systems (namely, *interface automata* [1,2], introduced in Sect. 2), and on previous formalisations of adaptive systems (discussed in Sect. 5) we distill in Sect. 3 a core model of adaptive systems called *adaptable interface automata* (AIA). The key feature of AIA are *control propositions* evaluated on states, the formal counterpart of control data. The choice of control propositions is arbitrary but it imposes a clear separation between ordinary, functional behaviours and adaptive ones. We then discuss in Sect. 4 how control propositions can be exploited in the specification and analysis of adaptive systems, focusing on various notions proposed in the literature, like *adaptability, feedback control loops*, and *control synthesis*. The approach based on control propositions can be applied to other computational models, yielding other instances of adaptable transition systems. The choice of interface automata is due to their simple and elegant theory.

Fig. 2. Three interface automata: Mac (left), Exe (centre), and Que (right)

2 Background

Interface automata were introduced in [2] as a flexible framework for component-based design and verification. We recall here the main concepts from [1].

Definition 1 (interface automaton). *An interface automaton P is a tuple $\langle V, V^i, \mathcal{A}^I, \mathcal{A}^O, \mathcal{T} \rangle$, where V is a set of states; $V^i \subseteq V$ is the set of initial states, which contains at most one element (if V^i is empty then P is called* empty*); $\mathcal{A}^I$ and $\mathcal{A}^O$ are two disjoint sets of* input *and* output *actions (we denote by $\mathcal{A} = \mathcal{A}^I \cup \mathcal{A}^O$ the set of all actions); and $\mathcal{T} \subseteq V \times \mathcal{A} \times V$ is a deterministic set of* steps *(i.e. $(u, a, v) \in \mathcal{T}$, $(u, a, v') \in \mathcal{T}$ implies $v = v'$).*

Example 1. Figure 2 presents three interface automata modelling respectively a machine Mac (left), an execution queue Exe (centre), and a task queue Que (right). Intuitively, each automaton models one component of our running example (cf. Fig. 1). The format of the states is as in our running example. The initial states are not depicted on purpose, because we will consider several cases. Here we assume that they are U, {} and [], respectively. The actions of the automata have been described in Sect. 1. The *interface* of each automaton is implicitly denoted by the action annotation: ? for inputs and ! for outputs.

Given $\mathcal{B} \subseteq \mathcal{A}$, we sometimes use $P_{|\mathcal{B}}$ to denote the automaton obtained by restricting the set of steps to those whose action is in $\mathcal{B}$. Similarly, the set of actions in $\mathcal{B}$ labelling the outgoing transitions of a state u is denoted by $\mathcal{B}(u)$. A *computation* ρ of an interface automaton P is a finite or infinite sequence of consecutive *steps* (or *transitions*) $\{(u_i, a_i, u_{i+1})\}_{i<n}$ from $\mathcal{T}$ (thus n can be ω).

A partial composition operator is defined for automata: in order for two automata to be composable their interface must satisfy certain conditions.

Definition 2 (composability). *Let P and Q be two interface automata. Then, P and Q are composable if $\mathcal{A}_P^O \cap \mathcal{A}_Q^O = \emptyset$.*

Let $shared(P, Q) = \mathcal{A}_P \cap \mathcal{A}_Q$ and $comm(P, Q) = (\mathcal{A}_P^O \cap \mathcal{A}_Q^I) \cup (\mathcal{A}_P^I \cap \mathcal{A}_Q^O)$ be the set of *shared* and *communication* actions, respectively. Thus, two interface automata can be composed if they share input or communication actions only.

Two composable interface automata can be combined in a *product* as follows.

Definition 3 (product). *Let P and Q be two composable interface automata. Then the product $P \otimes Q$ is the interface automaton $\langle V, V^i, \mathcal{A}^I, \mathcal{A}^O, \mathcal{T} \rangle$ such that $V = V_P \times V_Q$; $V^i = V_P^i \times V_Q^i$; $\mathcal{A}^I = (\mathcal{A}_P^I \cup \mathcal{A}_Q^I) \setminus comm(P, Q)$; $\mathcal{A}^O = \mathcal{A}_P^O \cup \mathcal{A}_Q^O$; and*

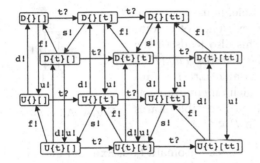 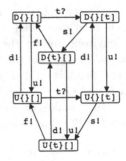

Fig. 3. The product Mac ⊗ Exe ⊗ Que (left) and the composition Mac | Exe | Que (right)

$\mathcal{T}$ *is the union of* $\{((v, u), a, (v', u)) \mid (v, a, v') \in \mathcal{T}_P \wedge a \notin shared(P, Q) \wedge u \in V_Q\}$
(i.e. P steps), $\{((v, u), a, (v, u')) \mid (u, a, u') \in \mathcal{T}_Q \wedge a \notin shared(P, Q) \wedge v \in V_P\}$
(i.e. Q steps), and $\{((v, u), a, (v', u')) \mid (v, a, v') \in \mathcal{T}_P \wedge (u, a, u') \in \mathcal{T}_Q \wedge a \in shared(P, Q)\}$ *(i.e. steps where P and Q synchronise over shared actions).*

In words, the product is a commutative and associative operation (up to isomorphism) that interleaves non-shared actions, while shared actions are synchronised in broadcast fashion, in such a way that shared input actions become inputs, communication actions become outputs.

Example 2. Consider the interface automata Mac, Exe and Que of Fig. 2. They are all pairwise composable and, moreover, the product of any two of them is composable with the remaining one. The result of applying the product of all three automata is depicted in Fig. 3 (left).

States in $P \otimes Q$ where a communication action is output by one automaton but cannot be accepted as input by the other are called *incompatible* or *illegal*.

Definition 4 (incompatible states). *Let P and Q be two composable interface automata. The set incompatible$(P, Q) \subseteq V_P \times V_Q$ of incompatible states of $P \otimes Q$ is defined as* $\{(u, v) \in V_P \times V_Q \mid \exists a \in comm(P, Q) . (a \in \mathcal{A}_P^O(u) \wedge a \notin \mathcal{A}_Q^I(v)) \vee (a \in \mathcal{A}_Q^O(v) \wedge a \notin \mathcal{A}_P^I(u))\}$.

Example 3. In our example, the product Mac ⊗ Exe ⊗ Que depicted in Fig. 3 (left) has several incompatible states, namely all those of the form "s{t}[t]" or "s{t}[tt]". Indeed, in those states, Que is willing to perform the output action s! but Exe is not able to perform the dual input action s?.

The presence of incompatible states does not forbid to compose interface automata. In an open system, compatibility can be ensured by a third automata called the *environment* which may e.g. represent the context of execution or an adaptation manager. Technically, an environment for an automaton R is a nonempty automaton E which is composable with R, synchronises with all output actions of R (i.e. $\mathcal{A}_E^I = \mathcal{A}_R^O$) and whose product with R does not have incompatible states. Interesting is the case when R is $P \otimes Q$ and E is a *compatible environment*, i.e. when the set *incompatible*$(P, Q) \times V_E$ is not reachable in $R \otimes E$.

Compatibility of two (composable, non-empty) automata is then expressed as the existence of a compatible environment for them. This also leads to the concept of *compatible* (or *usable*) states $cmp(P \otimes Q)$ in the product of two composable interface automata P and Q, i.e. those for which an environment E exists that makes the set of incompatible states $incompatible(P, Q)$ unreachable in $P \otimes Q \otimes E$.

Fig. 4. An environment

Example 4. Consider again the interface automata Mac, Exe and Que of Fig. 2. Automata Mac and Exe are trivially compatible, and so are Mac and Que. Exe and Que are compatible as well, despite of the incompatible states $\{t\}[t]$ and $\{t\}[tt]$ in their product Exe $\otimes$ Que. Indeed an environment that does not issue a second task execution requests $t!$ without first waiting for a termination notification (like the one in Fig. 4) can avoid reaching the incompatible states.

We are finally ready to define the composition of interface automata.

Definition 5 (composition). *Let P and Q be two composable interface automata. The composition $P \mid Q$ is an interface automaton $\langle V, V^i, \mathcal{A}^I_{P \otimes Q}, \mathcal{A}^O_{P \otimes Q}, \mathcal{T} \rangle$ such that $V = cmp(P \otimes Q)$; $V^i = V^i_{P \otimes Q} \cap V$; and $\mathcal{T} = \mathcal{T}_{P \otimes Q} \cap (V \times \mathcal{A} \times V)$.*

Example 5. Consider the product Mac $\otimes$ Exe $\otimes$ Que depicted in Fig. 3 (left). All states of the form $s\{t\}[t]$ and $s\{t\}[tt]$ are incompatible and states $D\{\}[tt]$ and $U\{\}[tt]$ are not compatible, since no environment can prevent them to enter the incompatible states. The remaining states are all compatible. The composition Mac $\mid$ Exe $\mid$ Que is the interface automaton depicted in Fig. 3 (right).

3   Adaptable Interface Automata

Adaptable interface automata extend interface automata with atomic propositions (state observations) a subset of which is called *control propositions* and play the role of the control data of [5].

Definition 6 (adaptable interface automata). *An adaptable interface automaton (AIA) is a tuple $\langle P, \Phi, l, \Phi^c \rangle$ such that $P = \langle V, V^i, \mathcal{A}^I, \mathcal{A}^O, \mathcal{T} \rangle$ is an interface automaton; Φ is a set of atomic propositions, $l : V \to 2^\Phi$ is a labelling function mapping states to sets of propositions; and $\Phi^c \subseteq \Phi$ is a distinguished subset of* control propositions.

Abusing the notation we sometimes call P an AIA with underlying interface automaton P, whenever this introduces no ambiguity. A transition $(u, a, u') \in T$ is called an *adaptation* if it changes the control data, i.e. if there exists a proposition $\phi \in \Phi^c$ such that either $\phi \in l(u)$ and $\phi \notin l(u')$, or vice versa. Otherwise, it is called a *basic* transition. An action $a \in A$ is called a *control action* if it labels at least one adaptation. The set of all control actions of an AIA P is denoted by $\mathcal{A}^C_P$.

Example 6. Recall the example introduced in Sect. 1. We raised the question whether the interface automaton S of Fig. 1 is (self-)adaptive or not. Two arguments were given. The first argument was *"the server schedules tasks only when it is up"*. That is, we identify two different behaviours of the server (when it is up or down, respectively), interpreting a change of behaviour as an adaptation. We can capture this interpretation by introducing a control proposition that records the state of the server. More precisely, we define the AIA Switch(S) in the following manner. The underlying interface automaton is S; the only (control) proposition is *up*, and the labelling function maps states of the form U{...}[...] into {*up*} and those of the form D{...}[...] into ∅. The control actions are then u and d. The second argument was *"the system starts the server up only when there is at least one task to schedule, and shuts it down only when no task has to be processed"*. In this case the change of behaviour (adaptation) is triggered either by the arrival of a task in the waiting queue, or by the removal of the last task scheduled for execution. Therefore we can define the control data as the state of both queues. That is, one can define an AIA Scheduler(S) having as underlying interface automaton the one of Fig. 1, as control propositions all those of the form $queues\_status\_q\_r$ (with $q \in \{\_, t\}$, and $r \in \{\_, t, tt\}$), and a labelling function that maps states of the form $s\{q\}[r]$ to the set $\{queues\_status\_q\_r\}$. In this case the control actions are s, f and t.

Computations. The computations of an AIA (i.e. those of the underlying interface automata) can be classified according to the presence of adaptation transitions. For example, a computation is *basic* if it contains no adaptive step, and it is *adaptive* otherwise. We will also use the concepts of *basic computation* starting at a state u and of *adaptation phase*, i.e. a maximal computation made of adaptive steps only.

Coherent Control. It is worth to remark that what distinguishes adaptive computations and adaptation phases are not the actions, because control actions may also label transitions that are not adaptations. However, very often an AIA has *coherent control*, meaning that the choice of control propositions is coherent with the induced set of control actions, in the sense that all the transitions labelled with control actions are adaptations.

Composition. The properties of composability and compatibility for AIA, as well as product and composition operators, are lifted from interface automata.

Definition 7 (composition). *Let P and Q be two AIA whose underlying interface automata P′, Q′ are composable. The composition P | Q is the AIA $\langle P' | Q', \Phi, l, \Phi^c \rangle$ such that the underlying interface automaton is the composition of P′ and Q′; $\Phi = \Phi_P \uplus \Phi_Q$ (i.e. the set of atomic propositions is the disjoint union of the atomic propositions of P and Q); $\Phi^c = \Phi_P^c \uplus \Phi_Q^c$; and l is such that $l((u,v)) = l_P(u) \cup l_Q(v)$ for all $(u,v) \in V$ (i.e. a proposition holds in a composed state if it holds in its original local state).*

Since the control propositions of the composed system are the disjoint union of those of the components, one easily derives that control coherence is preserved by composition, and that the set of control actions of the product is obtained as the union of those of the components.

4   Exploiting Control Data

We explain here how the distinguishing features of AIA (i.e. control propositions and actions) can be exploited in the design and analysis of self-adaptive systems. For the sake of simplicity we will focus on AIA with coherent control, as it is the case of all of our examples. Thus, all the various definitions/operators that we are going to define on AIA may rely on the manipulation of control actions only.

4.1   Design

Well-Formed Interfaces. The relationship between the set of control actions A_P^C and the alphabets A_P^I and A_P^O is arbitrary in general, but it could satisfy some pretty obvious constraints for specific classes of systems.

Definition 8 (adaptable, controllable and self-adaptive ATSs). *Let P be an AIA. We say that P is* adaptable *if $A_P^C \neq \emptyset$;* controllable *if $A_P^C \cap A_P^I \neq \emptyset$;* self-adaptive *if $A_P^C \cap A_P^O \neq \emptyset$.*

Intuitively, an AIA is *adaptable* if it has at least one control action, which means that at least one transition is an adaptation. An adaptable AIA is *controllable* if control actions include some input actions, or *self-adaptive* if control actions include some output actions (which are under control of the AIA).

From these notions we can derive others. For instance, we can say that an adaptable AIA is *fully self-adaptive* if $A_P^C \cap A_P^I = \emptyset$ (the AIA has full control over adaptations). Note that hybrid situations are possible as well, when control actions include both input actions (i.e. actions in A_P^I) and output actions (i.e. actions in A_P^O). In this case we have that P is both *self-adaptive* and *controllable*.

Example 7. Consider the AIA Scheduler(S) and Switch(S) described in Example 6, whose underlying automaton (S) is depicted in Fig. 1. Switch(S) is fully self-adaptive and not controllable, since its control actions do not include input actions, and therefore the environment cannot force the execution of control actions directly. On the other hand, Scheduler(S) is self-adaptive and controllable, since some of its control actions are outputs and some are inputs.

Consider instead the interface automaton A in the left of Fig. 5, which is very much like the automaton Mac ⊗ Exe ⊗ Que of Fig. 3, except that all actions but f have been turned into input actions and states of the form s{t}[tt] have been removed. The automaton can also be seen as the composition of the two automata on the right of Fig. 5. And let us call Scheduler(A) and Switch(A) the AIA obtained by applying the control data criteria of Scheduler(S) and Switch(S), respectively. Both Scheduler(A) and Switch(A) are adaptable and controllable, but only Scheduler(A) is self-adaptive, since it has at least one control output action (i.e. f!).

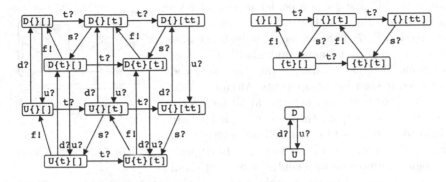

Fig. 5. An adaptable server (left) and its components (right)

Composition. As discussed in Sect. 3, the composition operation of interface automata can be extended seamlessly to AIA. Composition can be used, for example, to combine an adaptable basic component B and an adaptation manager M in a way that reflects a specific adaptation logic. In this case, natural well-formedness constraints can be expressed as suitable relations among sets of actions. For example, we can define when a component M controls another component B as follows.

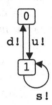

Fig. 6. A controller

Definition 9 (controlled composition). *Let B and M be two composable* AIA. *We say that M controls B in $B \mid M$ if $\mathcal{A}_B^C \cap \mathcal{A}_M^O \neq \emptyset$. In addition, we say that M controls completely B in $B \mid M$ if $\mathcal{A}_B^C \subseteq \mathcal{A}_M^O$.*

This definition can be used, for instance, to allow or to forbid mutual control. For example, if a manager M is itself at least partly controllable (i.e. $\mathcal{A}_M^C \cap \mathcal{A}_M^I \neq \emptyset$), a natural requirement to avoid mutual control would be that the managed component B and M are such the $\mathcal{A}_B^O \cap \mathcal{A}_M^C = \emptyset$, i.e. that B cannot control M.

Example 8. Consider the adaptable server depicted on the left of Fig. 5 as the basic component whose control actions are d, u and s. Consider further the controller of Fig. 6 as the manager, which controls completely the basic component. A superficial look at the server and the controller may lead to think that their composition yields the adaptive server of Fig. 1, yet this not the case. Indeed, the underlying interface automata are not compatible due to the existence of (unavoidable) incompatible states.

Control Loops and Action Classification. The distinction between input, output and control actions is suitable to model some basic interactions and well-formedness criteria as we explained above. More sophisticated cases such as control loops are better modelled if further classes of actions are distinguished.

As a paradigmatic example, let us consider the control loop of the MAPE-K reference model [9], illustrated in Fig. 7. This reference model is the most influential one for autonomic and adaptive systems. The name MAPE-K is due to the main activities of autonomic manager components (Monitor, Analyse, Plan, Execute) and the fact that all such activities operate and exploit the same Knowledge base.

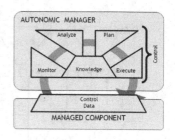

Fig. 7. MAPE-K loop

According to this model, a self-adaptive system is made of a component implementing the application logic, equipped with a control loop that monitors the execution through suitable sensors, analyses the collected data, plans an adaptation strategy, and finally executes the adaptation of the managed component through some effectors. The managed component is considered to be an adaptable component, and the system made of both the component and the manager implementing the control loop is considered as a self-adaptive component.

AIA can be composed so to adhere to the MAPE-K reference model as schematised in Fig. 8. First, the autonomic manager component M and the managed component B have their *functional* input and output actions, respectively $I \subseteq \mathcal{A}_M^I$, $O \subseteq \mathcal{A}_M^O$, $I' \subseteq \mathcal{A}_B^I$, $O' \subseteq \mathcal{A}_B^O$ such that no dual action is shared (i.e. $comm(B, M) \cap (I \cup I') = \emptyset$) but inputs may be shared (i.e. possibly $I \cap I' \neq \emptyset$). The autonomic manager is controllable and has hence a distinguished set of control actions $C = \mathcal{A}_B^C$. The dual of such control actions, i.e. the output actions of M that synchronise with the input control actions B can be regarded as *effectors*

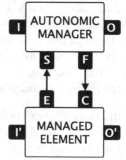

Fig. 8. MAPE-K actions

$F \subseteq \mathcal{A}_M^O$, i.e. output actions used to trigger adaptation. In addition, M will also have *sensor* input actions $S \subseteq \mathcal{A}_M^I$ to sense the status of B, notified via *emit* output actions $E \subseteq \mathcal{A}_M^O$. Clearly, the introduced sets partition inputs and outputs, i.e. $I \uplus S = \mathcal{A}_M^I$, $O \uplus F = \mathcal{A}_M^O$, $E \uplus I' = \mathcal{A}_B^I$ and $O' \uplus C = \mathcal{A}_M^O$.

4.2 Analysis and Verification

Property Classes. By the very nature of adaptive systems, properties that one is interested to verify on them can be classified according to the kind of computations that are concerned with, so that the usual verification (e.g. model checking problem) $P \models \psi$ (i.e. *"does the* AIA *P satisfy property ψ?"*) is instantiated in some of the computations of P depending of the class of ψ.

For example, some authors (e.g. [20,19,10]) distinguish the following three kinds of properties. *Local* properties are *"properties of one [behavioral] mode"*, i.e. properties that must be satisfied by basic computations only. *Adaptation* properties are to be *"satisfied on interval states when adapting from one behavioral mode to another"*, i.e. properties of adaptation phases. *Global* properties *"regard program behavior and adaptations as a whole. They should be satisfied by the adaptive program throughout its execution, regardless of the adaptations."*, i.e. properties about the overall behaviour of the system.

To these we add the class of *adaptability* properties, i.e. properties that may fail for local (i.e. basic) computations, and that need the adapting capability of the system to be satisfied.

Definition 10 (adaptability property). *Let P be an* AIA. *A property ψ is an* adaptability *property for P if $P \models \psi$ and $P_{|_{\mathcal{A}_P \setminus \mathcal{A}_P^C}} \not\models \psi$.*

Example 9. Consider the adaptive server of Fig. 1 and the AIA Scheduler(S) and Switch(S), with initial state U{}[]. Consider further the property *"whenever a task is received, the server can finish it"*. This is an adaptability property for Scheduler(S) but not for Switch(S). The main reason is that in order to finish a task it first has to be received (t) and scheduled (s), which is part of the adaptation logic in Scheduler(S) but not in Switch(S). In the latter, indeed, the basic computations starting from state U{}[] are able to satisfy the property.

Weak and Strong Adaptability. AIA are also amenable for the analysis of the computations of interface automata in terms of adaptability. For instance, the concepts of *weak* and *strong* adaptability from [13] can be very easily rephrased in our setting. According to [13] a system is *weakly adaptable* if *"for all paths, it always holds that as soon as adaptation starts, there exists at least one path for which the system eventually ends the adaptation phase"*, while a system is *strongly adaptable* if *"for all paths, it always holds that as soon as adaptation starts, all paths eventually end the adaptation phase"*.

Strong and weak adaptability can also be characterised by formulae in some temporal logic [13], ACTL [7] in our setting.

Definition 11 (weak and strong adaptability). *Let P be an* AIA. *We say that P is* weakly adaptable *if $P \models \mathbf{AG}\ \mathbf{EF}\ \mathbf{EX}\{\mathcal{A}_P \setminus \mathcal{A}_P^C\}true$, and* strongly adaptable *if $P \models \mathbf{AG}\ \mathbf{AF}\ (\mathbf{EX}\{\mathcal{A}_P\}true \wedge \mathbf{AX}\{\mathcal{A}_P \setminus \mathcal{A}_P^C\}true)$.*

The formula characterising weak adaptability states that along all paths (**A**) it always (**G**) holds that there is a path (**E**) where eventually (**F**) a state will be reached where a basic step can be executed ($\mathbf{EX}\{\mathcal{A}_P \setminus \mathcal{A}_P^C\}true$). Similarly, the formula characterising strong adaptability states that along all paths (**A**) it always (**G**) holds that along all paths (**A**) eventually (**F**) a state will be reached where at least one step can be fired ($\mathbf{EX}\{\mathcal{A}_P\}true$) and all fireable actions are basic steps ($\mathbf{AX}\{\mathcal{A}_P \setminus \mathcal{A}_P^C\}true$). Apart from its conciseness, such characterisations enable the use of model checking techniques to verify them.

Example 10. The AIA Switch(S) (cf. Fig. 1) is strongly adaptable, since it does not have any infinite adaptation phase. Indeed every control action (u or d) leads to a state where only basic actions (t, f or s) can be fired. On the other hand, Scheduler(S) is weakly adaptable due to the presence of loops made of adaptive transitions only (namely, t, s and f), which introduce an infinite adaptation phase. Consider now the AIA Scheduler(A) and Switch(A) (cf. Fig. 5). Both are weakly adaptable due to the loops made of adaptive transitions only: e.g. in Switch(A) there are cyclic behaviours made of the control actions u and d.

4.3   Reverse Engineering and Control Synthesis

Control data can also guide reverse engineering activities. For instance, is it possible to decompose an AIA S into a basic adaptable component B and a suitable controller M? We answer in the positive, by presenting first a trivial solution and then a more sophisticated one based on control synthesis.

Basic Decomposition. In order to present the basic decomposition we need some definitions. Let $P^{\perp_\mathcal{B}}$ denote the operation that given an automaton P results in an automaton $P^{\perp_\mathcal{B}}$ which is like P but where actions in $\mathcal{B} \subseteq \mathcal{A}$ have been complemented (inputs become outputs and vice versa). Formally, $P^{\perp_\mathcal{B}} = \langle V, V^i, ((\mathcal{A}^I \setminus \mathcal{B}) \cup (\mathcal{A}^O \cap \mathcal{B})), ((\mathcal{A}^O \setminus \mathcal{B}) \cup (\mathcal{A}^I \cap \mathcal{B})), \mathcal{T} \rangle$. This operation can be trivially lifted to AIA by preserving the set of control actions.

It is easy to see that interface automata have the following property. If P is an interface automaton and O_1, O_2 are sets of actions that partition $\mathcal{A}_P^O$ (i.e. $\mathcal{A}_P^O = O_1 \uplus O_2$), then P is isomorphic to $P^{\perp_{O_1}} \mid P^{\perp_{O_2}}$. This property can be exploited to decompose an AIA P as $M \mid B$ by choosing $M = P^{\perp_{\mathcal{A}_P^O \setminus \mathcal{A}_P^C}}$ and $B = P^{\perp_{\mathcal{A}_P^O \cap \mathcal{A}_P^C}}$. Intuitively, the manager and the base component are identical to the original system and only differ in their interface. All output control actions are governed by the manager M and become inputs in the base component B. Outputs that are not control actions become inputs in the manager. This decomposition has some interesting properties: B is fully controllable and, if P is fully self-adaptive, then M completely controls B.

Example 11. Consider the server Scheduler(S) (cf. Fig. 1). The basic decomposition provides the manager with underlying automata depicted in Fig. 9 (left) and the basic component depicted in Fig. 9 (right). Vice versa, if the server Switch(S) (cf. Fig. 1) is considered, then the basic decomposition provides the manager with underlying automata depicted in Fig. 9 (right) and the basic component depicted in Fig. 9 (left).

Decomposition as Control Synthesis. In the basic decomposition both M and B are isomorphic (and hence of equal size) to the original AIA S, modulo the complementation of some actions. It is however possible to apply heuristics in order to obtain smaller non-trivial managers and base components. One possibility is to reduce the set of actions that M needs to observe (its input actions). Intuitively, one can make the choice of ignoring some input actions and collapse the corresponding transitions. Of course, the resulting manager M must be checked for the absence of non-determinism (possibly introduced by the identification of states) but will be a smaller manager candidate. Once a candidate M is chosen we can resort to solutions to the control synthesis problem.

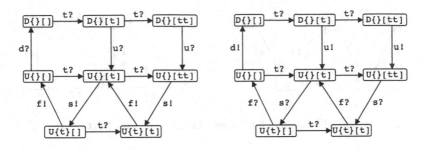

Fig. 9. A basic decomposition

We recall that the synthesis of controllers for interface automata [4] is the problem of solving the equation $P \mid Y \preceq Q$, for a given system Q and component P, i.e. finding a component Y such that, when composed with P, results in a system which *refines* Q. An interface automaton R *refines* an interface automaton S if (i) $\mathcal{A}_R^I \subseteq \mathcal{A}_S^I$, (ii) $\mathcal{A}_R^O \subseteq \mathcal{A}_S^O$, and (iii) there is an *alternating simulation* relation ϱ from R to S, and two states $u \in V_R^i$, $v \in V_S^i$ such that $(u, v) \in \varrho$ [1]. An *alternating simulation* relation ϱ from an interface automaton R to an interface automaton S is a relation $\varrho \subseteq V_R \times V_S$ such that for all $(u, v) \in \varrho$ and all $a \in \mathcal{A}_R^O(u) \cup \mathcal{A}_S^I(v)$ we have (i) $\mathcal{A}_S^I(v) \subseteq \mathcal{A}_R^I(u)$ (ii) $\mathcal{A}_R^O(u) \subseteq \mathcal{A}_S^O(v)$ (iii) there are $u' \in V_R, v' \in V_S$ such that $(u, a, u') \in \mathcal{T}_R$, $(v, a, v') \in \mathcal{T}_S$ and $(u', v') \in \varrho$.

The control synthesis solution of [4] can be lifted to AIA in the obvious way. The equation under study in our case will be $B \mid M \preceq P$. The usual case is when B is known and M is to be synthesised, but it may also happen that M is given and B is to be synthesised. The solution of [4] can be applied in both cases since the composition of interface automata is commutative. Our methodology is illustrated with the latter case, i.e. we first fix a candidate M derived from P. Then, the synthesis method of [4] is used to obtain B. Our procedure is not always successful: it may be the case that no decomposition is found.

Extracting the Adaptation Logic. In order to extract a less trivial manager from an AIA P we can proceed as follows. We define the bypassing of an action set $\mathcal{B} \subseteq \mathcal{A}$ in P as $P_{|\mathcal{B}, \equiv}$, which is obtained by $P_{|\mathcal{B}}$ (that is, the AIA obtained from P by deleting those transitions whose action belong to $\mathcal{B}$) collapsing the states via the equivalence relation induced by $\{u \equiv v \mid (u, a, v) \in \mathcal{T}_P \wedge a \in \mathcal{B}\}$.

The idea is then to choose a subset $\mathcal{B}$ of $\mathcal{A}_P \setminus \mathcal{A}_P^C$ (i.e. it contains no control action) that the manager M needs not to observe. The candidate manager M is then $P_{|\mathcal{B}, \equiv}^{\perp_{\mathcal{A}_P^O \setminus \mathcal{A}_P^C}}$. Of course, if the result is not deterministic, this candidate must be discarded: more observations may be needed.

Extracting the Application Logic. We are left with the problem of solving the equation $B \mid M \preceq P$ for given P and M. It is now sufficient to use the solution of [4] which defines B to be $(M \mid P^\perp)^\perp$, where $P^\perp$ abbreviates $P^{\perp_{\mathcal{A}_P}}$. If the obtained B and M are compatible, the reverse engineering problem has

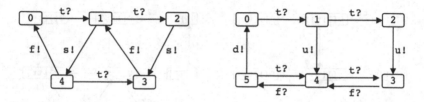

Fig. 10. Bypassed managers for `Scheduler(S)` (left) and `Switch(S)` (right)

been solved. Otherwise we are guaranteed that no suitable managed component B exists for the candidate manager M since the solution of [4] is sound and complete. A different choice of control data or hidden actions should be done.

Example 12. The manager $\texttt{Scheduler(S)}_{\{u,d\},\equiv}^{\perp\{u,d\}}$ (see Fig. 10, left) and the other manager $\texttt{Switch(S)}_{\{s\},\equiv}^{\perp\{f,s\}}$ (see Fig. 10, right) are obtained by removing some observations. For the former we obtain no solution, while for the latter we obtain the same base component of the basic decomposition (Fig. 9 left).

5 Related Works

Our proposal for the formalisation of self-adaptive systems takes inspiration by many former works in the literature. Due to lack of space we focus our discussion on the most relevant related works only.

S[B] systems [13] are a model for adaptive systems based on 2-layered transitions systems. The base transition system B defines the ordinary (and adaptable) behaviour of the system, while S is the adaptation manager, which imposes some regions (subsets of states) and transitions between them (adaptations). Further constraints are imposed by S via adaptation invariants. Adaptations are triggered to change region (in case of local deadlock). Weak and strong adaptability formalisations (casted in our setting in Sect. 4.2) are introduced.

Mode automata [12] have been also advocated as a suitable model for adaptive systems. For example, the approach of [20] represents adaptive systems with two layers: *functional layer*, which implements the application logic and is represented by state machines called *adaptable automata*, and *adaptation layer*, which implements the adaptation logic and is represented with a mode automata. Adaptation here is the change of mode. The approach considers three different kinds of specification properties (cf. 4.2): *local*, *adaptation*, and *global*. An extension of linear-time temporal logic (LTL) called *m*LTL is used to express them.

The most relevant difference between AIA and S[B] system or Mode automata is that our approach does not impose a two-layered asymmetric structure: AIA can be composed at will, possibly forming towers of adaptation [5] in the spirit of the MAPE-K reference architecture, or mutual adaptation structures. In addition, each component of an adaptive system (be it a manager or a managed component, or both) is represented with the same mathematical object, essentially a well-studied one (i.e. interface automata) decorated with some additional information (i.e. control propositions).

Adaptive Featured Transition Systems (A-FTS) have been introduced in [6] for the purpose of model checking adaptive software (with a focus on software product lines). A-FTS are a sort of transition systems where states are composed by the local state of the system, its configuration (set of active features) and the configuration of the environment. Transitions are decorated with executability conditions that regard the valid configurations. Adaptation corresponds to reconfigurations (changing the system's features). Hence, in terms of our white-box approach, system features play the role of control data. They introduce the notion of *resilience* as the ability of the system to satisfy properties despite of environmental changes (which essentially coincides with the notion of black-box adaptability of [8]). Properties are expressed in AdaCTL, a variant of the computation-tree temporal logic CTL. Contrary to AIA which are equipped with suitable composition operations, A-FTS are seen in [6] as monolithic systems.

6   Concluding Remarks

We presented a novel approach for the formalisation of self-adaptive systems, which is based on the notion of control propositions (and control actions). Our proposal has been presented by instantiating it to a well-known model for component-based system, interface automata. However, it is amenable to be applied to other foundational formalisms as well. In particular, we would like to verify its suitability for basic specification formalisms of concurrent and distributed systems such as process calculi. Among future works, we envision the investigation of more specific notions of refinement, taking into account the possibility of relating systems with different kinds of adaptability and general mechanisms for control synthesis that are able to account also for non-deterministic systems. Furthermore, our formalisation can be the basis to conciliate white- and black-box perspectives adaptation under the same hood, since models of the latter are usually based on variants of transition systems or automata. For instance, control synthesis techniques such as those used to modularise a self-adaptive system (white-box adaptation) or model checking techniques for game models (e.g. [3]) can be used to decide if and to which extent a system is able to adapt so to satisfy its requirements despite of the environment (black-box adaptation).

References

1. de Alfaro, L.: Game models for open systems. In: Dershowitz, N. (ed.) Verification: Theory and Practice. LNCS, vol. 2772, pp. 269–289. Springer, Heidelberg (2004)
2. de Alfaro, L., Henzinger, T.A.: Interface automata. In: ESEC/SIGSOFT FSE 2001. ACM SIGSOFT Software Engineering Notes, vol. 26(5), pp. 109–120. ACM (2001)
3. Alur, R., Henzinger, T.A., Kupferman, O.: Alternating-time temporal logic. Journal of the ACM 49(5), 672–713 (2002)

4. Bhaduri, P., Ramesh, S.: Interface synthesis and protocol conversion. Formal Aspects of Computing 20(2), 205–224 (2008)
5. Bruni, R., Corradini, A., Gadducci, F., Lluch Lafuente, A., Vandin, A.: A conceptual framework for adaptation. In: de Lara, J., Zisman, A. (eds.) FASE 2012. LNCS, vol. 7212, pp. 240–254. Springer, Heidelberg (2012)
6. Cordy, M., Classen, A., Heymans, P., Legay, A., Schobbens, P.-Y.: Model Checking Adaptive Software with Featured Transition Systems. In: Cámara, J., de Lemos, R., Ghezzi, C., Lopes, A. (eds.) Assurances for Self-Adaptive Systems. LNCS, vol. 7740, pp. 1–29. Springer, Heidelberg (2013)
7. De Nicola, R., Vaandrager, F.W.: Action versus state based logics for transition systems. In: Guessarian, I. (ed.) LITP 1990. LNCS, vol. 469, pp. 407–419. Springer, Heidelberg (1990)
8. Hölzl, M., Wirsing, M.: Towards a system model for ensembles. In: Agha, G., Danvy, O., Meseguer, J. (eds.) Formal Modeling: Actors, Open Systems, Biological Systems. LNCS, vol. 7000, pp. 241–261. Springer, Heidelberg (2011)
9. IBM Corporation: An Architectural Blueprint for Autonomic Computing (2006)
10. Kulkarni, S.S., Biyani, K.N.: Correctness of component-based adaptation. In: Crnković, I., Stafford, J.A., Schmidt, H.W., Wallnau, K. (eds.) CBSE 2004. LNCS, vol. 3054, pp. 48–58. Springer, Heidelberg (2004)
11. Lints, T.: The essentials in defining adaptation. Aerospace and Electronic Systems 27(1), 37–41 (2012)
12. Maraninchi, F., Rémond, Y.: Mode-automata: About modes and states for reactive systems. In: Hankin, C. (ed.) ESOP 1998. LNCS, vol. 1381, pp. 185–199. Springer, Heidelberg (1998)
13. Merelli, E., Paoletti, N., Tesei, L.: A multi-level model for self-adaptive systems. In: Kokash, N., Ravara, A. (eds.) FOCLASA 2012. EPTCS, vol. 91, pp. 112–126 (2012)
14. Oreizy, P., Gorlick, M.M., Taylor, R.N., Heimbigner, D., Johnson, G., Medvidovic, N., Quilici, A., Rosenblum, D.S., Wolf, A.L.: An architecture-based approach to self-adaptive software. Intelligent Systems and their Applications 14(3), 54–62 (1999)
15. Laddaga, R., Robertson, P., Shrobe, H.: Introduction to self-adaptive software: Applications. In: Laddaga, R., Robertson, P., Shrobe, H. (eds.) IWSAS 2001. LNCS, vol. 2614, pp. 1–5. Springer, Heidelberg (2003)
16. Salehie, M., Tahvildari, L.: Self-adaptive software: Landscape and research challenges. ACM Transactions on Autonomous and Adaptive Systems 4(2), 1–42 (2009)
17. Salvaneschi, G., Ghezzi, C., Pradella, M.: Context-oriented programming: A programming paradigm for autonomic systems (v2). CoRR abs/1105.0069 (2012)
18. Zadeh, L.A.: On the definition of adaptivity. Proceedings of the IEEE 51(3), 469–470 (1963)
19. Zhang, J., Goldsby, H., Cheng, B.H.C.: Modular verification of dynamically adaptive systems. In: Moreira, A., Schwanninger, C., Baillargeon, R., Grechanik, M. (eds.) AOSD, pp. 161–172. ACM (2009)
20. Zhao, Y., Ma, D., Li, J., Li, Z.: Model checking of adaptive programs with mode-extended linear temporal logic. In: EASe, pp. 40–48. IEEE Computer Society (2011)

Compiling Logics

Mihai Codescu[1], Fulya Horozal[3], Aivaras Jakubauskas[3],
Till Mossakowski[2], and Florian Rabe[3]

[1] Friedrich-Alexander University, Erlangen-Nürnberg, Germany
[2] DFKI GmbH Bremen, Germany
[3] Computer Science, Jacobs University Bremen, Germany

Abstract. We present an architecture that permits compiling declarative logic specifications (given in some type theory like LF) into implementations of that logic within the Heterogeneous Tool Set Hets. The central contributions are the use of declaration patterns for singling out a suitable subset of signatures for a particular logic, and the automatic generation of datatypes and functions for parsing and static analysis of declaratively specified logics.

1 Introduction

In [5], we presented an extension of the Heterogeneous Tool Set HETS [14] with a framework for representing logics independently of their foundational assumptions. The key idea [17] is that a graph of theories in a type theoretical logical framework like LF [9] can fully represent a model theoretic logic. Our integration used this construction to make the process of extending HETS with a new logic more declarative, on one side, and fully formal, on the other side.

However, the new logic in HETS inherits the syntax of the underlying logical framework. This is undesirable for multiple reasons. Firstly, a logical framework unifies many concepts that are distinguished in individual logics. Examples are binding and application (unified by higher-order abstract syntax), declarations and axioms (unified by the Curry-Howard correspondence), and different kinds of declarations (unified by LFP – LF with declaration patterns). Therefore, users of a particular logic may find it unintuitive to use the concrete syntax (and the associated error messages) of the logical framework.

Secondly, only a small fragment of the syntax of the logical framework is used in a particular logic. For example, first-order logic only requires two base types *term* for terms and *form* for formulas and not the whole dependent type theory of LF. Therefore, it is unnecessarily complicated if implementers of additional services for a particular logic have to work with the whole abstract syntax of the logical framework. Such services include in particular logic translations from logics defined in LF to logics implemented by theorem provers.

Therefore, we introduce an architecture that permits compiling logics defined in LF into custom definitions in arbitrary programming languages. This is similar to parser generators, which provide implementations of parsers based on a language definition in a context-free grammar. Our work provides implementations

N. Martí-Oliet and M. Palomino (Eds.): WADT 2012, LNCS 7841, pp. 111–126, 2013.
© IFIP International Federation for Information Processing 2013

of a parser and a type-checker based on a context-sensitive language definition in
the recent extension of LF with *declaration patterns* [11] (LFP). Here declaration
patterns give a formal specification of the syntactic shape of the declarations in
the theories of a logic.

For realising this approach, we will use the MMT framework, which provides
a scalable Module system for Mathematical Theories [19] independently of the
logic and foundation. The MMT tool provides an API for parsing, checking, and
flattening MMT theories. We will build our implementation on this, and use it
in particular for modular theories in LFP.

2   Preliminaries

2.1   Institutions and Hets

The Heterogeneous Tool Set (Hets, [14]) is a set of tools for multi-logic specifications, which combines parsers, static analyzers, and theorem provers. Hets
provides a heterogeneous specification language built on top of CASL [1] and
uses the development graph calculus [13] as a proof management component.

Hets formalizes the logics and their translations using the abstract model
theory notions of institutions and institution comorphisms (see [6] and [7]).

Definition 1. *An institution is a quadruple $I = (\mathbf{Sig}, \mathbf{Sen}, \mathbf{Mod}, \models)$ where:*

- **Sig** *is a category of* signatures*;*
- **Sen** : **Sig** → *Set is a functor to the category Set of small sets and func-
  tions, giving for each signature Σ its set of* sentences **Sen**(Σ) *and for
  each signature morphism* $\varphi : \Sigma \rightarrow \Sigma'$ *the* sentence translation *function*
  Sen(φ) : **Sen**(Σ) → **Sen**(Σ') *(denoted by a slight abuse also* φ*);*
- **Mod** : **Sig**^{op} → *Cat is a functor to the category of categories and functors
  Cat* [1] *giving for each signature Σ its category of models* **Mod**(Σ) *and for
  each signature morphism* $\varphi : \Sigma \rightarrow \Sigma'$ *the* model reduct *functor* **Mod**(φ) :
  Mod(Σ') → **Mod**(Σ) *(denoted* $\_|_\varphi$*);*
- *a satisfaction relation* $\models_\Sigma \subseteq |\mathbf{Mod}(\Sigma)| \times \mathbf{Sen}(\Sigma)$ *for each signature Σ*

such that the following satisfaction condition holds:

$$M'|_\varphi \models_{\Sigma'} e \Leftrightarrow M' \models_\Sigma \varphi(e)$$

*for each $M' \in |\mathbf{Mod}(\Sigma')|$ and $e \in \mathbf{Sen}(\Sigma)$, expressing that truth is invariant
under change of notation and context.*

For example, the institution of propositional logic *PL* has signatures consisting
of a set of propositional symbols, and signature morphisms are just functions
between those sets. Models are functions from the signature to the set of truth

---

[1] We disregard here the foundational issues, but notice however that *Cat* is actually
a so-called quasi-category.

values $\{true, false\}$ giving the interpretation of each proposition, and sentences are defined inductively, starting with the propositional symbols and applying a finite number of Boolean connectives. Sentence translation means replacement of the translated symbols. The reduct of a Σ'-model $m : \Sigma' \to \{true, false\}$ along a signature morphism $\varphi : \Sigma \to \Sigma'$ is just the composition $\varphi; m : \Sigma \to \{true, false\}$. Finally, satisfaction is given by the standard truth-tables semantics and it is straightforward to see that the satisfaction condition holds.

Definition 2. *Given two institutions I_1, I_2 with $I_i = (\mathbf{Sig}_i, \mathbf{Sen}_i, \mathbf{Mod}_i, \models^i)$, an institution comorphism from I_1 to I_2 consists of a functor $\Phi : \mathbf{Sig}_1 \to \mathbf{Sig}_2$ and natural transformations $\beta : \Phi; \mathbf{Mod}_2 \Rightarrow \mathbf{Mod}_1$ and $\alpha : \mathbf{Sen}_1 \Rightarrow \Phi; \mathbf{Sen}_2$, such that the following satisfaction condition holds:*

$$M' \models^2_{\Phi(\Sigma)} \alpha_\Sigma(e) \Leftrightarrow \beta_\Sigma(M') \models^1_\Sigma e,$$

where Σ is an I_1-signature, e is a Σ-sentence in I_1 and M' is a $\Phi(\Sigma)$-model in I_2.

Hets has been designed as an extensible tool: new institutions can be plugged in without having to modify the institution-independent parts. Hets implements institutions in Haskell using a multiparameter type class [12] with functional dependencies. Functional dependencies are needed because no operation will involve all types of the multiparameter type class; hence we need a method to derive the missing types.

```
class Logic lid sublogics sign mor sen basic_spec symb_map
      | lid -> sublogics sign mor sen basic_spec symb_map
   where
      logic_name :: lid -> String
      id :: lid -> sign -> mor
      comp :: lid ->  mor -> mor -> mor
      parse_basic_spec :: lid -> String -> basic_spec
      parse_symb_map :: lid -> String -> symb_map
      map_sen :: lid -> mor -> sen -> sen
      basic_analysis :: lid -> sign -> basic_spec
                     -> (sign, [sen])
      stat_symb_map :: lid -> symb_map -> sign -> mor
      minSublogic :: lid -> sublogics -> sign -> Bool
      minSublogic :: lid -> sublogics -> sen -> Bool
```

Fig. 1. The basic components of a logic in Hets

The Logic class of Hets is presented in Fig. 1, in a very simplified form (e.g., error handling is omitted). For each logic, we introduce a new singleton type lid that gives the name (logic_name), or constitutes the identity of the logic. All other parameters of the type class depend on this type, and all operations take it as first argument. The types basic_spec and symb_map serve as a more user-friendly syntax for signatures and morphisms, respectively. The methods of the type class `Logic` give then the definition of the category of signatures and signature morphisms. Parsers for basic specifications and symbol maps must be provided, as well as functions for static analysis, that transforms basic specifications into theories of the logic and symbol maps into signature morphisms, respectively. We also have a function that gives the translation of a sentence along a signature morphism. Finally, Hets includes a sublogic analysis mechanism, which is important because it reduces the number of logic instances. A sublogic can use the operations of the main logic and functions are available that give the minimal sublogic of an item (methods `minSublogic`).

2.2   The LATIN Meta-Framework

Logical Frameworks in LATIN. The Logic Atlas and Integrator (LATIN) project [4] develops a foundationally unconstrained framework for the representation of institutions and comorphism [17]. It abstract from individual logical frameworks such as LF [9] or Isabelle [15] by giving a general definition of a logical framework, called the LATIN meta-framework [5].

The central component of a logical framework $\mathbb{F}$ in the sense of LATIN is a category (whose components are called signatures and signature morphisms) with inclusions. LATIN follows a "logics-as-theories and translations-as-morphisms" approach, providing a general construction of an institution from a $\mathbb{F}$-signature and of an institution comorphism from a $\mathbb{F}$-morphism [5].

The basic idea is that, given an $\mathbb{F}$-signature L, the signatures of the institution $\mathbb{F}(L)$ are the extensions $L \hookrightarrow \Sigma$ in $\mathbb{F}$. Similarly morphisms from $L \hookrightarrow \Sigma$ to $L \hookrightarrow \Sigma'$ are the commuting triangles formed by morphism $\sigma : \Sigma \to \Sigma'$. The $\mathbb{F}(L)$-sentences over Σ are obtained as certain Σ-objects in $\mathbb{F}$, and the $\mathbb{F}(L)$-models of Σ are obtained as certain $\mathbb{F}$-morphisms out of Σ. We refer to [5] for the details.

The Logical Framework LFP. For the purposes of this paper, we will use a logical framework $\mathbb{F} = \text{LFP}$ based on an extension of modular LF [9,20] with declaration patterns. Here we will briefly introduce LFP and direct the reader to [11] for the details on declaration patterns.

Figure 2 gives the fragment of the grammar for LFP-theories that is sufficient for this paper. Here the parts pertaining to declaration patterns are underlined. In particular, LFP-signatures and morphisms are produced from Σ and σ, respectively.

Let us first consider the language without declaration patterns. Modules are the toplevel declarations. Their semantics is defined in terms of the category of LF *signatures* and signature morphisms (see, e.g., [10]); the latter are called *views* in modular LF.

A non-modular signature Σ declares a list of typed constants c. Correspondingly, views from a signature T_1 to a signature T_2 consist of assignments $c := E$, which map T_1-constants to T_2-expressions.

$$
\begin{array}{ll}
\text{Modules} & M ::= \%\text{sig } T = \{\varSigma\} \mid \%\text{view } v : T_1 \to T_2 = \{\sigma\} \\
\text{Theories} & \varSigma ::= c : E \mid \%\text{include } T \mid \%\text{pattern } p = P \\
\text{Morphisms } \sigma & ::= c := E \mid \%\text{include } v \mid \%\text{pattern } p := P \\
\text{Expressions } E & ::= \text{type} \mid c \mid x \mid \{x : E\}\, E \mid [x : E]\, E \mid E\, E \\
& \quad \mid \quad \underline{E, E} \mid \underline{[E]_{x=1}^{E}} \mid \underline{E_E} \mid \underline{Nat} \mid \underline{0} \mid \underline{succ(E)} \\
\underline{\text{Patterns}} & \underline{P} ::= \underline{p} \mid \underline{\{\varSigma\}} \mid \underline{[x : E]\, P} \mid \underline{P\, E}
\end{array}
$$

Fig. 2. Grammar of LFP

Expressions are formed from the universe of types **type**, constants c, bound variables x, dependent function types ($\varPi$-types) $\{x : E\}\, E$, λ-abstraction $[x : E]\, E$, and application $E\, E$. As usual, we write $E_1 \to E_2$ instead of $\{x : E_1\}\, E_2$ whenever x does not occur in E_2. A valid view extends homomorphically to a (type-preserving) map of all T_1-expressions to T_2-expressions.

To this, the module system adds the ability for signatures *include* other signatures. Morphisms out of such signatures must correspondingly include a morphism.

Example 1 (First-order logic in LF). The LF signatures below encode propositional logic (PL) and first-order logic (FOL) in LF. We will use these encodings as our running example in this paper.

```
%sig Base = {
    form : type
    ded  : form → type
}
%sig PL^Syn = {
    %include Base
    false : form
    imp   : form → form → form
}
%sig FOL^Syn = {
    %include PL^Syn
    term : type
    ∀    : (term → form) → form
}
```

The signature *Base* declares an LF-type *form* of propositional formulas and a *form*-indexed type family *ded*. This type family exemplifies how logic encodings in LF follow the Curry-Howard correspondence to represent judgments as types and proofs as terms: Terms of type *ded F* represent derivations of the judgment "F is true". Furthermore, PL^{Syn} encodes the syntax of PL by declaring propositional connectives (here only falsehood and implication). Finally, we obtain the FOL syntax in the signature FOL^{Syn} by adding to PL^{Syn} a constant *term* for the type of terms and the universal quantifier $\forall$. The latter is a binder and thus is represented in LF using higher-order abstract syntax.

Then the LF signature PL^{Pf} below encodes the proof theory of PL by providing the inference rules associated with each symbol declared in PL^{Syn}:

```
%sig PL^Pf = {
    falseE : ded false → {F} ded F
    impI   : (ded A → ded B) → ded (A imp B)
    impE   : ded (A imp B) → ded A → ded B
}
```

We obtain the proof theory of first-order logic accordingly.

Declaration Patterns. Let us now consider the extension of modular LF with declaration patterns. Declaration patterns formalize what it means to be an arbitrary theory of a logic L: A declaration pattern gives a formal specification of the syntactic shape of a class of L-declarations, and in a legal L-theory, each declaration must match one of the L-patterns.

Declaration patterns P are formed from pattern constants p, signatures $\{\Sigma\}$, λ-abstractions $[x : E]\, P$, and applications $P\, E$ of patterns P to expressions E.

Example 2 (Declaration patterns for PL). Consider the following version of the signature PL^{Syn} with declaration patterns for legal PL-signatures.

```
%sig PL^Syn = {
    %include Base
    false : form
    imp  : form → form → form
    %pattern props = {
        q : form
    }
    %pattern axiom = [F : form] {
        m : ded F
    }
}
```

The declaration pattern *props* allows for the declaration of propositional variables of the form $q : form$ in PL-signatures. And the declaration pattern *axiom* formalizes the shape of axiom declarations. Each axiom declaration must be of the form $m : ded\, F$ for some proposition F.

More technically, consider an LFP-signature $L = L_S, L_P$ where L_S is a signature of modular LF and L_P is a list of pattern declarations. Then we define a morphism $L \hookrightarrow \Sigma$ to exist if $L_S \hookrightarrow \Sigma$ is an inclusion in modular LF and every declaration in $\Sigma \setminus L_S$ matches one of the patterns in L_P. Consequently, the institution LFP(L) contains exactly the "well-patterned" signatures.

Sequences. Even though most logics do not use sequences, it turns out that sequences are usually necessary to write down declaration patterns. For example, in theories of typed first-order logic, function symbol declarations use a sequence of types – the argument types of the function symbol. More precisely, in the case of FOL, the signatures Σ should contain only declarations of the form $f : term \rightarrow \ldots \rightarrow term \rightarrow term$ (n-ary function symbols) and $p : term \rightarrow \ldots \rightarrow term \rightarrow form$ (n-ary predicate symbols). Therefore, our language also uses *expression sequences* and *natural numbers*. These are formed by the underlined productions for expressions:

- E_1, E_2 for the concatenation of two sequences,
- E_n for the n-th element of E,
- $[E(x)]_{x=1}^{n}$ for the sequence $E(1), \ldots, E(n)$ where n has type Nat and $E(x)$ denotes an expression E with a free variable $x : Nat$; we write this sequence as E^n whenever x does not occur free in E,
- Nat for the type of natural numbers,
- 0 and $succ(n)$ for zero and the successor of a given natural number n.

Example 3 (Declaration patterns for FOL). Using sequences, we can give the following version of the signature FOL^{Syn} that declares appropriate declaration patterns:

```
%sig FOL^{Syn} = {
    %include PL^{Syn}
    term : type
    ∀     : (term → form) → form
    %pattern ops = [n : Nat] {
        f : term^n → term
    }
    %pattern preds = [n : Nat] {
        q : term^n → form
    }
}
```

The declaration pattern *ops* allows for the declaration of function symbols of the form $f : term^n \rightarrow term$, which take a sequence of first-order terms of length n and return a first-order term for any natural number n. Similarly, the declaration pattern *preds* allows for the declaration of predicate symbols of the form $q : term^n \rightarrow form$ for any natural number n.

More complex patterns arise if we consider sorted first-order logic (SFOL):

Example 4 (Declaration patterns for Sorted FOL). $SFOL^{Syn}$ is similar to FOL^{Syn} except that we use a type *sort* to encode the set of sorts and an LF type family tm indexed by *sort* that provides the type $tm\,S$ of terms of sort S (i.e., $t : tm\,S$ is a declaration of a term t of sort S). Universal quantification $\forall$ is sorted, i.e., it first takes a sort argument S and then binds a variables of type $tm\,S$.

Now the declaration pattern *sorts* allows for the declaration of sorts $s : sort$. The declaration patterns *sortedOps* and *sortedPreds* formalize the

shape of declarations of sorted function and predicate symbols of the form
$f : tm\, s_1 \to \ldots \to tm\, s_n \to tm\, t$ and $q : tm\, s_1 \to \ldots tm\, s_n \to form$, respectively, for any sort $s_1 : sort, \ldots, s_n : sort$ and $t : sort$. Note that we can extend all left or right associative infix operators to sequences: The sequence $[tm\, s_i]_{i=1}^n$ normalizes to $tm\, s_1, \ldots, tm\, s_n$ and the type $[tm\, s_i]_{i=1}^n \to tm\, t$ normalizes to $tm\, s_1 \to \ldots \to tm\, s_n \to tm\, t$. Correspondingly, for a function f of that type and a sequence E that normalizes to $E_1, \ldots, E_n$, the expressions $f\, E$ normalizes to $f\, E_1 \ldots E_n$.

```
%sig SFOL^{Syn} = {
    %include PL^{Syn}
    sort : type
    tm   : sort → type
    ∀    : {S : sort} (tm S → form) → form
    %pattern sorts = {
        s : sort
    }
    %pattern sortedOps = [n : Nat] [s : sort^n] [t : sort] {
        f : [tm s_i]_{i=1}^n → tm t
    }
    %pattern sortedPreds = [n : Nat] [s : sort^n] {
        q : [tm s_i]_{i=1}^n → form
    }
}
```

We avoid giving the type system for this extension of LF and refer to [11] for the details. Intuitively, natural numbers and sequences occur only in pattern expressions, and fully applied pattern expressions normalize to expressions of the form $\{\Sigma\}$ where Σ is a plain LF signature.

3   Simple Signatures

In principle, the category of $LFP(L)$ is already defined by giving an LFP-signature L such as FOL^{Syn}. However, the resulting $LFP(L)$-signatures are defined in terms of LFP-signatures, which is often more complex than desirable in practice.

Therefore, we introduce simple LFP-signatures below. These form a subclass of LFP-signatures that covers all the typical cases for the syntax of logics. At the same time, they are so restricted that they can be described in a way that is closer to what would appear in a stand-alone definition of a logic (i.e., in a setting without a logical framework).

First we need an auxiliary definition about LF types:

Definition 3. *A type is called* atomic *if it is of the form* $t\, a_1 \ldots, a_n$ *and* composed *if it is of the form* $\{x : A\}\, B$.

The dependency erasure T^- *of a type* T *is defined by* $(t\, a_1 \ldots, a_n)^- = t$ *and* $(\{x : A\}\, B)^- = A^- \to B^-$.

Intuitively, dependency erasure turns all dependent type families $t : A \to$ type into simple types $t :$ type and all dependent function types into simple function types. For example, the dependency erasure of the $SFOL^{Syn}$ declarations tm and $\forall$ yields $tm :$ type and $\forall : sort \to (tm \to form) \to form$, respectively. In particular, applying the dependency erasure everywhere turns an LF signature into a simply-typed signature.

Then we define simple signatures and their constituents as follows:

Definition 4. *Let N be the set of identifiers declared in an* LFP-*signature L. Let $c \in N$ be the identifier of a type family.*

A declaration $n : T$ in L is called

- *a connective for c with arguments $a_1, \ldots, a_n \in N$ if $T^- = a_1 \to \ldots \to a_n \to c$.*
- *an untyped quantifier for c quantifying over $b \in N$ with scope $s \in N$ if $T = T^- = (b \to s) \to c$.*
- *a typed quantifier for c with argument $a \in N$ quantifying over $b \in N$ with scope $s \in N$ if $T^- = a \to (b \to s) \to c$ and $T = \{x : A\} (b\,x \to C) \to S$ (i.e., A, C, and S must be atomic and $A^- = a$, $C^- = c$, and $S^- = s$).*

A pattern declaration is called simple *if it is of the form* %pattern $p := [x_1 : E_1] \ldots [x_n : E_n] \{\Sigma\}$ *such that (i) each E_i is either atomic or of the form $[E(x)]_{x=1}^n$ for atomic E and (ii) Σ declares only connectives. Then the connectives in Σ are called p-connectives.*

Finally, L is called simple *if it declares only connectives, untyped and typed quantifiers, and simple patterns.*

In particular, our running examples are simple:

Example 5. PL^{Syn}, FOL^{Syn} *and* $SFOL^{Syn}$ *are simple signatures. In* PL^{Syn}, *imp is a connective for* $form$ *with arguments* $form, form$. *In* FOL^{Syn}, $\forall$ *is an untyped quantifier for* $form$ *quantifying over* $term$ *with scope* $form$. *In* $SFOL^{Syn}$, $\forall$ *is a typed quantifier for* $form$ *with argument* $sort$ *quantifying over* tm *with scope* $form$. *Also in* $SFOL^{Syn}$, *sortedOps is a simple pattern, and f is an sortedOps-connective with arguments* $tm, \ldots, tm$.

4 Compiling Simple Signatures to Hets

In [5], we introduced an extension of Hets with a component that generates an instance of the type class Logic from its representation in a logical framework.

This generation is conceptually straightforward by reducing all necessary operations to the ones of LF. As described in the introduction, this reduction to the logical framework leads to the counter-intuitive identification of many notions. For example, for the syntax of FOL^{Syn}, it generates a logic with *i)* a single Haskell type for expressions (not distinguishing between formulas and terms), *ii)* one Haskell constructor for each logical symbol (not distinguishing between connectives and quantifiers) and *iii)* a single type of FOL-declarations (not distinguishing function and predicate symbols).

By restricting attention to simple signatures, we can obtain a – much more complex – algorithm that generates Haskell classes that avoid these identifications. In the following, we describe this algorithm using our running examples. Throughout the following, we assume a fixed simple signature L with one distinguished identifier o (designating the type of formulas).

Expressions. For every declaration of a type family with name C in L, we generate one inductive data type named C whose constructors are obtained from L as follows:

- for every connective n for C with arguments $a_1, \ldots, a_n$: a constructor n with arguments $a_1, \ldots, a_n$,
- for every p-connective n for C with arguments $a_1, \ldots, a_n$: a constructor p_n with arguments $\texttt{String}$ (intuition: the name of the matching declaration), $a_1, \ldots, a_n$,
- for every untyped quantifier n for C quantifying over b with scope s: a constructor n with arguments $\texttt{String}$ (intuition: the name of the bound variable) and s,
- for every typed quantifier n for C with argument a quantifying over b with scope s: a constructor n with arguments a, $\texttt{String}$ (intuition: the name of the bound variable), and s,
- if there is any quantifier over C: a constructor $\texttt{C\_var}$ with argument $\texttt{String}$ (intuition: references to bound variables).

For example, in the case of FOL^{Syn}, this yields the following Haskell types[2,3]

```
data Form =
    False |
    Imp Form Form |
    Forall String Form |
    Preds_q [Term]

data Term =
    Ops_f [Term]
```

As expected, sentences are mapped along a signature morphism homomorphically on the structure of the sentence, by replacing each symbol according to the corresponding declaration map that composes the morphism.

Declarations. For every declaration of a pattern, we generate one record type with the following selectors: *i)* a selector returning $\texttt{String}$ (intuition: the name

---

[2] Here and below, we ignore minor syntactical issues, e.g., when compiling into Haskell, an additional step makes all identifiers begin with an upper case letter.

[3] [T] is Haskell's notation for the type of lists over T.

of the matching declaration), *ii)* for each bound variable $x : E$ of the pattern: one selector of the same name returning *i)* E^- if E is atomic or *ii)* $[E'^-]$ if $E = [E'(x)]_{x=1}^n$.

For example, %pattern $props = \{p : form\}$ from PL^{Syn} generates the Haskell type

```
data Props_decl = Props_d {name :: String}
```

The declaration patterns %pattern $ops = \lambda n : Nat.\ \{f : term^n \to term\}$ and %pattern $preds = \lambda n : Nat.\ \{p : term^n \to form\}$ from FOL^{Syn} generate:

```
data Ops_decl = Ops_decl {
   name :: String,
   n    :: Int
}
data Preds_decl = Preds_decl {
   name :: String,
   n    :: Int
}
```

At the specification level, we get a new type of declarations for axioms, using the pattern in the Base signature:

```
data Axiom_decl = Axiom_d {
    name :: String,
    formula :: Form
}
```

Signatures. Finally, we define a type of declarations as the disjoint union of the types generated for the declaration patterns and a type of signatures as lists of declarations. For example, for FOL^{Syn}, this yields the Haskell types:

```
data Decl = Ops_d Ops_decl | Preds_d Preds_decl
data Sign = Sign {decls :: [Decl]}
```

The assumption that signatures are lists of declarations is typical of logical frameworks like LFP. This is not a substantial restriction because many logics directly have such signatures, and even relations like a subsort relation can be coded as special declarations. Hets makes extensive use of equality tests between signatures. In our setting, we generate code that reports two signatures to be equal if they both have the same declarations, perhaps in a different order.

Similarly, signature morphisms have a source and a target signature and maps between declarations of the same kind. In the case of first-order logic, this amounts to:

```
data Morphism = Morphism {
  ssign, tsign :: Sign,
  ops_map :: Map Ops_decl Ops_decl,
  preds_map :: Map Preds_decl Preds_decl
}
```

This construction of signature morphisms can be applied for logics whose signatures are collections of sets of symbols of the same kind. In particular, this is the case for the logics in the logic graph of Hets. As a more general criterion, the construction covers signatures that can be represented as tuple sets [8]. There are however institutions that fall outside this framework. For example, the signatures of institutions for UML [3] are based on graphs rather than sets, and therefore are difficult to represent in LF.

External Syntax. The above steps of the compilation have generated the internal (abstract) syntax. The generation of data types for external (concrete) syntax and for parsing function is substantially more complex. Therefore, we use a logic-independent representation that can be implemented (in our case, as a part of Hets) once and for all.

Expressions are represented as syntax trees, which are inspired by OpenMath objects [2]:

```
data Tree =
            Var String |
            App String [Tree] |
            Bind String String Tree |
            TBind String String Tree Tree
```

The constructors represent references to bound variables, application of a connective to a list of arguments, untyped quantifier applications, and typed quantifier applications. This data type differs crucially from the above in that it is untyped, i.e., all expressions have the same type, which makes it logic-independent. At the same time this lack of typing does not impede sophisticated parsing methods (which often do not depend on type information), e.g., by using mixfix notations for connectives or complex notations for binders. As a default notation, our implementation assumes that connectives are applied as $(c\,t_1 \ldots t_n)$, and quantifiers as $q\,x : t.\,t'$ where $: t$ is omitted if q is untyped.

The logic-independent representation of declarations is as follows:

```
data ParseDecl = ParseDecl {
    pattern    :: String,
    name       :: String,
    arguments  :: [Tree]
}
```

Here **name** is the name of the declaration, **pattern** is the name of the instantiated declaration pattern, and **arguments** gives the list of arguments to that declaration pattern. The external syntax for signatures is a list of such **ParseDecl**.

In concrete syntax, we use the name of the declaration pattern as a keyword for the declaration. Thus, when defining a logic, we choose the names of the declaration patterns accordingly, e.g., *ops* and *preds* in FOL^{Syn} are chosen to mimic the Hets syntax for CASL. For example, the SFOL specification of a magma would look like as follows:

```
logic SFOL
spec Magma =
 sorts u
 ops comp 2 u u u
```

Similarly, we might use • instead of *axiom* as the name of the declaration pattern for axioms in order to mimic the Hets syntax of axioms.

Parsing functions for these data types are easy to write in Hets. Thus, the compilation process only generates logic-specific functions that translate from external to internal syntax. Comprehensive type-checking is performed separately by the MMT framework as described in Section 5.

Sublogics. For the sublogic analysis, we make use of the modular structure of the logic representation in LFP. Ideally, every feature of the logical language is introduced in a separate module (say L_i^{Syn}) that can be later re-used. As a result, the signature representing the syntax of the logic contains a number of imports, e.g.:

$$\%\text{sig } L^{Syn} = \{$$
$$\quad \%\text{include } L_1^{Syn}$$
$$\quad \vdots$$
$$\quad \%\text{include } L_n^{Syn}$$
$$\}$$

We can then generate a data type for sublogics with Boolean flags that indicate whether a feature is present in the sublogic or not:

```
data L_SL = L_SL {
   has_L_1 :: Bool,
   ...,
   has_L_n :: Bool
}
```

Thus each of the possible combinations of signatures included in L_syn gives rise to a lattice of sublogics of the logic. Note that with this approach we cannot represent sublogics based on certain restrictions made on the structure of sentences, e.g. the Horn fragment of a logic.

5   Workflow

For parsing and static analysis, we use the MMT tool [18,16], which already provides logic-independent parsing and type checking. In fact, the logic compilation itself is implemented as a part of MMT already.

Thus, we obtain two workflows. Firstly, the logic-level workflow occurs when Hets reads a new logic definition, e.g.,

```
logic LFP
spec SFOL^Syn = ...

newlogic SFOL =
  meta LFP
  syntax SFOL^Syn
  proofs ...
  models ...
  foundation ...
end
```

Here $SFOL^{Syn}$ is defined in Hets using the logical framework LFP, and then $SFOL^{Syn}$ it is used as a part of the definition of the logic SFOL. Such logic definitions have the advantage that the logic becomes a fully formal object, verified within the logical framework, and users of the tool can specify their favorite logic without having to understand the implementation details behind Hets. In this case, Hets invokes MMT to compile $SFOL^{Syn}$ into Haskell code, which is then compiled as a part of Hets resulting in a new Hets logic.

Secondly, the specification-level work flow occurs when the logic is used to write a new specification such as in Magma in Section 4. In this case, Hets invokes MMT for parsing and static analysis of the specification Magma. The result is exported by MMT in terms of the logic-independent ParseDecl and Tree data types, which are then read by Hets into the logic-specific data types produced by the compilation during the first workflow.

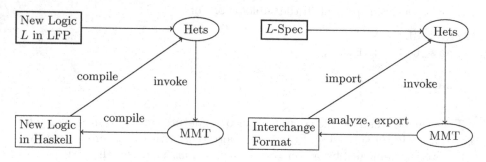

Fig. 3. Logic-Level (left) and Specification-Level (right) Workflows

6 Conclusions and Future Work

We have presented a milestone on the way of making the specification of new logics in Hets fully declarative, while maintaining the good and diverse tool support that logics in Hets enjoy. Declaration patterns allow the specification of signatures and syntax of logics in a declarative way that simultaneously further automatizes the integration of logics in Hets.

Future work will strengthen the implementation of the approach, in particular improving the customizability of the notations for concrete syntax of expressions (`Tree`) and declarations (`ParseDecl`). Moreover, we want to integrate more features of logics into the compilation workflow. For example, it is relatively easy to extend the workflow to proof terms so that Hets can read and verify user-written proofs. It is also interesting but more difficult to generate data types that represent models, model reduction, and model morphisms. Then, colimits, amalgamability checks, and institution comorphisms can in principle be integrated into both the declarative specification in LFP and the compilation to to Hets, but finding concise declarative representations with good properties will be a challenge. Last but not least, we aim at the declarative integration of proof tools into Hets. At the language level, such an integration can already be achieved now, either via adding the prover's logic declaratively to Hets, or by looking for a logic already integrated to Hets that is close to the prover's logic, and provide a syntax translation to adapt this to the prover's input language(s). At the level of interaction with the prover, one could think of grammars for interpreting the prover's output in terms of a status ontology, and more complex reactive specifications of interfaces of interactive provers.

Acknowledgement. This work was funded by the German Federal Ministry of Education and Research under grant 01 IW 10002 (project SHIP) and by the Deutsche Forschungsgemeinschaft (DFG) within grants KO 2428/9-1 and MO 971/2-1 (project LATIN).

References

1. Astesiano, E., Bidoit, M., Kirchner, H., Krieg-Brückner, B., Mosses, P., Sannella, D., Tarlecki, A.: CASL: The Common Algebraic Specification Language. Theoretical Computer Science 286(2), 153–196 (2002)
2. Buswell, S., Caprotti, O., Carlisle, D., Dewar, M., Gaetano, M., Kohlhase, M.: The Open Math Standard, Version 2.0. Technical report, The Open Math Society (2004), http://www.openmath.org/standard/om20
3. Cengarle, M.V., Knapp, A., Tarlecki, A., Wirsing, M.: A Heterogeneous Approach to UML Semantics. In: Degano, P., De Nicola, R., Meseguer, J. (eds.) Concurrency, Graphs and Models. LNCS, vol. 5065, pp. 383–402. Springer, Heidelberg (2008)
4. Codescu, M., Horozal, F., Kohlhase, M., Mossakowski, T., Rabe, F.: Project Abstract: Logic Atlas and Integrator (LATIN). In: Davenport, J.H., Farmer, W.M., Rabe, F., Urban, J. (eds.) Calculemus/MKM 2011. LNCS (LNAI), vol. 6824, pp. 289–291. Springer, Heidelberg (2011)
5. Codescu, M., Horozal, F., Kohlhase, M., Mossakowski, T., Rabe, F., Sojakova, K.: Towards Logical Frameworks in the Heterogeneous Tool Set Hets. In: Mossakowski, T., Kreowski, H.-J. (eds.) WADT 2010. LNCS, vol. 7137, pp. 139–159. Springer, Heidelberg (2012)
6. Goguen, J., Burstall, R.: Institutions: Abstract model theory for specification and programming. Journal of the Association for Computing Machinery 39(1), 95–146 (1992)
7. Goguen, J., Rosu, G.: Institution morphisms. Formal Aspects of Computing 13, 274–307 (2002)

8. Goguen, J.A., Tracz, W.: An Implementation-Oriented Semantics for Module Composition. In: Leavens, G.T., Sitaraman, M. (eds.) Foundations of Component-Based Systems, ch. 11, pp. 231–263. Cambridge University Press, New York (2000)
9. Harper, R., Honsell, F., Plotkin, G.: A framework for defining logics. Journal of the Association for Computing Machinery 40(1), 143–184 (1993)
10. Harper, R., Sannella, D., Tarlecki, A.: Structured presentations and logic representations. Annals of Pure and Applied Logic 67, 113–160 (1994)
11. Horozal, F.: Logic translations with declaration patterns (2012), https://svn.kwarc.info/repos/fhorozal/pubs/patterns.pdf
12. Jones, S.P., Jones, M., Meijer, E.: Type classes: exploring the design space. In: Proceedings of the ACM Haskell Workshop (1997)
13. Mossakowski, T., Autexier, S., Hutter, D.: Development Graphs - Proof Management for Structured Specifications. Journal of Logic and Algebraic Programming 67(1-2), 114–145 (2006)
14. Mossakowski, T., Maeder, C., Lüttich, K.: The Heterogeneous Tool Set, HETS. In: Grumberg, O., Huth, M. (eds.) TACAS 2007. LNCS, vol. 4424, pp. 519–522. Springer, Heidelberg (2007)
15. Paulson, L.C.: Isabelle: A Generic Theorem Prover. LNCS, vol. 828. Springer, Heidelberg (1994)
16. Rabe, F.: The MMT System (2008), https://trac.kwarc.info/MMT/
17. Rabe, F.: A Logical Framework Combining Model and Proof Theory. Mathematical Structures in Computer Science (to appear, 2013), http://kwarc.info/frabe/Research/rabe_combining_10.pdf
18. Rabe, F., Kohlhase, M.: A Scalable Module System (2011), http://arxiv.org/abs/1105.0548
19. Rabe, F., Kohlhase, M.: A Web-Scalable Module System for Mathematical Theories (2011) (under review), http://kwarc.info/frabe/Research/mmt.pdf
20. Rabe, F., Schürmann, C.: A Practical Module System for LF. In: Cheney, J., Felty, A. (eds.) Proceedings of the Workshop on Logical Frameworks: Meta-Theory and Practice (LFMTP), pp. 40–48. ACM Press (2009)

Transformation Systems with Incremental Negative Application Conditions

Andrea Corradini[1], Reiko Heckel[2], Frank Hermann[3,*], Susann Gottmann[3,*], and Nico Nachtigall[3,*]

[1] Dipartimento di Informatica, Università di Pisa, Italy
`andrea@di.unipi.it`
[2] University of Leicester, UK
`reiko@mcs.le.ac.uk`
[3] Interdisciplinary Center for Security, Reliability and Trust,
Université du Luxembourg, Luxembourg
`{frank.hermann,susann.gottmann,nico.nachtigall}@uni.lu`

Abstract. In several application areas, Graph Transformation Systems (GTSs) are equipped with Negative Application Conditions (NACs) that specify "forbidden contexts", in which the rules shall not be applied. The extension to NACs, however, introduces inhibiting effects among transformation steps that are not local in general, causing a severe problem for a concurrent semantics. In fact, the relation of *sequential independence* among derivation steps is not invariant under switching, as we illustrate with an example. We first show that this problem disappears if the NACs are restricted to be *incremental*. Next we present an algorithm that transforms a GTS with arbitrary NACs into one with incremental NACs only, able to simulate the original GTS. We also show that the two systems are actually equivalent, under certain assumptions on NACs.

Keywords: graph transformation, concurrent semantics, negative application conditions, switch equivalence.

1 Introduction

Graph Transformation Systems (GTSs) are an integrated formal specification framework for modelling and analysing structural and behavioural aspects of systems. The evolution of a system is modelled by the application of rules to the graphs representing its states and, since typically such rules have local effects, GTSs are particularly suitable for modelling concurrent and distributed systems where several rules can be applied in parallel. Thus, it is no surprise that a large body of literature is dedicated to the study of the concurrent semantics of graph transformation systems [6,1,2].

The classical results include – among others – the definitions of parallel production and shift equivalence [15], exploited in the Church-Rosser and Parallelism theorems [7]: briefly, derivations that differ only in the order in which independent steps are applied are considered to be equivalent. Several years

---

* Supported by the Fonds National de la Recherche, Luxembourg (3968135, 4895603).

N. Martí-Oliet and M. Palomino (Eds.): WADT 2012, LNCS 7841, pp. 127–142, 2013.

later, taking inspiration from the theory of Petri nets, deterministic processes were introduced [6], which are a special kind of GTSs, endowed with a partial order, and can be considered as canonical representatives of shift-equivalence classes of derivations. Next, the unfolding of a GTS was defined as a typically infinite non-deterministic process which summarises all the possible derivations of a GTS [4]. Recently, all these concepts have been generalised to transformation systems based on ($\mathcal{M}$-)adhesive categories [8,5,3].

In this paper, we consider the concurrent semantics of GTSs that use the concept of Negative Application Conditions (NACs) for rules [11], which is widely used in applied scenarios. A NAC allows one to describe a sort of "forbidden context", whose presence around a match inhibits the application of the rule. These inhibiting effects introduce several dependencies among transformation steps that require a shift of perspective from a purely local to a more global point of view when analysing such systems.

Existing contributions that generalise the concurrent semantics of GTSs to the case with NACs [17,10] are not always satisfactory. While the lifted Parallelism and Concurrency Theorems provide adequate constructions for composed rules specifying the effect of concurrent steps, a detailed analysis of possible interleavings of a transformation sequence leads to problematic effects caused by the NACs. As shown in [12], unlike the case without NACs, the notion of *sequential independence* among derivation steps is not stable under switching. More precisely, it is possible to find a derivation made of three direct transformations $s = (s_1; s_2; s_3)$ where s_2 and s_3 are sequentially independent and to find a derivation $s' = (s_2'; s_3'; s_1')$ that is shift equivalent to s (obtained with the switchings $(1 \leftrightarrow 2; 2 \leftrightarrow 3)$), but where s_2' and s_3' are sequentially dependent on each other. This is a serious problem from the concurrent semantics point of view, because for example the standard colimit technique [6] used to generate the process associated with a derivation does not work properly, since the causalities between steps do not form a partial order in general.

In order to address this problem, we introduce a restricted kind of NACs, based on incremental morphisms [12]. We first show that sequential independence is invariant under shift equivalence if all NACs are incremental. Next we analyse to which extent systems with general NACs can be transformed into systems with incremental NACs. For this purpose, we provide an algorithmic construction *INC* that takes as input a GTS and yields a corresponding GTS with incremental NACs only. We show that the transformation system obtained via *INC* simulates the original one, i.e., each original transformation sequence induces one in the derived system. Thus, this construction provides an over-approximation of the original system. We also show that this simulation is even a bisimulation, if the NACs of the original system are obtained as colimits of incremental NACs.

In the next section we review main concepts for graph transformation systems. Sect. 3 discusses shift equivalence and the problem that sequential independence with NACs is not stable in general. Thereafter, Sect. 4 presents incremental NACs and shows the main result on preservation of independence. Sect. 5 presents the algorithm for transforming systems with general NACs into those with incremental ones and shows under which conditions the resulting system is equivalent. Finally, Sect. 6 provides a conclusion and sketches future developments. The proofs of the main theorems are included in the paper.

2   Basic Definitions

In this paper, we use the double-pushout approach [9] to (typed) graph transformation, occasionally with negative application conditions [11]. However, we will state all definitions and results at the level of adhesive categories [16]. A category is *adhesive* if it is closed under pushouts along monomorphisms (hereafter *monos*) as well as under pullbacks, and if all pushouts along a mono enjoy the van Kampen property. That means, when such a pushout is the bottom face of a commutative cube such as in the left of Fig. 1, whose rear faces are pullbacks, the top face is a pushout if and only if the front faces are pullbacks. In any adhesive category we have uniqueness of pushout complements along monos, monos are preserved by pushouts and pushouts along monos are also pullbacks. As an example, the category of typed graphs for a fixed type graph TG is adhesive [8].

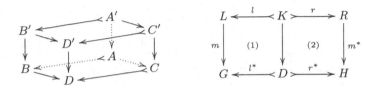

Fig. 1. van Kampen condition (left) and DPO diagram (right)

In the rest of the paper, unless differently stated, all objects and arrows live in an arbitrary but fixed adhesive category **C**.

A *rule* $p = (L \xleftarrow{l} K \xrightarrow{r} R)$ consists of a span of two monos l and r. Given a morphism $m : L \to G$ called the *match*, a *direct transformation* (or *step*) $G \xRightarrow{p,m} H$ from G to a H exists if a double-pushout (DPO) diagram can be constructed as in the right of Fig. 1, where (1) and (2) are pushouts.

The applicability of rules can be restricted by specifying negative conditions requiring the non-existence of certain structures in the context of the match. A *(negative) constraint* on an object L is a morphism $n : L \to \hat{L}$. A morphism $m : L \to G$ satisfies n (written $m \models n$) iff there is no mono $q : \hat{L} \rightarrowtail G$ such that $n; q = m$. A negative application condition (NAC) on L is a set of constraints N. A morphism $m : L \to G$ *satisfies* N (written $m \models N$) if and only if m satisfies every constraint in N, i.e., $\forall n \in N : m \models n$.

All along the paper we shall consider only monic matches and monic constraints: possible generalisations are discussed in the concluding section.

A graph transformation system (GTS) $\mathcal{G}$ consists of a set of rules, possibly with NACs. A derivation in $\mathcal{G}$ is a sequence of direct transformations $s = (G_0 \xRightarrow{p_1,m_1} G_1 \xRightarrow{p_2,m_2} \cdots \xRightarrow{p_n,m_n} G_n)$ such that all p_i are in $\mathcal{G}$; we denote it also as $s = s_1; s_2; \ldots; s_n$, where $s_k = (G_{k-1} \xRightarrow{p_k,m_k} G_k)$ for $k \in \{1, \ldots, n\}$.

3 Independence and Shift Equivalence

Based on the general framework of adhesive categories, this section recalls the relevant notions for sequential independence and shift equivalence and illustrates the problem that independence is not stable under switching in presence of NACs. In the DPO approach, two consecutive direct transformations $s_1 = G_0 \overset{p_1,m_1}{\Longrightarrow} G_1$ and $s_2 = G_1 \overset{p_2,m_2}{\Longrightarrow} G_2$ as in Fig. 2 are *sequentially independent* if there exist morphisms $i : R_1 \to D_2$ and $j : L_2 \to D_1$ such that $j; r_1^* = m_2$ and $i; l_2^* = m_1^*$. In this case, using the local Church-Rosser theorem [8] it is possible to construct a derivation $s' = G_0 \overset{p_2,m_2'}{\Longrightarrow} G_1' \overset{p_1,m_1'}{\Longrightarrow} G_2$ where the two rules are applied in the opposite order. We write $s_1; s_2 \sim_{sh} s'$ to denote this relation.

Given a derivation $s = s_1; s_2; \ldots s_i; s_{i+1}; \ldots; s_n$ containing sequentially independent steps s_i and s_{i+1}, we denote by $s' = switch(s, i, i+1)$ the equivalent derivation $s' = s_1; s_2; \ldots s_i'; s_{i+1}'; \ldots; s_n$, where $s_i; s_{i+1} \sim_{sh} s_i'; s_{i+1}'$. Shift equivalence $\equiv_{sh}$ over derivations of $\mathcal{G}$ is defined as the transitive and "context" closure of $\sim_{sh}$, i.e., the least equivalence relation containing $\sim_{sh}$ and such that if $s \equiv_{sh} s'$ then $s_1; s; s_2 \equiv_{sh} s_1; s'; s_2$ for all derivations s_1 and s_2.

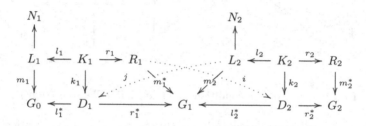

Fig. 2. Sequential independence

The definitions of independence and shift equivalence carry over to transformations with NACs [18] by requiring that the match for p_2 in G_0 given by $m_2' = j; l_1^*$ satisfies the NAC of p_2 and the induced match of p_1 into graph G_1' obtained by $G_0 \overset{p_2,m_2'}{\Longrightarrow} G_1'$ satisfies the NAC of p_1.

Throughout the paper, we use a short notation for transformation rules in our examples as for instance in Ex. 1 below. A rule p is depicted as $(L \overset{p}{\Rightarrow} R)$ showing the left and right hand sides of p. The intermediate interface graph K containing all preserved elements can be obtained as intersection of L and R. Numbers and positions of elements indicate the mappings. If a rule has a NAC, then we depict it inside the left hand side and indicate the NAC-only elements by dotted line style. If a NAC contains more than one constraint, then they are marked by different numbers. However, this situation does not appear in any figure of this paper.

Example 1 (context-dependency of independence with NACs). Fig. 3 presents three transformation sequences starting with graph G_0 via rules $p1, p2$ and $p3$. Rule $p3$ has a NAC, which is indicated by dotted lines (one node and two edges).

In the first sequence $s = G_0 \xrightarrow{p_1, m_1} G_1 \xrightarrow{p_2, m_2} G_2 \xrightarrow{p_3, m_3} G_3 = (s_1; s_2; s_3)$ shown in the top of Fig. 3, steps s_1 and s_2 are sequentially independent, and so are s_2 and s_3. After switching the first and the second step we derive $s' = switch(s, 1, 2) = (s'_2; s'_1; s_3)$ (middle of Fig. 3) so that both sequences are shift equivalent ($s \equiv_{sh} s'$). Since s'_1 and s_3 are independent, we can perform a further switch $s'' = switch(s', 2, 3) = (s'_2; s'_3; s''_1)$ shown in the bottom sequence in Fig. 3. However, steps s'_2 and s'_3 are dependent from each other in s'', because the match for rule $p3$ will not satisfy the corresponding NAC for a match into G_0. Hence, independence can change depending on the derivation providing the context, even if derivations are shift equivalent.

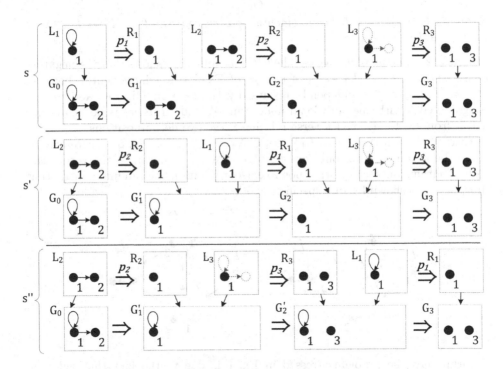

Fig. 3. Independence of p_2 and p_3 is not preserved by switching with p_1

4   Restricting to Incremental NACs

In this section we show that under certain assumptions on the NACs of the rules, the problem identified in Ex. 1 does not occur. Intuitively, for each constraint $n : L \to \hat{L}$ in a NAC we will require that it is *incremental*, i.e., that $\hat{L}$ does not extend L in two (or more) independent ways. Therefore, if there are two different ways to decompose n, one has to be an extension of the other. Incremental arrows have been considered in [12] for a related problem: here we present the definition for monic arrows only, because along the paper we stick to monic NACs.

Definition 1 (incremental monos and NACs). *A mono $f : A \rightarrowtail B$ is called incremental, if for any pair of decompositions $g_1; g_2 = f = h_1; h_2$ as in the diagram below where all morphisms are monos, there is either a mediating morphism $o : O \rightarrowtail O'$ or $o' : O' \rightarrowtail O$, such that the resulting triangles commute.*

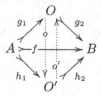

A monic NAC N over L is incremental *if each constraint $n : L \rightarrowtail \hat{L} \in N$ is incremental.*

Example 2 (Incremental NACs). The left diagram below shows that the negative constraint $n_3 : L_3 \to \hat{L}_3 \in N_3$ of rule p_3 of Ex. 1 is not incremental, because $\hat{L}_3$ extends L_3 in two independent ways: by the loop on 1 in O_3, and by the outgoing edge with one additional node 2 in O'_3. Indeed, there is no mediating arrow from O_3 to O'_3 or vice versa relating these two decompositions.

Instead the constraint $n_4 : L_4 \to \hat{L}_4 \in N_4$ of rule p_4 of Fig. 5 is incremental: it can be decomposed in only one non-trivial way, as shown in the top of the right diagram, and for any other possible decomposition one can find a mediating morphism (as shown for one specific case).

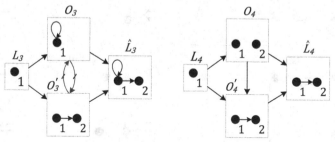

Intuitively, the problem stressed in Ex. 1 is due to the fact that rules p_1 and p_2 delete from G_0 two independent parts of the forbidden context for p_3. Therefore p_3 depends on the firing of p_1 *or* on the firing of p_2, while p_1 and p_2 are independent. This form of *or-causality* from sets of *independent* events is known to be a source of ambiguities in the identification of a reasonable causal ordering among the involved events, as discussed in [19]. The restriction to incremental NACs that we consider here is sufficient to avoid such problematic situations (as proved in the main result of this section) essentially because if both p_1 and p_2 delete from G_0 part of an incremental NAC, then they cannot be independent, since the NAC cannot be factorized in two independent ways.

Incrementality of monos enjoys some nice properties: it is preserved by decomposition of arrows, and it is both preserved and reflected by pushouts along monos, as stated in the next propositions.

Proposition 1 (decomposition of monos preserve incrementality). *Let* $f: A \rightarrowtail B$ *be an incremental arrow and* $f = g; h$ *with monos* $g: A \rightarrowtail C$ *and* $h: C \rightarrowtail B$. *Then both* h *and* g *are incremental.*

Proposition 2 (preservation and reflection of incrementality by POs).

In the diagram to the right, let $B \xrightarrow{f^*} D \xleftarrow{g^*} C$ *be the pushout of the monic arrows* $B \xleftarrow{g} A \xrightarrow{f} C$. *Then* f *is incremental if and only if* f^* *is incremental.*

$$\begin{array}{ccc} A & \overset{g}{\rightarrowtail} & B \\ {\scriptstyle f}\downarrow & & \downarrow{\scriptstyle f^*} \\ C & \underset{g^*}{\rightarrowtail} & D \end{array}$$

We come now to the main result of this section: if all NACs are incremental, then sequential independence of direct transformations is invariant with respect to the switch of independent steps.

Theorem 1 (invariance of independence under shift equivalence). *Assume transformation sequences* $s = G_0 \overset{p_1,m_1}{\Longrightarrow} G_1 \overset{p_2,m_2}{\Longrightarrow} G_2 \overset{p_3,m_3}{\Longrightarrow} G_3$ *and* $s' = G_0 \overset{p_2,m_2'}{\Longrightarrow} G_1' \overset{p_3,m_3'}{\Longrightarrow} G_2' \overset{p_1,m_1''}{\Longrightarrow} G_3$ *using rules* p_1, p_2, p_3 *with incremental NACs only as in the diagram below, such that* $s \equiv_{sh} s'$ *with* $s' = switch(switch(s, 1, 2), 2, 3)$.

$$\begin{array}{ccccc} G_0 & \overset{p_2,m_2'}{=\!=\!\Longrightarrow} & G_1' & \overset{p_3,m_3'}{=\!=\!\Longrightarrow} & G_2' \\ {\scriptstyle p_1,m_1}\Big\Downarrow & & \Big\| {\scriptstyle p_1,m_1'} & & \Big\Downarrow {\scriptstyle p_1,m_1''} \\ G_1 & \overset{p_2,m_2}{=\!=\!\Longrightarrow} & G_2 & \overset{p_3,m_3}{=\!=\!\Longrightarrow} & G_3 \end{array}$$

Then, $G_1 \overset{p_2,m_2}{\Longrightarrow} G_2$ *and* $G_2 \overset{p_3,m_3}{\Longrightarrow} G_3$ *are sequentially independent if and only if* $G_0 \overset{p_2,m_2'}{\Longrightarrow} G_1'$ *and* $G_1' \overset{p_3,m_3'}{\Longrightarrow} G_2'$ *are.*

Proof. Let N_1, N_2 and N_3 be the NACs of p_1, p_2 and p_3, respectively. Due to sequential independence of $G_1 \overset{p_2,m_2}{\Longrightarrow} G_2$ and $G_2 \overset{p_3,m_3}{\Longrightarrow} G_3$, match $m_3 : L_3 \to G_2$ extends to a match $m_3^* : L_3 \to G_1$ satisfying N_3. Using that both m_3 and m_3^* satisfy N_3, we show below that the match $m_3'' : L_3 \to G_0$, that exists by the classical local Church-Rosser, satisfies N_3, too. This provides one half of the independence of $G_0 \overset{p_2,m_2'}{\Longrightarrow} G_1'$ and $G_1' \overset{p_3,m_3'}{\Longrightarrow} G_2'$.

By reversing the two horizontal sequences in the diagram above with the same argument we obtain the proof for the other half, i.e., that the comatch of p_2 into G_2' satisfies the equivalent right-sided NAC of N_2, which is still incremental thanks to Prop. 2. Finally reversing the vertical steps yields the reverse implication, that independence of the upper sequence implies independence of the lower.

The diagram in Fig. 4(a) shows a decomposition of the transformations $G_0 \overset{p_1,m_1}{\Longrightarrow} G_1 \overset{p_2,m_2}{\Longrightarrow} G_2$ and $G_0 \overset{p_2,m_2'}{\Longrightarrow} G_1' \overset{p_1,m_1'}{\Longrightarrow} G_2$ according to the proof of the

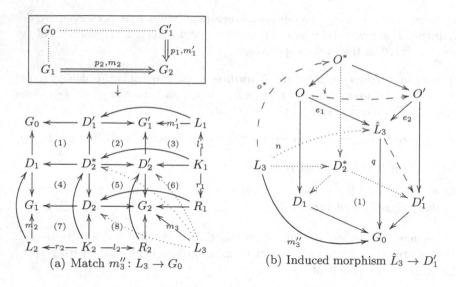

(a) Match $m_3'' : L_3 \to G_0$ (b) Induced morphism $\hat{L}_3 \to D_1'$

Fig. 4. Constructions for proof of Thm. 1

local Church-Rosser theorem ([8], Thm. 3.20). Hence $D_2' G_2 D_2 D_2^*$ is a pullback while all other squares are pushouts, and all morphisms are monos.

The match m_3 is also shown. Let us assume that $G_1 \overset{p_2,m_2}{\Longrightarrow} G_2$ and $G_2 \overset{p_3,m_3}{\Longrightarrow} G_3$ are independent. Then, there exists $L_3 \to D_2$ commuting with m_3 such that $m_3^* = L_3 \to D_2 \to G_1$ satisfies N_3. Also, $G_1' \overset{p_1,m_1'}{\Longrightarrow} G_2$ and $G_2 \overset{p_3,m_3}{\Longrightarrow} G_3$ are independent because equivalence of s and s' requires to switch them, so there exists $L_3 \to D_2'$ commuting with m_3 such that $m_3' = L_3 \to D_2' \to G_1'$ satisfies N_3.

There exists a morphism $L_3 \to D_2^*$ commuting the resulting triangles induced by pullback (5). Matches $L_3 \to D_2^* \to D_1$ and $L_3 \to D_2^* \to D_1'$ satisfy N_3 because they are prefixes of matches m_3^* and m_3', respectively; indeed, it is easy to show that $m; m' \models n \Rightarrow m \models n$ for injective matches m, m' and constraint n.

To show that $m_3'' = L_3 \to D_2^* \to D_1 \to G_0 = L_3 \to D_2^* \to D_1' \to G_0$ satisfies N_3, by way of contradiction, assume $n : L_3 \to \hat{L}_3 \in N_3$ with morphism $q : \hat{L}_3 \to G_0$ commuting with m_3''. We can construct the cube in Fig. 4(b) as follows. The bottom face is pushout (1), faces front left (FL), front right (FR) and top (TOP) are constructed as pullbacks. The commutativity induces unique morphism $O^* \to D_2^*$ making the back faces commuting and thus, all faces in the cube commute. Back left face (BL) is a pullback by pullback decomposition of pullback (TOP+FR) via (BL+(1)) and back right face (BR) is a pullback by pullback decomposition of pullback (TOP+FL) via (BR+(1)). We obtain $o^* : L_3 \to O^*$ as induced morphism from pullback (BL+FL) and using the assumption $m_3'' = n; q$. Further, by the van Kampen property, the top face is a pushout. Since the constraint is incremental and $L_3 \to O \to \hat{L}_3 = L_3 \to O' \to \hat{L}_3$, without loss of generality we have a morphism $i : O \to O'$ commuting the triangles.

We show that $e_2 \colon O' \leftrightarrow \hat{L}_3$ is an isomorphism. First of all, e_2 is a mono by pullback (FR) and mono $D_1' \to G_0$. Pushout (TOP) implies that the morphism pair (e_1, e_2) with $e_1 \colon O \to \hat{L}_3$ and $e_2 \colon O' \to \hat{L}_3$ is jointly epimorphic. By commutativity of $i; e_2 = e_1$, we derive that also $(i; e_2, e_2)$ is jointly epi. By definition of jointly epi, we have that for arbitrary (f, g) it holds that $i; e_2; f = i; e_2; g$ and $e_2; f = e_2; g$ implies $f = g$. This is equivalent to $e_2; f = e_2; g$ implies $f = g$. Thus, e_2 is an epimorphism. Together with e_2 being a mono (see above) we conclude that e_2 is an isomorphism, because adhesive categories are balanced [16]. This means, there exists a mediating morphism $\hat{L}_3 \to O' \to D_1'$ which contradicts the earlier assumption that $L_3 \to D_2^* \to D_1'$ satisfies N_3.   □

Example 3. If in Fig. 3 we replace rule p_3 by rule p_4 of Fig. 5 that has an incremental NAC, so that $s = G_0 \xrightarrow{p_1, m_1} G_1 \xrightarrow{p_2, m_2} G_2 \xrightarrow{p_4, m_4} G_3 = (s_1; s_2; s_4)$, then the problem described in Ex. 1 does not hold anymore, because s_2 and s_4 are not sequentially independent, and they remain dependent in the sequence $s'' = s_2'; s_4'; s_1''$.

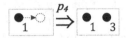

Fig. 5. Rule p_4 with incremental NAC

5   Transforming General NACs into Incremental NACs

In this section we show how to compile a set of rules P with arbitrary NACs into a (usually much larger) set of rules $INC(P)$ having incremental NACs only. The construction guarantees that every derivation using rules in P can be transformed into a derivation over $INC(P)$. Additionally, we show that P and $INC(P)$ are actually equivalent if all constraints in P are obtained as colimits of incremental constraints.

The example shown in Fig. 6 can help getting an intuition about the transformation. It shows one possible outcome (indeed, the algorithm we shall present is non-deterministic) of the application of the transformation to rule p_3, namely the set $INC(\{p_3\}) = \{p_{31}, p_{32}\}$ containing rules with incremental NACs only. It is not difficult to see that p_3 can be applied to a match if and only if either p_{31} or p_{32} can be applied to the same match (determined by the image of node 1), and the effect of the rules is the same (adding a new node). In fact, if either p_{31} or p_{32} can be applied, then also p_3 can be applied to the same match, because at least one part of its NAC is missing (the loop if p_{31} was applied, otherwise the edge). Viceversa, if p_3 can be applied, then either the loop on 1 is missing, and p_{31} is applicable, or the loop is present but there is no non-looping edge from 1, and thus p_{32} can be applied. As a side remark, notice that the NACs p_{31} or p_{32} "cover" the non-incremental NAC of p_3, which is possible because the left-hand side of p_{32} is larger.

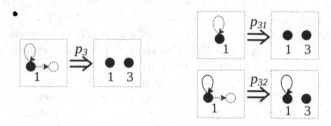

Fig. 6. Rule p_3 (left) and the set $INC(\{p_3\}) = \{p_{31}, p_{32}\}$ (right)

Let us start with some auxiliary technical facts that hold in adhesive categories and that will be exploited to show that the compilation algorithm terminates, which requires some ingenuity because sometimes a single constraint can be compiled into several ones.

Definition 2 (finitely decomposable monos). *A mono $A \xrightarrow{f} B$ is called at most k-decomposable, with $k \geq 0$, if for any sequence of arrows $f_1; f_2; \cdots ; f_h = f$ where for all $1 \leq i \leq h$ arrow f_i is a mono and it is not an iso, it holds $h \leq k$. Mono f is called k-decomposable if it is at most k-decomposable and either $k = 0$ and f is an iso, or there is a mono-decomposition like the above with $h = k$. A mono is* finitely decomposable *if it is k-decomposable for some $k \in \mathbb{N}$. A 1-decomposable mono is called* atomic.

From the definition it follows that all and only the isos are 0-decomposable. Furthermore, any atomic (1-decomposable) mono is incremental, but the converse is false in general. For example, in **Graph** the mono $\{\bullet\} \rightarrowtail \{\bullet \to \bullet\}$ is incremental but not atomic. Actually, it can be shown that in **Graph** all incremental monos are at most 2-decomposable, but there exist adhesive categories with k-decomposable incremental monos for any $k \in \mathbb{N}$.

Furthermore, every finitely decomposable mono $f : A \rightarrowtail B$ can be factorized as $A \rightarrowtail K \xrightarrow{g} B$ where g is incremental and "maximal" in a suitable sense.

Proposition 3 (decomposition and incrementality). *Let $f : A \rightarrowtail B$ be finitely decomposable. Then there is a factorization $A \rightarrowtail K \xrightarrow{g} B$ of f such that g is incremental and there is no K' such that $f = A \rightarrowtail K' \rightarrowtail K \xrightarrow{g} B$, where $K' \rightarrowtail K$ is not an iso and $K' \rightarrowtail K \xrightarrow{g} B$ is incremental. In this case we call g* maximally incremental w.r.t. f.

Proposition 4 (preservation and reflection of k-decomposability).
Let the square to the right be a pushout and a be a mono. Then b is a k-decomposable mono if and only if d is a k-decomposable mono.

$$\begin{array}{ccc} A & \xrightarrow{a} & B \\ b\downarrow & (1) & \downarrow d \\ C & \xrightarrow{c} & D \end{array}$$

In the following construction of incremental NACs starting from general ones, we will need to consider objects that are obtained starting from a span of monos, like pushout objects, but that are characterised by weaker properties.

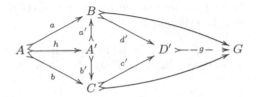

Fig. 7. Quasi-pushout of monos in an adhesive category

Definition 3 (quasi-pushouts of monos). *Let* $B \leftarrowtail A \rightarrowtail C$ *be a span of monos as in Fig. 7, and let* $A \rightarrowtail A'$ *be a mono such that there are monos* $A' \rightarrowtail B$ *and* $A' \rightarrowtail C$ *making the triangles commute. Let* $B \rightarrowtail D' \leftarrowtail C$ *be the pushout of* $B \leftarrowtail A' \rightarrowtail C$. *Then* $B \rightarrowtail D' \leftarrowtail C$ *is a* quasi-pushout (based on A') *of* $B \leftarrowtail A \rightarrowtail C$, *and* D' *is the* quasi-pushout object. *If* $A \rightarrowtail A'$ *is not an iso, the quasi-pushout is called* proper.

Proposition 5 (properties of quasi-pushouts)

1. *Let* $B \leftarrowtail A \rightarrowtail C$ *be a span of monos, and let* $B \rightarrowtail G \leftarrowtail C$ *be a cospan of monos such that the square* $B \leftarrowtail A \rightarrowtail C \rightarrowtail G \leftarrowtail B$ *commutes. Then there is a quasi-pushout* $B \rightarrowtail D' \leftarrowtail C$ *of* $B \leftarrowtail A \rightarrowtail C$ *such that the mediating morphism* $g : D' \to G$ *is mono.*
2. *Let* $B \leftarrowtail A \rightarrowtail C$ *be a span of monos. If objects* B *and* C *are finite (i.e., they have a finite number of subobjects), then the number of non-isomorphic distinct quasi-pushouts of the span is finite.*
3. *In span* $B \leftarrowtail A \overset{b}{\rightarrowtail} C$, *suppose that mono* b *is* k-decomposable, *and that* $B \overset{d'}{\rightarrowtail} D' \leftarrowtail C$ *is a quasi-pushout based on* A', *where* $h : A \rightarrowtail A'$ *is not an iso. Then mono* $d' : B \rightarrowtail D'$ *is at most* $(k-1)$-decomposable.
4. *Quasi-pushouts preserve incrementality: if* $B \overset{d'}{\rightarrowtail} D' \leftarrowtail C$ *is a quasi-pushout of* $B \leftarrowtail A \overset{b}{\rightarrowtail} C$ *and* b *is incremental, then also* $d' : B \rightarrowtail D'$ *is incremental.*

We describe now how to transform a rule p with arbitrary finitely decomposable constraints into a set of rules with simpler constraints: this will be the basic step of the algorithm that will compile a set of rules with finitely decomposable NACs into a set of rules with incremental NACs only.

Definition 4 (compiling a rule with NAC). *Let* $p = \langle L \leftarrowtail K \rightarrowtail R, N \rangle$ *be a rule with NAC, where the NAC* $N = \{n_i : L \rightarrowtail L_i \mid i \in [1, s]\}$ *is a finite set of finitely decomposable monic constraints and at least one constraint, say* n_j, *is not incremental. Then we define the set of rules with NACs* $INC(p, n_j)$ *in the following way.*

(a) If $K \rightarrowtail L \overset{n_j}{\rightarrowtail} L_j$ has no pushout complement, then $INC(p, n_j) = \{p'\}$, where p' if obtained from p by dropping constraint n_j.

(b) Otherwise, let $L \overset{n'_j}{\rightarrowtail} M_j \overset{k}{\rightarrowtail} L_j$ be a decomposition of n_j such that k is maximally incremental w.r.t. n_j (see Prop. 3). Then $INC(p, n_j) = \{p', p_j\}$, where:

　1. p' is obtained from p by replacing constraint $n_j : L \rightarrowtail L_j$ with constraint $n'_j : L \rightarrowtail M_j$.

　2. $p_j = \langle M_j \hookleftarrow K' \rightarrowtail R', N' \rangle$, where $M_j \hookleftarrow K' \rightarrowtail R'$ is obtained by applying rule $\langle L \hookleftarrow K \rightarrowtail R \rangle$ to match $n'_j : L \rightarrowtail M_j$, as in the next diagram.

$$
\begin{array}{ccccc}
L & \overset{l}{\longleftarrow} & K & \overset{r}{\longrightarrow} & R \\
{\scriptstyle n'_j}\downarrow & (1) & \downarrow & (2) & \downarrow \\
M_j & \overset{l^*}{\longleftarrow} & K' & \overset{r^*}{\longrightarrow} & R'
\end{array}
$$

Furthermore, N' is a set of constraints $N' = N'_1 \cup \cdots \cup N'_s$ obtained as follows. (1) $N'_j = \{k : M_j \rightarrowtail L_j\}$. (2) For all $i \in [1, s] \setminus \{j\}$, $N'_i = \{n_{ih} : M_j \rightarrowtail L_{ih} \mid L_i \rightarrowtail L_{ih} \hookleftarrow M_j$ is a quasi-pushout of $L_i \hookleftarrow L \rightarrowtail M_j\}$.

Before exploring the relationship between p and $INC(p, n_j)$ let us show that the definition is well given, i.e., that in Def. 4(b).2 the applicability of $\langle L \hookleftarrow K \rightarrowtail R \rangle$ to match $n'_j : L \rightarrowtail M_j$ is guaranteed.

In fact, by the existence of a pushout comple-ment of $K \rightarrowtail L \overset{n_j}{\rightarrowtail} L_j$ we can build a pushout that is the external square of the diagram on the right; next we build the pullback (2) and obtain $K \rightarrowtail X$ as mediating morphism. Since $(1) + (2)$

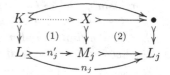

is a pushout, (2) is a pullback and all arrows are mono, from Lemma 4.6 of [16] we have that (1) is a pushout, showing that $K \rightarrowtail L \rightarrowtail M_j$ has a pushout complement.

The following result shows that $INC(p, n_j)$ can simulate p, and that if the decomposition of constraint n_j has a pushout complement, then also the converse is true.

$$
\begin{array}{ccccccc}
\{L_i\} & \overset{\{n_i\}}{\longleftarrow} & L & \overset{l}{\longleftarrow} & K & \overset{r}{\longrightarrow} & R \\
& {\scriptstyle q}\searrow & {\scriptstyle m}\downarrow & (1) & \downarrow & (2) & \downarrow {\scriptstyle m^*} \\
& & G & \overset{l^*}{\longleftarrow} & D & \overset{r^*}{\longrightarrow} & H
\end{array}
$$

Fig. 8. DPO diagram with NAC

Proposition 6 (relationship between p and $INC(p, n_j)$). In the hypotheses of Def. 4, if $G \overset{p}{\Longrightarrow} H$ then $G \overset{INC(p,n_j)}{\Longrightarrow} H$. Furthermore, if the decomposition of n_j $(L \overset{n'_j}{\rightarrowtail} M_j \overset{k}{\rightarrowtail} L_j)$ has a pushout complement, then $G \overset{INC(p,n_j)}{\Longrightarrow} H$ implies $G \overset{p}{\Longrightarrow} H$.

Example 4. Fig. 6 shows one possible outcome of $INC(\{p_3\})$, as discussed at the beginning of this section. As the NAC of p_3 is a colimit of incremental arrows, $\{p_3\}$ and $INC(\{p_3\})$ are equivalent.

Instead, let $p = \langle id_{\{\bullet^1\}}, n : \{\bullet^1\} \rightarrowtail \{\bullet^1 \rightarrow \bullet^2 \rightarrow \bullet^3\}\rangle$ be a rule (the identity rule on graph $\{\bullet^1\}$) with a single negative constraint n, which is not incremental. Then according to Def. 4 we obtain $INC(p, n) = \{p', p_1\}$ where $p' = \langle id_{\{\bullet^1\}}, n' : \{\bullet^1\} \rightarrowtail \{\bullet^1 \rightarrow \bullet^2\}\rangle$ and $p_1 = \langle id_{\{\bullet^1 \rightarrow \bullet^2\}}, n'' : \{\bullet^1 \rightarrow \bullet^2\} \rightarrowtail \{\bullet^1 \rightarrow \bullet^2 \rightarrow \bullet^3\}\rangle$. Note that all constraints in $INC(p, n)$ are incremental, but the splitting of n as $n'; n''$ does not have a pushout complement. Indeed, we can find a graph to which p_1 is applicable but p is not, showing that the condition we imposed on NACs to prove that p and $INC(p, n_j)$ are equivalent is necessary. In fact, let $G = \{\bullet^2 \leftarrow \bullet^1 \rightarrow \bullet^{2'} \rightarrow \bullet^{3'}\}$, and let x be the inclusion morphism from $\{\bullet^1 \rightarrow \bullet^2\}$ to G. Then $G \xLongrightarrow{p_1,x} G$, but the induced inclusion match $m : \{\bullet^1\} \rightarrow G$ does not satisfy constraint n.

Starting with a set of rules with arbitrary (but finitely decomposable) NACs, the construction of Def. 4 can be iterated in order to get a set of rules with incremental NACs only, that we shall denote $INC(P)$. As expected, $INC(P)$ simulates P, and they are equivalent if all NACs are obtained as colimits of incremental constraints.

Definition 5 (compiling a set of rules). *Let P be a finite set of rules with NACs, such that all constraints in all NACs are finitely decomposable. Then the set $INC(P)$ is obtained by the following procedure.*

$INC(P) := P$
while *(there is a rule in $INC(P)$ with a non-incremental constraint)* **do**
    let $\hat{k} = max\{k \mid$ *there is a k-decomposable non-incremental*
                 constraint in $INC(P)\}$
    let $\hat{n}$ be a $\hat{k}$-*decomposable non-incremental constraint of $\hat{p} \in INC(P)$*
    Set $INC(P) := (INC(P) \setminus \{\hat{p}\}) \cup INC(\hat{p}, \hat{n})$
endwhile
return $INC(P)$

Theorem 2 (correctness and conditional completeness of compilation)

1. *The algorithm of Def. 5 terminates.*
2. *$INC(P)$ contains rules with incremental NACs only.*
3. *$INC(P)$ simulates P, i.e., $G \xLongrightarrow{P} H$ implies $G \xLongrightarrow{INC(P)} H$ for all G.*
4. *Suppose that each constraint of each rule in P is the colimit of incremental monos, i.e., for each constraint $L \rightarrowtail L'$, L' is the colimit object of a finite diagram $\{L \rightarrowtail L_i\}_{i \in I}$ of incremental monos. Then P and $INC(P)$ are equivalent, i.e., we also have that $G \xLongrightarrow{INC(P)} H$ implies $G \xLongrightarrow{P} H$.*

Proof. Point 2 is obvious, given the guard of the **while** loop, provided that it terminates. Also the proofs of points 3 and 4 are pretty straightforward, as they follow by repeated applications of Prop. 6. The only non-trivial proof is that of termination.

To this aim, let us use the following lexicographic ordering, denoted $\langle \mathbb{N}^k, \sqsubseteq \rangle$ for a fixed $k \in \mathbb{N}$, that is obviously well-founded. The elements of $\mathbb{N}^k$ are sequences of natural numbers of length k, like $\sigma = \sigma_1 \sigma_2 \ldots \sigma_k$. The ordering is defined as $\sigma \sqsubset \sigma'$ iff $\sigma_h < \sigma'_h$, where $h \in [1, k]$ is the *highest* position at which σ and σ' differ.

Now, let k be the minimal number such that all non-incremental constraints in P are at most k-decomposable, and define the *degree* of a rule p, $deg(p)$, as the sequence $\sigma \in \mathbb{N}^k$ given by

$$\sigma_i = |\{n \mid n \text{ is an } i\text{-decomposable non-incremental constraint of } p\}|$$

Define $deg(Q)$ for a finite set of rules as the componentwise sum of the degrees of all the rules in Q.

Next we conclude by showing that at each iteration of the loop of Def. 5 the degree $deg(INC(P))$ decreases strictly. Let $\hat{p}$ be a rule and $\hat{n}$ be a non-incremental constraint, $\hat{k}$-decomposable for a maximal $\hat{k}$. The statement follows by showing that $INC(\hat{p}, \hat{n})$ has at least one $\hat{k}$-decomposable non-incremental constraint less than $\hat{p}$, while all other constraints are at most $(\hat{k} - 1)$-decomposable.

This is obvious if $INC(\hat{p}, \hat{n})$ is obtained according to point (a) of Def. 4. Otherwise, let $INC(\hat{p}, \hat{n}) = \{p', p_j\}$ using the notation of point (b). In this case rule p' is obtained from $\hat{p}$ by replacing the selected constraint with one that is at most $(\hat{k} - 1)$-decomposable. Furthermore, each other constraint n_i is replaced by a set of constraints, obtained as quasi-pushouts of n_i and n'_j. If n_i is incremental, so are all the new constraints obtained as quasi-pushouts, by Prop. 5(4), and thus they don't contribute to the degree. If instead n_i is non-incremental, then it is h-decomposable for $h \leq \hat{k}$, by definition of $\hat{k}$. Then by Prop. 5(3) all constraints obtained as proper quasi-pushouts are at most $(h - 1)$-decomposable, and only one (obtained as a pushout) will be h-decomposable.   □

6 Discussion and Conclusion

In our quest for a stable notion of independence for conditional transformations, we have defined a restriction to *incremental NACs* that guarantees this property (Thm. 1). Incremental NACs turn out to be quite powerful, as they are sufficient for several case studies of GTSs. In particular, the well studied model transformation from class diagrams to relational data base models [14] uses incremental NACs only. In an industrial application for translating satellite software (pages 14-15 in [20]), we used a GTS with more than 400 rules, where only 2 of them have non-incremental NACs. Moreover, the non-incremental NACs could also have been avoided by some modifications of the GTS. Incremental morphisms have been considered recently in [12], in a framework different but related to ours, where requiring that matches are *open maps* one can restrict the applicability of transformation rules without using NACs.

We have also presented a construction that compiles as set of rules with general (finitely-decomposable) NACs into a set of rules with incremental NACs only. For NACs that are obtained as colimits of incremental ones, this compilation yields an equivalent system, i.e., for every transformation in the original GTS there exists one compatible step in the compiled one and vice versa (Thm. 2), and therefore the rewrite relation on graphs is still the same. In the general case, the compiled system provides an overapproximation of the original GTS, which nevertheless can still be used to analyse the original system.

In fact our intention is to define a stable notion of independence on transformations with general NACs. Using the compilation, we can declare a two-step sequence independent if this is the case for all of its compilations, or more liberally, for at least one of them. Both relations should lead to notions of equivalence that are finer than the standard shift equivalence, but that behave well thanks to Thm. 1. Moreover, independence should be expressed directly on the original system, rather than via compilation. Such a revised relation will be the starting point for developing a more advanced theory of concurrency for conditional graph transformations, including processes and unfoldings of GTSs.

The main results in this paper can be applied for arbitrary adhesive transformation systems with monic matches. However, in some cases (like for attributed graph transformation system) the restriction to injective matches is too strict (rules contain terms that may be mapped by the match to equal values). As shown in [13], the concept of NAC-schema provides a sound and intuitive basis for the handling of non-injective matches for systems with NACs. We are confident that an extension of our results to general matches is possible based on the concept of NAC-schema.

Another intersting topic that we intend to study is the complexity of the algorithm of Def. 5, and the size of the set of rules with incremental constraints, $INC(P)$, that it generates. Furthermore, we plan to extend the presented results for shift equivalence to the notion of permutation equivalence, which is coarser and still sound according to [13]. Finally, we also intend to address the problem identified in Ex. 1 at a more abstract level, by exploiting the event structures with or-causality of events that are discussed in depth in [19].

References

1. Baldan, P., Corradini, A., Montanari, U., Rossi, F., Ehrig, H., Löwe, M.: Concurrent Semantics of Algebraic Graph Transformations. In: Ehrig, H., Kreowski, H.J., Montanari, U., Rozenberg, G. (eds.) The Handbook of Graph Grammars and Computing by Graph Transformations. Concurrency, Parallelism and Distribution, vol. 3, pp. 107–188. World Scientific (1999)
2. Baldan, P., Corradini, A., Heindel, T., König, B., Sobociński, P.: Processes for adhesive rewriting systems. In: Aceto, L., Ingólfsdóttir, A. (eds.) FOSSACS 2006. LNCS, vol. 3921, pp. 202–216. Springer, Heidelberg (2006)

3. Baldan, P., Corradini, A., Heindel, T., König, B., Sobociński, P.: Unfolding grammars in adhesive categories. In: Kurz, A., Lenisa, M., Tarlecki, A. (eds.) CALCO 2009. LNCS, vol. 5728, pp. 350–366. Springer, Heidelberg (2009)
4. Baldan, P., Corradini, A., Montanari, U., Ribeiro, L.: Unfolding semantics of graph transformation. Inf. Comput. 205(5), 733–782 (2007)
5. Corradini, A., Hermann, F., Sobociński, P.: Subobject Transformation Systems. Applied Categorical Structures 16(3), 389–419 (2008)
6. Corradini, A., Montanari, U., Rossi, F.: Graph processes. Fundamenta Informaticae 26(3/4), 241–265 (1996)
7. Ehrig, H.: Introduction to the Algebraic Theory of Graph Grammars (A Survey). In: Claus, V., Ehrig, H., Rozenberg, G. (eds.) Graph Grammars 1978. LNCS, vol. 73, pp. 1–69. Springer, Heidelberg (1979)
8. Ehrig, H., Ehrig, K., Prange, U., Taentzer, G.: Fundamentals of Algebraic Graph Transformation. EATCS Monographs in Theor. Comp. Science. Springer (2006)
9. Ehrig, H., Pfender, M., Schneider, H.: Graph grammars: an algebraic approach. In: 14th Annual IEEE Symposium on Switching and Automata Theory, pp. 167–180. IEEE (1973)
10. Ehrig, H., Habel, A., Lambers, L.: Parallelism and Concurrency Theorems for Rules with Nested Application Conditions. EC-EASST 26, 1–24 (2010)
11. Habel, A., Heckel, R., Taentzer, G.: Graph Grammars with Negative Application Conditions. Fundamenta Informaticae 26(3,4), 287–313 (1996)
12. Heckel, R.: DPO Transformation with Open Maps. In: Ehrig, H., Engels, G., Kreowski, H.-J., Rozenberg, G. (eds.) ICGT 2012. LNCS, vol. 7562, pp. 203–217. Springer, Heidelberg (2012)
13. Hermann, F., Corradini, A., Ehrig, H.: Analysis of Permutation Equivalence in M-adhesive Transformation Systems with Negative Application Conditions. In: MSCS (to appear, 2013)
14. Hermann, F., Ehrig, H., Golas, U., Orejas, F.: Efficient Analysis and Execution of Correct and Complete Model Transformations Based on Triple Graph Grammars. In: Bézivin, J., Soley, R., Vallecillo, A. (eds.) Proc. Int. Workshop on Model Driven Interoperability (MDI 2010), pp. 22–31. ACM (2010)
15. Kreowski, H.J.: Is parallelism already concurrency? Part 1: Derivations in graph grammars. In: Ehrig, H., Nagl, M., Rozenberg, G., Rosenfeld, A. (eds.) Graph Grammars 1986. LNCS, vol. 291, pp. 343–360. Springer, Heidelberg (1987)
16. Lack, S., Sobocinski, P.: Adhesive and quasiadhesive categories. ITA 39(3), 511–545 (2005)
17. Lambers, L., Ehrig, H., Orejas, F., Prange, U.: Parallelism and Concurrency in Adhesive High-Level Replacement Systems with Negative Application Conditions. In: Ehrig, H., Pfalzgraf, J., Prange, U. (eds.) Proceedings of the ACCAT Workshop at ETAPS 2007, vol. 203/6, pp. 43–66. Elsevier (2008)
18. Lambers, L.: Certifying Rule-Based Models using Graph Transformation. Ph.D. thesis, Technische Universität Berlin (2009)
19. Langerak, R., Brinksma, E., Katoen, J.P.: Causal ambiguity and partial orders in event structures. In: Mazurkiewicz, A., Winkowski, J. (eds.) CONCUR 1997. LNCS, vol. 1243, pp. 317–331. Springer, Heidelberg (1997)
20. Ottersten, B., Engel, T.: Interdisciplinary Centre for Security, Reliability and Trust - Annual Report 2011. University of Luxembourg, Interdisciplinary Centre for Security, Reliability and Trust (SnT), http://www.uni.lu/content/download/52106/624943/version/1/file/SnT_AR2011_final_web.pdf

Statistical Model Checking
for Composite Actor Systems*

Jonas Eckhardt[1], Tobias Mühlbauer[1], José Meseguer[2], and Martin Wirsing[3]

[1] Technische Universität München
[2] University of Illinois at Urbana-Champaign
[3] Ludwig-Maximilians-Universität München

Abstract. In this paper we propose the so-called composite actor model for specifying composed entities such as the Internet. This model extends the actor model of concurrent computation so that it follows the "Reflective Russian Dolls" pattern and supports an arbitrary hierarchical composition of entities. To enable statistical model checking we introduce a new scheduling approach for composite actor models which guarantees the absence of unquantified nondeterminism. The underlying executable specification formalism we use is the rewriting logic-based semantic framework Maude, its probabilistic extension PMaude, and the statistical model checker PVESTA. We formalize a model transformation which—given certain formal requirements—generates a scheduled specification. We prove the correctness of the scheduling approach and the soundness of the transformation by introducing the notions of strong zero-time rule confluence and time-passing bisimulation and by showing that the transformation is a time-passing bisimulation for strongly zero-time rule confluent composite actor specifications.

Keywords: actor system, rewriting logic, Maude, composite actor, statistical model checking.

1 Introduction

The actor model is a classical model for concurrent computation [16] which witnessed revived interest with the advent of multi-core programming and Cloud-scale computing. Several modern programming languages such as Erlang and Scala base their concurrency models on actors [7,14]. An actor is a concurrent object which operates asynchronously and interacts with other actors by sending asynchronous messages [3]. Temporal logic properties of actor-based models can be automatically verified either by exact model checking algorithms [13] or, in an approximate but more scalable way, by statistical model checking (see e.g., [4]).

---

* This work has been partially sponsored by the EU-funded projects FP7-257414 ASCENS and FP7-256980 NESSoS, and AFOSR Grant FA8750-11-2-0084. Tobias Mühlbauer is also partially supported by the Google Europe Fellowship in Structured Data Analysis. We thank Mirco Tribastone for his helpful comments on this paper. We further thank all reviewers for their valuable feedback.

N. Martí-Oliet and M. Palomino (Eds.): WADT 2012, LNCS 7841, pp. 143–160, 2013.
© IFIP International Federation for Information Processing 2013

These approaches are all based on the original "flat" actor model but many interesting applications such as the Internet and Cloud systems are not flat, as they are composed of various participants and systems and are hierarchically structured into different layers and networks. Such composed entities are often safety- and security-critical, and have strong qualitative and quantitative formal requirements. The above mentioned analysis approaches, however, rely on flat actor models and cannot handle and model check composite models in a direct way.

In this paper, we extend the actor model to a so-called *composite actor model* that directly addresses hierarchical concurrent systems and present a model transformation which makes statistical model checking usable for composite actor model specifications. The composite actor model follows the so-called "Reflective Russian Dolls" model [19] and supports an arbitrary hierarchical composition of entities. As underlying executable specification formalism we use the rewriting logic language Maude and its real-time and probabilistic extensions.

Current statistical model checking methods require that the system is purely probabilistic, i.e., that there is no unquantified nondeterminism in the choice of transitions. This is nontrivial to achieve for distributed systems where many different components may perform local transitions concurrently. There are two complementary ways for guaranteeing the absence of nondeterminism: either by associating continuous probability distributions with message delays and computation time and by relying on the fact, that for continuous distributions the probability of sampling the same real number twice is zero (see e.g., [13]), or by introducing a scheduler that provides a deterministic ordering of messages (see e.g., [6]). We follow the latter approach and propose a new scheduling method for well-formed composite actor models that guarantees the absence of nondeterminism. We formalize the approach in Maude and study its soundness by proving the correctness of the scheduling approach, termination and confluence of the underlying equational specification, and by showing the absence of unquantified nondeterminism from any scheduled well-formed composite actor specification. We further formalize a model transformation which—given a composite actor specification that adheres to certain formal requirements—generates a scheduled specification. To prove the soundness of the transformation we introduce the notions of strong zero-time rule confluence and time-passing bisimulation and then show that the transformation, which is only a simulation by itself, is indeed a time-passing bisimulation for strongly zero-time rule confluent composite actor specifications. To the best of our knowledge, our solution is the first one making it possible to analyze such systems in a faithful way by statistical model checking. We have applied our method to several complex case studies (see [12,21,11,20]); for reasons of space we only illustrate a simple example.

Outline. The paper is structured as follows: In Sect. 2 we explicate shortly the (flat) actor model and the "Reflective Russian Dolls" model, and show how to build the composite actor model in Maude by applying the Russian Dolls approach to the flat actor model. In Sect. 3 we present our model transformation together with the scheduling approach, their formalization in Maude and prove the soundness properties. Sect. 4 then gives a short presentation of our methodology for the statistical model checking analysis of actor model-based specifications. We conclude by discussing related work, summarizing our results, and sketching further work.

2   The Composite Actor Model

2.1   The Actor Model of Computation

Our specifications are based on the *actor model of computation* [16,15,2], a mathematical model of concurrent computation in distributed systems. Similar to the *object-oriented programming paradigm*, in which the philosophy that *everything is an object* is adopted, the *actor model* follows the paradigm that *everything is an actor*. An *actor* is a concurrent object that encapsulates a state and can be addressed using a unique name. Actors communicate with each other using asynchronous messages. Upon receiving a message, an actor can change its state and can send messages to other actors. Actors can be used to model and reason about distributed and concurrent systems in a natural way [2].

2.2   The *Reflective Russian Dolls* Model

In rare situations, the state of a distributed system can be thought of as a *flat configuration* which contains objects and messages. Such a *flat configuration* can be modelled as a *flat soup* (i.e., a multiset) that consists of actors and messages. However, as a distributed system becomes more complex, hierarchies are introduced to better represent the structure of the system and its communication patterns. A flat model does not reflect boundaries in a hierarchical system which impose conditional communication barriers. In a flat model, every participant can communicate with everybody else. However, some concepts, like a firewall, rely on the existence of physical boundaries that messages from the outside have to cross in order to reach destinations within that boundary.

In [19], Meseguer and Talcott present the *Reflective Russian Dolls* (RRD) model which extends and formalizes previous work on actor reflection and provides a generic formal model of distributed object reflection. The rewriting logic-based model combines logical reflection and hierarchical structuring. In their model, the state of a distributed system is not represented by a *flat soup*, but rather as a *soup of soups*, each enclosed within specific boundaries. As with traditional Russian dolls, soups can be nested up to an arbitrary depth.

Figure 1 illustrates the basic idea using a system that is guarded by a firewall. Each of the boxes represents a system. The firewall consists of a subsystem which itself is composed of several components $C_1 \ldots C_n$. Message M is addressed to the innermost component C_n and as such has to pass the boundary of the firewall. The firewall possibly transforms the message to M' (e.g., tags a message with a security clearance). After that, the boundary of the sub-system has to be crossed which, respectively, can also alter the message to M''.

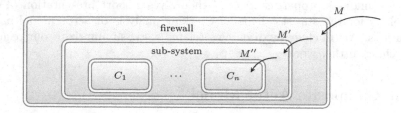

Fig. 1. Example of a Russian Dolls model of a system with boundaries

Mathematically, this can be modelled by boundary operators of the form

$$b : s_1, \ldots, s_n, Configuration \to Configuration$$

where $s_1, \ldots, s_n$ are additional sorts. These sorts are called the *parameters* of the boundary operator. Boundary operators encapsulate a configuration together with several parameters, and as with Russian dolls, they can be nested arbitrarily. Using the Russian Dolls model, sophisticated distributed systems, that rely on system boundaries, can be modeled [19].

2.3   The Composite Actor Model

We extend the original *actor model of computation* for flat configurations with the *RRD model* to allow the specification of hierarchically structured composite actor systems. Actors are now allowed to contain a soup, that is a multiset, of actors and messages in their state where each of the actors in the sub-configuration can again contain another soup of actors and messages, and so on. Referring to the aforementioned firewall example, the firewall actor can be represented by the term

```
<0 :Firewall | config: <0.0 :Subsystem | config: <0.0.0 :C₁> ... <0.0.N-1 :C_N>>>
```

Thereby the firewall actor has address 0 and contains a sub-configuration in its state denoted by the `config` attribute. In this sub-configuration there is a subsystem actor with address 0.0 which again contains a sub-configuration which consists of the components (C_i) with adresses 0.0.0,...,0.0.N-1.

The naming of actors follows an hierarchical naming scheme. This naming scheme is comparable to Internet domain names, which are structured in top-level (e.g., "de"), second- (e.g., "google") and third-level (e.g., "www") domain names. The hierarchical naming scheme for the actor model basically builds a hierarchical name tree, in which children addresses are composed of their parents address and an appended number which uniquely identifies them among their siblings (e.g., a name could be 4.0.8.2). Further information on the naming scheme can be found in [11,20].

The analysis of real world systems often requires the system model to include a notion of time (e.g., as for quality of service properties in our case studies [6,21]). Thus, we introduce activation times for messages, i.e., times at which messages are intended to be processed.

We model composite actor models as hierarchically structured soups of messages and actors. In Maude, we represent a soup as a term of sort `Config`. A configuration is associative and commutative and is constructed by the operators `op null : -> Config` and `op __ : Config Config -> Config [assoc comm id: null]`. The sorts for actors and messages are both subsorts of `Config`. Additionally, we introduce the sort `Address`, which represents an actor's address, and which is a subsort of `Nat`. Addresses can be concatenated using the constructor `op _._ : Address Address -> Address [assoc]`. Messages are terms of sort `Msg` and are created by the constructor `(_,_<-_): Float Address Content -> Msg`, which takes the message's "activation" time, the receiver address and the actual contents of the message as arguments. Finally, actors can be created by the constructor `op <_:_|_> : Address ActorType AttributeSet -> Actor`, which takes the address of the actor, the type of the actor, and an additional set of attributes (a term of sort `AttributeSet`, which represents an associative, commutative set of terms of sort `Attribute`) as arguments.

The hierarchical nature of the composite actor model is made explicit by the dedicated attribute `config`, which contains the inner soup of an actor. The constructor `op config:_ : Config -> Attribute [gather(&)]` creates this inner soup for an arbitrary term of sort `Config`. [1]

Following the idea of the actor model, actors in the composite actor model communicate only via asynchronous message passing. This convention allows the local specification of actors, i.e., the specification of an actor's semantics does not need to contain knowledge about the structure of the composed system and rewrite rules that capture the semantics do not have to include the receiving actor. As such, a specification of an actor only requires local transition rules (message handling) and boundary crossing rules to be specified. Thereby, boundary crossing rules are rules that insert a message into an actor's sub-configuration and rules that move a message from an actor's sub-configuration to the configuration the actor is in. More precisely, composite actor model specifications only contain (possibly probabilistic) rewrite rules of the following type:

*C*onsume: one actor consumes a message at time t and may emit timed $(t_1,...,t_n)$ messages or spawn new actors in its sub-configuration.

```
< A : T | config: C, AS> (t, A<-M) =>
    < A : T | config: C', AS'> (t1, A1<-M1) ... (tn, An<-Mn)
```

*B*oundary-Down: one actor consumes a message at time t and inserts it into its subconfiguration.

```
< A : T | config: C, AS> (t, X<-M) => < A : T | config: C, (t', X'<-M'), AS'>
```

*B*oundary-Up: one actor consumes a message from its subconfiguration at time t and emits it at its level in the composite actor hierarchy.

```
< A : T | config: (t, X<-M) C, AS> => < A : T | config: C, AS'> (t', X'<-M')
```

A, $A_1,...,A_n$, X, X' are thereby terms of sort `Address`, T of sort `ActorType`, C, C' of sort `Config`, t, $t_1,...,t_n$, t' of sort `Float`, M, $M_1,...,M_n$, M' of sort `Content`

---

[1] In general, it is possible to allow an actor to have its inner soup(s) in arbitrarily named attributes. The convention of having a single sub-configuration in a predefined attribute, however, simplifies scheduling approaches for statistical model checking.

and AS and AS' of sort AttributeSet. Any of these rules may have added probabilistic information [4] or be subjected to a condition.

Local specifications of actors provide modularity which is a key technique
to tackle the complexity of large distributed systems. In previous work [12] we
describe how meta objects, i.e., distributed objects that mediate and control the
behavior of one or more distributed objects, can be made highly reusable as
formal patterns and how a distributed system can be modelled as a composition
of such formal patterns.

Note that even though messages in the composite actor model are timed, the
execution of composite actor model specifications in pure Maude would neither
take activation times nor the concept of a global time into account. Rather all
possible executions regardless of activation times would be executed.

Definition (Well-Timedness). In the following we are only interested in "well-
timed" executions where messages are processed according to their activation
time and global time is advancing monotonously. More precisely, we call a run
$u_0 \xrightarrow{r_0} u_1 \xrightarrow{r_1} \ldots$ of a composite actor specification well-timed if for any i, the
transition $u_i \xrightarrow{r_i} u_{i+1}$ is triggered by the application of rule r_i at time t_i, and
$t_i \leq t_{i+1}$ where t_i is the smallest activation time occurring in u_i.

3   Scheduling Approach for Composite Actor Models

The absence of unquantified nondeterminism is a requirement for statistical
model checking using Maude/PMaude and PVESTA [4]. Currently, there are
two approaches for assuring this requirement: (i) by associating continuous probability distributions with message delays and computation time [4] and (ii) by
introducing a scheduler which guarantees a deterministic execution order of messages in the actor system [6]. Both approaches however rely on a flat soup of
actors and cannot, in their current state, handle composite actor models. In this
work we adapt the scheduling approach (ii) for composite actor models. In order
to promote modularity and to make the scheduling approach transparent, we
propose a model transformation approach that does not require the specification of an actor to have knowledge about the composition of the system and
about the scheduling approach as such. Given a composite actor specification
M—that adheres to certain formal requirements—we generate a scheduled composite actor specification SM by a model transformation $M \rightarrow SM$ such that
SM guarantees the absence of unquantified nondeterminism.

3.1   Well-Formedness Requirements

To enable statistical model checking we require the following well-formedness
conditions on the original composite actor specification. We call a composite
actor specification M well-formed if an initial state is defined and the following
two formal requirements are fulfilled:

(1.) The specification must adhere to the composite actor model, i.e., entities must be specified as actors which communicate only via asynchronous message passing and there are only rewrite rules of type *consume, boundary-down,* or *boundary-up* (see Sect. 2.3).

(2.) The specification must have no unquantified nondeterminism in the choice of rewrite rules, i.e., for each message there is at most one matching rewrite rule.

Moreover, we assume w.l.o.g. that (3.) any message m = (t, A<-m') is executed at its activation time t (i.e. there exists a matching rule which at time t is applied to m and the actor <A : T | C>). This is not a restriction since e.g. the loss of m can easily be modeled by the rule (t, A<-M)<A : T | C> => <A : T| C>.

If a specification M fulfills these requirements, it is still nondeterministic (e.g. if several messages occur in a configuration there may be a nondeterministic choice to which message a rewrite rule will be applied), but our scheduling approach eliminates all unquantified nondeterminism in the generated scheduled composite actor model specification SM. Note that we introduce requirement (2.) as the scheduling approach introduced by the model transformation only eliminates unquantified nondeterminism in consumption of messages, but does not in the choice of rewrite rules. Together with the scheduling approach, requirement (2.) ensures no unquantified nondeterminism in the whole specification, which is a requirement for statistical model checking with PVeStA.

3.2 Model Transformation $M \to SM$

Given a module M that specifies a well-formed composite actor model specification, the transformation $M \to SM$ creates a module SM, for which the transformation preserves all sort declarations, all operators, and all equations. Moreover, we introduce new sorts that define an explicit scheduler and new message types to represent scheduled messages as well as two auxiliary sorts of messages where active and scheduled messages are annotated by the address of the sending actor:

```
sort Scheduler ScheduleList ScheduleMsg LocActiveMsg LocScheduleMsg .
subsorts Scheduler ScheduleMsg LocScheduleMsg LocActiveMsg < Config .
subsort LocScheduleMsg < ScheduleList .
```

where terms of sort ScheduleMsg are created by the constructor [_] : Msg -> ScheduleMsg and represent messages that are emitted by rewrite rules and that are to be inserted in the scheduler. The constructors {_,_} : Address Msg -> LocActiveMsg and [_,_] : Address ScheduleMsg -> LocScheduleMsg, generate terms of sort LocActiveMsg and LocScheduleMsg respectively. Both contain either a term of sort Msg (in case of a LocActiveMsg) or ScheduleMsg (in case of a LocScheduleMsg) and a term of sort Address [2].Terms of sort ScheduleListrepresent

---

[2] Terms of sort LocScheduleMsg are used to store the address of the actor in whose configuration a scheduled message was emitted (or the topmost address if the message was emitted at the top-most level) and the scheduled message itself. Similarly, terms of sort LocActiveMsg are used to store the same address but of a scheduled message that has been made active by the scheduler. These auxiliary messages are used in our scheduling approach.

a list of `ScheduleMsg` using the constructor `op _;_ : ScheduleList ScheduleList -> ScheduleList [assoc id: nil]`. A term of sort `Scheduler` is created by the constructor `op {_|_} : Float ScheduleList -> Scheduler`, which contains the current time and a list of messages that are to be scheduled.

For each rewrite rule r (which is either of type *consume, boundary-down, or boundary-up*) in M, a rewrite rule r' is added to SM, where each term `MSG` of sort `Msg` on the right side of r is transformed to a term `[MSG]` of sort `ScheduleMsg`. We also add the specification of our scheduling approach to the new specification (for details see Sect. 3.3).

Moreover, any initial state st of M is transformed to a state st' of SM as follows: (i) every message `MSG` of sort `Msg` is transformed to a term `[MSG]` of sort `ScheduleMsg` and (ii) an empty scheduler `{0.0 | nil}` is added. Then the result of the model transformation is the canonical form $[st']$ of st', which is of the form `AC {0.0 | SL}`, where `AC` represents the message-free actor hierarchy of st and `SL` is the message list in the scheduler that now contains all messages of st as scheduled messages.

3.3   The Scheduling Approach

On a high level of abstraction, in order to remove all unquantified nondeterminism, our scheduling approach takes control of when a rewrite rule is executed. Our well-formedness requirements specify that only three types of rewrite rules (*consume, boundary-down,* and *boundary-up*) are allowed; and that there is no unquantified nondeterminism in the choice of these rules. Each of these rewrite rules requires an active message, i.e., non-scheduled message, to be present either at the hierarchy level of the actor or in its sub-configuration. Thus, a rewrite can occur only if an active message is present in the composite actor hierarchy. Furthermore, since we require that one single message is consumed by exactly one actor, this rewrite can be determined up to probabilistic choices. In order to ensure that only one message at a time is emitted we introduce an operation `step` and, similar to the special tick rule that is defined in [4], define a one-step computation of a model written in the scheduled composite actor model as a transition of the form

$$u \xrightarrow{step} v \to w$$

where

(i) u is a canonical term, which represents the global state of a scheduled composite actor system and in which all messages are contained in the scheduler.

(ii) v is a canonical term obtained by removing the next addressed scheduled message `[A,[(T, A'<-M)]]` from the scheduler, and by inserting it, as a non-scheduled addressed message `{A,(T, A'<-M)}`, into the configuration of the actor at address `A`. Additionally, the global time in the scheduler is advanced to T. This operation is called `step`.

(iii) w is a canonical term obtained after zero or one rewrites the actor performs to forward (*boundary-down* or *boundary-up*) or *consume* the message.

More precisely, the scheduling approach for the composite actor model consists of the following steps:

(1.) If no active message is present at any level of the hierarchy, the next scheduled message is marked as active and inserted in the top-most configuration by the scheduler.

(2.) Then, the message is pushed down the hierarchy by equational simplification, until it eventually reaches the configuration where it has been emitted. The resulting term is in canonical form.

(3.) Since the specification adheres to the composite actor model, there exists at most one rewrite rule of type *consume*, *boundary-down*, or *boundary-up* which consumes the message and possibly produces several scheduled messages.

(4.) Finally, the scheduled messages are pulled up the hierarchy by equational simplification, until it eventually reaches the top-most level where the scheduled messages are inserted into the scheduler which stores the messages in a strict order, i.e., by the scheduled activation time and if equal by Maude's built-in term order. The resulting term is in canonical form.

In the following, we present the Maude specification of the scheduling approach which forms the final part of the model transformation $M \to SM$. To efficiently distinguish composite actor hierarchies that contain messages in any of its subconfigurations and those that do not, we use conditional sort memberships. The sort of term `ActorConfig`, a subsort of `Config`, thereby represents a configuration which contains no messages in any of its sub-configurations. In order to be able to make this distiction, we introduce the sort `InertActor`, which represents a term of sort `Actor`, that is either flat, i.e., that does not contain any subconfiguration, or whose subconfiguration is of sort `ActorConfig`. This is expressed by the following (conditional) sort membership axioms:

```
mb < A : T | config: AC, AS > : InertActor .
cmb ACT : InertActor if flatActor(ACT) .
op flatActor : Actor -> Bool .
eq flatActor(< A : T | config: C, AS >) = false .
eq flatActor(ACT) = true [owise] .
```

where `A` is a variable of sort `Address`, `T` of sort `ActorType`, `AC` of sort `ActorConfig`, `AS` of sort `AttributeSet`, `ACT` of sort `Actor`, and `C` of sort `Config`.

Having these (conditional) sort membership axioms, we can easily identify configurations that do not contain any active message: terms of sort `ActorConfig`. These are the configurations that can advance, i.e., the **step** operation can be called on these configurations. The **step** operation is defined as

```
op step : Config -> Config [iter] .
eq step(AC {gt | [A1, [(t1 , A <- M1)]]; SL}) = {A1, (t1 , A <- M1)} AC {t1 | SL} .
```

where `AC` is a term of sort `ActorConfig`, `gt` and `t1` of sort `Float`, `A1` and `A` of sort `Address`, `M1` of sort `Msg`, and `SL` of sort `ScheduleList`. The **step** equation can only be applied on a term of sort `Config` which consists of a term of sort `ActorConfig` and a term of sort `Scheduler`. It inserts the first message of the scheduler in the configuration and updates the time of the scheduler.

The remaining equations use the following variables: `A` and `A'` of sort `Address`, `T` of sort `ActorType`, `SM` of sort `ScheduleMsg`, `LSM` of sort `LocScheduleMsg`, `S` of sort

Scheduler, SL of sort `ScheduleList` C of sort `Config`, AS of sort `AttributeSet`, AM of sort `Msg`, and finally `gt`, `t1`, and `t2` of sort `Float`.

To put scheduled messages as active messages back into the configuration where they have been emitted, the address of the actor containing the scheduled message needs to be stored together with the scheduled message. The equations `create-loc-msg1` and `create-loc-msg2` create a `LocScheduleMsg` from a scheduled message that is either in an actor's sub-configuration or at the top-most level of the hierarchy.

```
eq [create-loc-msg1] : <A : T | config: SM C, AS> = [A,SM] <A : T | config: C, AS> .
eq [create-loc-msg2] : SM C S = [ toplevel, SM ] C S .
```

Then, terms of sort `LocScheduleMsg` are pulled up the hierarchy by the `pull-up` equation and finally inserted in the scheduler by the `insert-in-scheduler` equation.

```
eq [pull-up] : < A : T | config: LSM C, AS > = LSM < A : T | config: C, AS > .
eq [insert-in-scheduler] : LSM S = insert(S, LSM) .
```

Terms of sort `LocActiveMsg` are pushed down the hierarchy by the equation `push-down` and are finally inserted in the correct subconfiguration by the equation `insert-in-configuration` or by the equation `insert-toplevel`, if the scheduled message has been emitted in the top-most level.

```
eq [push-down] : < A : T | config: C, AS > {A . A', AM}
  = < A : T | config: C {A . A', AM}, AS > .
eq [insert-in-configuration] : < A : T | config: C, AS > {A , AM}
  = < A : T | config: C AM, AS > .
eq [insert-toplevel] : {toplevel, AM} S = AM S .
```

Finally, the operator `insert` inserts a term of sort `LocScheduleMsg` into the scheduler.

```
op insert : Scheduler LocScheduleMsg -> Scheduler .
op insert : ScheduleList LocScheduleMsg -> ScheduleList .
eq insert({gt | SL}, LSM) = {gt | insert(SL,LSM)} .
eq [insert-list] : insert([A1,[(t1, A<-M1)]]; SL, [A2,[(t2, A'<-M2)]]) =
  if (t1 < t2) or ((t1 == t2)) and lt(M1,M2) then
    [A1,[(t1, A<-M1)]] ; insert(SL,[A2,[(t2, A'<-M2)]])
  else
    [A2,[(t2, A'<-M2)]] ; [A1,[(t1, A<-M1)]] ; SL
  fi .
eq insert(nil, LSM) = LSM .
```

3.4   Correctness of the Scheduling Approach

In this section we analyse the correctness of the scheduling approach by proving that the equational specification is terminating and confluent modulo associativity and commutativity and by showing that the introduction of the scheduler eliminates all unquantified nondeterminism.

Proposition. The equational specification of the scheduling approach is terminating and confluent modulo associativity and commutativity (AC).

Proof sketch (termination). The equations that need to be discussed are the recursive ones: `insert-list`, `pull-up`, and `push-down`. `insert-list` terminates since the `ScheduleList` argument of the `insert-list` operator in the recursive call gets smaller. `pull-up`, for a specific scheduled message, terminates, since the distance of the message's location to the toplevel gets smaller with each level of recursion. `push-down`, for a specific addressed active message, terminates, since the distance between the message's address and the addresses of the actors in the sub-configuration the message is inserted into gets small with each level of recursion.   □

Proof sketch (confluence modulo AC). As the equational specification of the scheduling approach is terminating, it is sufficient to prove local confluence. For most equations local confluence is achieved by applying the AC property of the configurations. The only exception are the equations that insert a message into the scheduler. For the insertion equations local confluence results from the fact that the message order in the scheduler list is a total ordering. The messages are ordered by activation time and if equal by Maude's built-in term order.   □

Lemma 1. The scheduling approach emits at most one active message at a time (after a call to the `step` operation).

Proof sketch. The `step` operator takes an `ActorConfig` together with a `Scheduler` as its argument. By construction an `ActorConfig` does not contain any message and all messages are in the scheduler. `step` emits one addressed active message into the toplevel. If the message's address is the toplevel the message is then "unwrapped", i.e., converted into an active message; otherwise it is pushed down to the level it addresses and is "unwrapped" there. As only one addressed active message is emitted, the scheduling approach emits at most one active message after a call to the `step` operation.   □

Theorem 1. Let SM be a scheduled composite actor specification. If SM satisfies the requirements for the scheduling approach for scheduled composite actor models (see Sect. 3.1), then for any one-step computation $u \xrightarrow{step} v \to w$ of SM, there is no unquantified nondeterminism possible; however, there may be probabilistic choices in the application of an actor rewrite rule in v.

Proof sketch. We prove the theorem by reductio ad absurdum. Assume there exists unquantified nondeterminism in a one-step computation of SM. As a `step` operation is possible for u, the configuration in a state u has to be of sort `ActorConfig`, which means that it contains no active messages. As all the rewrite rules require an active message, no unquantified nondeterminism is possible in u. w is a state after a rewrite triggered by an active message. However, in a scheduled composite actor model specification all rewrite rules may only produce scheduled messages and thus no active messages. Since no active messages are present in the configuration of a state w, no rewrites are possible and thus no unquantified nondeterminism is possible in w. Thus there has to be unquantified nondeterminism in a state v. Because of the well-formedness of the original specification an actor can only react to an active message with rewrite rules that fulfill the unquantified nondeterminism requirement, there have to be at least two active messages in the configuration of a state v to get unquantified nondeterminism. Since the `step` operation works on terms of sort `ActorConfig` in which there are no active messages, this means that the scheduling approach has to emit more than one active message. Lemma 2 however states that the scheduling approach emits at most one active message after a call to the step operation. Thus there cannot be two active messages in the configuration of a state u and as a consequence, no unquantified nondeterminism is possible in u. As there is no unquantified nondeterminism possible in u, v, and w, there is no unquantified nondeterminism possible for any one-step computation $u \xrightarrow{step} v \rightarrow w$ of SM. □

3.5   Soundness of the Model Transformation

In this section we analyse the soundness of the model transformation. In particular, we show that SM is a simulation of M for any model transformation $M \rightarrow SM$ where M is well-formed. Correctness needs the additional assumption of strong rule confluence: if M is well-formed and strongly rule confluent, then M and SM are time-passing bisimilar.

First, we introduce some definitions for (timed) probabilistic labelled transition systems and recall the notions of simulation and bisimilation [18].

Definition (Timed Probabilistic Labelled Transition System). A timed probabilistic (tp) labelled transition system $\mathcal{A} = (A, L, \mu, \xrightarrow{l,t})$ consists of a set of states A, a family of probabilistic distributions $\mu : A \times L \rightarrow [0, 1]$ and a labelled transition relation $\xrightarrow{l,t} \subseteq A \times A$ where $l \in L$ and $t \in \mathbb{R}_{\geq 0}$. We write $u \xrightarrow{l,t} v$ for a transition from u to v where the label l indicates which rule r of the rewriting specification is applied to u, $\mu(u, l_r)$ is the probability, and t is the time delay, i.e., the difference of global time between u and v.

In the following, the specifications M and SM form the labelled transition systems $\mathcal{M} = (Can(M), L_M, \mu_M, \xrightarrow{L_M, t}_M)$ of M and $\mathcal{SM} = (Can(SM), L_{SM},$ $\mu_{SM}, \xrightarrow{L_{SM}, t}_{SM})$ of SM, where $L_{SM} = L_M$ is the set of rules in M, μ_M is a family of distribution functions, μ_{SM} is a family of distribution functions which coincides with μ_M on the image of the model transformation, and $Can(M)$ and $Can(SM)$ are the canonical terms of sort config of M and SM, respectively. The labelled transition systems $\mathcal{M}$ and $\mathcal{SM}$ describe the set of all well-timed one-step rewrites of the probabilistic rewrite theory of M and SM, respectively.

Definition (Simulation and Bisimulation). Given timed probabilistic (tp) labelled transition systems $\mathcal{A} = (A, L_A, \mu_A, \xrightarrow{l,t}_A)$ and $\mathcal{B} = (B, L_B, \mu_B, \xrightarrow{l,t}_B)$ and a bijection of the labels $L_A \leftrightarrow L_B$, where $\mu_A : A \times L_A \rightarrow [0, 1]$ and $\mu_B : B \times L_B \rightarrow [0, 1]$ are families of probability distributions. Then a simulation of tp-transition systems is a binary relation $H \subseteq A \times B$ such that if $a \xrightarrow{l_A, t}_A a'$ and aHb then there is b' such that $b \xrightarrow{l_B, t}_B b'$ and $a'Hb'$, $\mu_A(a, l_A) = \mu_B(b, l_B)$, and $l_A \leftrightarrow l_B$. If both H and H^{-1} are simulations, then we call H a bisimulation.

Remark (Relation between M and SM). We relate the scheduled composite actor specification SM with the composite actor specification M as follows: For any canonical term u of SM in form AC {t | SL} one can construct a term $\hat{u}$ of M by inserting all messages of SL into the right subconfiguration of AC: for each term [A , [(at1, A1 <- M)]] of sort LocScheduleMsg of SL, the message (at1, A1 <- M) is inserted into the subconfiguration AC at adress A (or into the toplevel configuration of AC if A = toplevel).

Theorem 2. Let M be a well-formed composite actor model specification and SM the scheduled composite actor model specification that is the result of the model transformation $M \to SM$. Then $\mathcal{SM}$ is a simulation of $\mathcal{M}$.

Proof sketch. $H : \mathcal{SM} \to \mathcal{M}$ is a simulation where $H = \{(u, \hat{u})|u = $ AC {gt | SL} for some AC, gt, SL}. $\qquad\qquad\qquad\qquad\qquad\qquad\qquad\qquad\qquad\qquad\qquad\square$

To show the existence of a time-passing bisimulation between $\mathcal{M}$ and $\mathcal{SM}$ we need some further definitions:

Definition (Delay and Zero-Time Transition). We distinguish between delay transitions $a \xrightarrow{l,t} a'$ with $t > 0$, which indicate the passing of time, and zero-time transitions $a \xrightarrow{l,0} a'$ which are executed without taking any time. By $A_{>0} = \{a \in A|$ there is no a', l such that $a \xrightarrow{l,0} a'\}$ we denote the set of all terms to which no zero-time transition is applicable.

Definition (Complete Zero-Time *-Transition). A zero-time *-transition $a \Rightarrow_L a'$ consists of a finite sequence of zero-time transitions; it is complete if it cannot be extended further:

For $n \in \mathbb{N}, L : \{1, \ldots, n\} \to L_{\mathcal{A}}$ we call $a \Rightarrow_L a'$ a zero-time transition if and only if $\exists a_1, \ldots, a_n$ such that $a \xrightarrow{L(1),0} a_1 \to \ldots \to a_{n-1} \xrightarrow{L(n),0} a', a_n = a'$. It is called complete if $a' \in A_{>0}$.

If $\pi : \{1, \ldots, n\} \to \{1, \ldots, n\}$ is a permutation, we write $a \Rightarrow_{\pi(L)} a'$ for the zero-time *-transition $a \Rightarrow_{L'} a'$, where $L'(i) = L(\pi(i))$ for $i = 1, \ldots, n$.

Definition (Time-passing Simulation). Given tp-labelled transition systems $\mathcal{A} = (A, L_{\mathcal{A}}, \mu_{\mathcal{A}}, \xrightarrow{l,t}_{\mathcal{A}})$ and $\mathcal{B} = (B, L_{\mathcal{B}}, \mu_{\mathcal{B}}, \xrightarrow{l,t}_{\mathcal{B}})$ and a bijection $L_{\mathcal{A}} \leftrightarrow L_{\mathcal{B}}$. Then $H \subseteq A \times B$ is called a time-passing simulation if $\forall a, b, l_{\mathcal{A}}, t > 0, a', L, a''$:

$$aHb \wedge a \xrightarrow{l_{\mathcal{A}},t}_{\mathcal{A}} a' \wedge a' \Rightarrow_L a'' \text{ complete}$$
$$\Rightarrow \exists \text{ a permutation } \pi : \{1, \ldots, n\} \to \{1, \ldots, n\}, b', b'', l_{\mathcal{B}}, \text{ s.t.}$$
$$b \xrightarrow{l_{\mathcal{B}},t}_{\mathcal{B}} b' \wedge b' \Rightarrow_{\pi(L)} b'' \wedge a'' H b'' \wedge l_{\mathcal{A}} \leftrightarrow l_{\mathcal{B}}$$

Corollary. Every simulation (of tp transition systems) is a time-passing simulation.

Definition (Strong Zero-Time Rule Confluence). An actor specification is called strongly zero-time rule confluent if the order of applying any two zero-time actor rules to the same term does not matter:

A specification M is strongly zero-time rule confluent, if for any two zero-time rules r_1, r_2 with labels l_{r_1}, l_{r_2} and for all canonical terms u, v, w of sort Config: $u \xrightarrow{l_{r_1},0} v$ and $u \xrightarrow{l_{r_2},0} w$ implies that there exist v', w' such that $v \xrightarrow{l_{r_2},0} v'$ and $w \xrightarrow{l_{r_1},0} w'$ and $v' = w'$.

Many composite actor specifications satisfy the condition of strong zero-time rule confluence; in particular, except for one [12], all our published case studies studies [21,20,11] are strongly zero-time rule confluent.

Lemma 2. Let M be a well-formed and strongly zero-time rule confluent composite actor specification. For any zero-time *-transition $v_0 \Rightarrow_L v_n$ of M of form $v_0 \xrightarrow{L(1),0} v_1 \ldots v_{n-1} \xrightarrow{L(n),0} v_n$ and any rule r which is applicable with label l to a message m in v_0, the following holds:

(1.) If r is not applied in any of the v_i then m occurs in all v_i and r stays applicable with label l in all $v_i, i = 0 \ldots n$.

(2.) If M is strongly zero-time rule confluent and $v_{n-1} \xrightarrow{L(n),0} v_n$ is the first rule application of r with label $L(n) = l$, then for appropriate v'_i , $v_0 \xrightarrow{L(n),0} v'_1 \xrightarrow{L(1),0} v'_2 \ldots v'_{n-1} \xrightarrow{L(n-1),0} v'_n$ and $v_n = v'_n$.

Proof. Proof by induction on the length n of the transition. □

Lemma 3. Let M be well-formed and strongly zero-time rule confluent and SM be the result of the model transformation $M \to SM$. For all canonical terms v_0 of sort `Config` of M , w_0 of SM with $\hat{w}_0 = v_0$, and any zero-time *-transition $v_0 \Rightarrow_L v'$ with $v' \in M_{>0}$ there exist $w' \in SM_{>0}$ and a permutation π such that $w_0 \Rightarrow_{\pi(L)} w'$ and $\hat{w}' = v$.

Proof. Proof by induction on the length of the transition using Lemma 2. □

Theorem 3. Let M be a well-formed and strongly zero-time rule confluent composite actor model specification and SM be the result of the model transformation $M \to SM$. Then M and SM are time-passing bisimilar.

Proof. Consider the relation H as in Theorem 2 and let $H_{>0} \subseteq SM_{>0} \times M_{>0}$. □

Remark. One can also solve the general case where interdependent messages have the same activation time and the specification is not zero-time rule commutative by assigning random delays to the scheduled activation times of the dependent messages. As in [4], we assume that the probability that a random number is sampled twice is 0. If this process is recursively applied to the dependent messages in the scheduler, a fix point is reached in which there exist no two dependent messages with the same scheduled activation time.

3.6 Example: Scheduling Approach for the Composite Actor Model

As an example of how the scheduling approach works, we model a forest of binary trees, where the leaf nodes send messages to each other while the intermediate nodes only forward and delay these messages. The leaf nodes in the forest are of type `Leaf` and the intermediate nodes of type `Node`. Message contents are created by the operator `cnt : -> Contents`. The following listing shows the specification of the behavior of the intermediate and leaf nodes, where the original composite actor specification M is shown on the left side and the modified rules of the specification SM after the transformation on the right side:

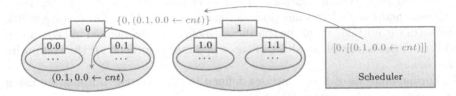

Fig. 2. The term in canonical form after one step

```
crl [intermediate-boundary-up1] :              crl [intermediate-boundary-up1] :
  <A : Node |config: (t, A'<-cnt) C> =>          <A : Node |config: (t, A'<-cnt) C> =>
  <A : Node |config: C> (t+0.1, A'<-cnt)         <A : Node |config: C> [(t+0.1, A'<-cnt)]
  if |A'| <= |A| .                               if |A'| <= |A| .
crl [intermediate-boundary-up2] :              crl [intermediate-boundary-up2] :
  <A : Node |config: (t, A'<-cnt) C> =>          <A : Node |config: (t, A'<-cnt) C> =>
  <A : Node |config: C> (t+0.1, A'<-cnt)         <A : Node |config: C> [(t+0.1, A'<-cnt)]
  if |A'| > |A| /\ pref(A', |A|) =/= A .         if |A'| > |A| /\ pref(A', |A|) =/= A .
rl [intermediate-boundary-down] :              rl [intermediate-boundary-down] :
  <A : Node |config: C> (t, A.A'<-cnt) =>        <A : Node |config: C> (t, A.A'<-cnt) =>
  <A : Node |config: C (t+0.1, A.A'<-cnt)> .     <A : Node |config:C [(t+0.1,A.A'<-cnt)]> .
rl [leaf-receive-and-send] :                   rl [leaf-receive-and-send] :
  <A : Leaf |nil> (t, A<-cnt) =>                 <A : Leaf |nil> (t, A<-cnt) =>
  <A : Leaf |nil> (t+0.1, rndA(|A|)<-cnt) .      <A : Leaf |nil> [(t+0.1, rndA(|A|)<-cnt)] .
```

In both specifications, A and A' are variables of sort Address, t of sort Float, C of sort Config, pref: Address Nat -> Address an operator returning a prefix of a given length, |_| : Address -> Nat an operator returning the length of a given address, and rndA: Nat -> Address an operator returning a random Address of a given length for a binary tree.

The initial configuration consists of two actor trees and the top-level scheduler which contains one message. The following listing shows this configuration, where the original one (M) is shown on the left, and SM on the right.

```
                                            <0: Node | config:
<0: Node | config: (0.1, 0.0 <- cnt)          <0.0 : Leaf | nil> <0.1 : Leaf | nil>>
  <0.0 : Leaf | nil> <0.1 : Leaf | nil>>    <1: Node | config:
<1: Node | config:                            <1.0 : Leaf | nil> <1.1 : Leaf | nil>>
  <1.0 : Leaf | nil> <1.1 : Leaf | nil>>    {0.0 | [0, [(0.1, 0.0 <- cnt)]]}
```

Figure 2 illustrates the resulting term in canonical form after one step, where the term in canonical form is depicted in solid black and in addition, the intermediate messages in gray. The arrows show the corresponding equational simplifications.

4   Statistical Model Checking Methodology

We propose the following methodology to verify hierarchically structured distributed systems:

(1.) Specification of the real-world system as a composite actor system in Maude/PMaude using the composite actor model as a foundation and fulfilling the formal requirements for the scheduling approach for composite actor models.

(2.) Definition of standard probabilistic temporal logic and quantitative temporal logic properties for the system.

(3.) (Automated) transformation of the model specified in (1.) to a scheduled composite actor model specification and specification of an initial state which consists of the initial state of the specification of the model defined in (1.) and an instance of the top-level scheduler of the scheduling approach for the composite actor model.

(4.) Formal analysis of the properties defined in (2.) over the initial state using the statistical model checker PVESTA.

Besides the formal requirements, we further require that the specification does not contain zero-time loops, i.e., a recurring series of messages is produced where each message has the same activation time.

More technically the statistical model checker PVESTA calls the operator run to start the execution of a sample of a composite actor model. The operator is repeatedly called by PVESTA until a specified amount (denoted by the variable LIMIT) of global time has passed. In Maude, the run operator is defined as follows:

```
op run : Config Float -> Config .
eq run(AC {gt | SL}, LIMIT) =
  if (gt <= LIMIT) then run(step(AC {gt | SL}), LIMIT) else AC {gt | SL} fi .
```

Several case studies have been conducted to validate this methodology. For example, the approach has been used to detect bugs in the design of a group key distribution service [21,11], to predict the performance of a distributed broker-based Publish/Subscribe service [21,20], and to improve the availability of service of Internet-based service architectures such as Clouds using a denial of service protection mechanism together with dynamic resource provisioning [12].

From a more practical point of view, the statistical model checking methodology for hierarchical specifications based on the composite actor model has proven itself effective in our case studies. Compared to LTL model checking, we were able to model check meaningful qualitative as well as quantitative properties of rather large instances of the specifications. E.g., during the work on [12], we model checked quantitative properties with a high confidence of 99% on specification instances with up to 500 individual actors. Thereby, the statistical model checking process was running for several hours on a cluster of 32 machines. Model checking just simple qualitative properties with LTL model checking on the same specification instances would have exceeded a graspable timeframe.

5   Related Work and Concluding Remarks

This work is mainly related to the ideas of actors [16,15,2], PMaude [4], the statistical model checker PVESTA [5], and the original scheduling approach for flat actor configurations mentioned in [6]. Only recently Ölveczky et al. [8] proposed a probabilistic strategy language for probabilistic rewrite theories that is implemented in Maude and offers the possibility of statistical model checking with a model checking algorithm implemented as a Maude meta-level functionality. Bruni et al. [9] have shown that a framework to describe adaptive behavior in multi-layered component hierarchies can naturally be realized in Maude based on the Reflective Russian Dolls Model and quantitatively analyzed using PVESTA.

In a broader sense this work is related to process calculi such as the Mobile Ambients calculus and its probabilistic extension [10,17]. While process calculi mainly focus on dynamic behavior, our approach also emphasizes the representation of data.

In this paper we have presented the composite actor model and argued that it is well-suited for specifying concurrent Cloud and Internet-based systems which are composed of various participants and subsystems and are hierarchically structured in different layers and networks. Our model extends the actor model of concurrent computation and supports an arbitrary hierarchical composition of entities. As a second main result we have defined a model transformation which extends a composite actor specification by a new scheduling approach that guarantees the absence of nondeterminism and, as a consequence, enables statistical model checking. To show the soundness of our approach we have proven termination and confluence of the (equational part of the) scheduler specification and shown the absence of unquantified nondeterminism in scheduled composite actor specifications. To prove the soundness of the model transformation we introduced the notions of strong zero-time rule confluence and time-passing bisimulation and showed that the transformation is a time-passing bisimulation for strongly zero-time rule confluent composite actor specifications.

Until today we have successfully applied the composite actor approach to three non-trivial Cloud case studies [12,21,20,11] by formally specifying and analyzing them using Maude and PVESTA.

In the future we plan to extend and complement our composite actor modeling approach by a correct-by-construction model-driven program synthesis. As a target system we choose the ØMQ (zeromq) socket library [1] to act as a concurrent framework for Cloud services. Currently, we have partially implemented one of our case studies [12] as a ØMQ application based on the Maude model and shown that large parts of the transformation process can be automated.

References

1. ØMQ: The Intelligent Transport Layer (August 07, 2012), http://www.zeromq.org/
2. Agha, G.: Actors: a model of concurrent computation in distributed systems. MIT Press (1986)
3. Agha, G., Hewitt, C.: Concurrent programming using actors. In: Object-Oriented Concurrent Programming, pp. 37–53. MIT Press (1988)
4. Agha, G., Meseguer, J., Sen, K.: PMaude: Rewrite-based Specification Language for Probabilistic Object Systems. ENTCS 153(2), 213–239 (2006)
5. AlTurki, M., Meseguer, J.: PVESTA: A parallel statistical model checking and quantitative analysis tool. In: Corradini, A., Klin, B., Cîrstea, C. (eds.) CALCO 2011. LNCS, vol. 6859, pp. 386–392. Springer, Heidelberg (2011)
6. AlTurki, M., Meseguer, J., Gunter, C.A.: Probabilistic Modeling and Analysis of DoS Protection for the ASV Protocol. ENTCS 234, 3–18 (2009)
7. Armstrong, J., Virding, R., Wikström, C., Williams, M.: Concurrent Programming in Erlang. Prentice Hall (1996)
8. Bentea, L., Ölveczky, P.C.: Probabilistic real-time rewrite theories and their expressive power. In: Fahrenberg, U., Tripakis, S. (eds.) FORMATS 2011. LNCS, vol. 6919, pp. 60–79. Springer, Heidelberg (2011)

9. Bruni, R., Corradini, A., Gadducci, F., Lluch Lafuente, A., Vandin, A.: Modelling and Analyzing Adaptive Self-assembly Strategies with Maude. In: Durán, F. (ed.) WRLA 2012. LNCS, vol. 7571, pp. 118–138. Springer, Heidelberg (2012)
10. Cardelli, L., Gordon, A.D.: Mobile ambients. In: Nivat, M. (ed.) FOSSACS 1998. LNCS, vol. 1378, pp. 140–155. Springer, Heidelberg (1998)
11. Eckhardt, J.: A Formal Analysis of Security Properties in Cloud Computing. Master's thesis, LMU Munich, TU Munich (2011)
12. Eckhardt, J., Mühlbauer, T., AlTurki, M., Meseguer, J., Wirsing, M.: Stable Availability under Denial of Service Attacks through Formal Patterns. In: de Lara, J., Zisman, A. (eds.) FASE 2012. LNCS, vol. 7212, pp. 78–93. Springer, Heidelberg (2012)
13. Eker, S., Meseguer, J., Sridharanarayanan, A.: The Maude LTL model checker. In: WRLA. ENTCS, vol. 71, pp. 162–187 (2002)
14. Haller, P., Sommers, F.: Actors in Scala. Artima Developer (2012)
15. Hewitt, C., Baker, H.G.: Laws for communicating parallel processes. In: IFIP Congress, pp. 987–992 (1977)
16. Hewitt, C., Bishop, P., Steiger, R.: A universal modular actor formalism for artificial intelligence. In: IJCAI, pp. 235–245 (1973)
17. Kwiatkowska, M., Norman, G., Parker, D., Vigliotti, M.G.: Probabilistic Mobile Ambients. TCS 410(12-13), 1272–1303 (2009)
18. Larsen, K.G., Skou, A.: Bisimulation through Probabilistic Testing. Inf. Comput. 94(1), 1–28 (1991)
19. Meseguer, J., Talcott, C.: Semantic Models for Distributed Object Reflection. In: Magnusson, B. (ed.) ECOOP 2002. LNCS, vol. 2374, pp. 1–36. Springer, Heidelberg (2002)
20. Mühlbauer, T.: Formal Specification and Analysis of Cloud Computing Management. Master's thesis, LMU Munich, TU Munich (2011)
21. Wirsing, M., Eckhardt, J., Mühlbauer, T., Meseguer, J.: Design and Analysis of Cloud-Based Architectures with KLAIM and Maude. In: Durán, F. (ed.) WRLA 2012. LNCS, vol. 7571, pp. 54–82. Springer, Heidelberg (2012)

Barbed Semantics for Open Reactive Systems*

Fabio Gadducci and Giacoma Valentina Monreale

Department of Informatics, University of Pisa, Italy

Abstract. Reactive systems (RSs) represent a meta-framework aimed at deriving labelled transition systems from unlabelled ones such that the induced bisimilarity is a congruence. Such a property is desirable, since it allows one to replace a subsystem with an equivalent one without changing the behaviour of the overall system. One of the main drawback of RSs is the restriction to the analysis of ground (i.e., completely specified) systems. Only recently the theory was extended to consider open systems (and rules) and an associated strong bisimulation equivalence. However, the resulting bisimilarity adopted for the formalism turns out to be a congruence only under very restrictive conditions, hindering the applicability of the framework. In this paper we suggest to consider (strong and weak) barbed equivalence as an alternative for open RSs. After proving that it is always a congruence, we instantiate our proposal by addressing the semantics of Asynchronous CCS and of Mobile Ambients.

Keywords: Open reactive systems, (saturated) barbed bisimilarities.

1 Introduction

The most commom technique for specifying the dynamics of a system is based on a reduction semantics: a set representing the states of the system, plus an unlabelled relation among these states, usually built out of a finite set of rules, denoting the evolutions. Despite the advantage of conveying the behaviour with relatively few rules, the main drawback of reduction semantics is that the dynamics of a system is described in a monolithic way, meaning that the evolution is obtained by closing the set of rules under all contexts. Thus, a subsystem can be interpreted only by inserting it inside the contexts where a reduction may take place. To make the analysis simpler, it is often necessary to consider descriptions accounting separately for the behaviour of each single sub-component, thus increasing modularity and enhancing the opportunities for verification.

The theory of reactive systems (RSs) [17] offers a framework that allows to distill labelled transition systems (LTSs), hence to obtain behavioural equivalences, for formalisms based on reduction semantics. The idea is simple: a system p has a labelled transition $p \xrightarrow{c} p'$ if the system obtained by inserting p inside the (unary) minimal context c may reduce to p'. The notion of "minimal" context is expressed via the categorical notion of relative pushout (RPO), which ensures

---

* Research partially supported by the EU FP7-ICT IP ASCEns (IP 257414).

N. Martí-Oliet and M. Palomino (Eds.): WADT 2012, LNCS 7841, pp. 161–177, 2013.

that bisimilarity is a congruence when enough RPOs exist. A grupoidal extension (GRSs) has been proposed [24], in order to deal with equationally specified systems (e.g. processes in a calculus with structural congruence).

Should all the possible contexts allowing a reduction be admitted as labels, the resulting bisimilarity would in general be coarser. Despite being better suited in some case studies, such an equivalence (called saturated bisimilarity) is usually intractable, since it has to tackle a potentially infinite set of contexts. The problem has been addressed in [6] by introducing an "efficient" characterisation of these semantics, where one avoids considering all contexts by using in a cunning way RPOs, at the price of modifying the standard, symmetric presentation of the (either strong or weak) bisimulation equivalences.

Despite their wide range of applicability, saturated and "minimal contexts" bisimilarities still fail to cover all the interesting possibilities: the former is sometimes too fine-grained, while the latter is too coarse. As for process calculi, the standard way out of the impasse is to consider barbs [20] (that is, predicates on the states of a system) and barbed equivalences (where the check of such predicates is added to the bisimulation game). The flexibility of the definition allows for recasting a variety of observational, bisimulation-based equivalences. Therefore, a suitable notion of barbed saturated semantics for RSs is introduced in [4], and it is further investigated in [3,5], where an efficient characterisation for both the strong and weak case is presented, avoiding to consider all contexts.

Even if these extensions witness the vitality of the formalism, one of the fiercest limits suffered by RSs was the restriction to ground rules for describing the dynamics of a system. We are aware of two different attempts to deal with parametric rules in the (G)RS framework. In [10], the notion of a category of second-order contexts is introduced and the RPO construction is carried out in this setting, to capture the lazy observational equivalence of λ-calculus. In [16], a general theory for open (G)RSs is developed, by considering open terms and rules. So, now the transitions are labelled not only with the minimal context but also with the most general instantiation allowing a reduction. More explicitly, $p \xrightarrow[x]{c} p'$ if p inserted into the context c and instantiated with the (possibly open) term x may evolve into a state p'. The notions of minimal context and most general instantiation are captured at once by the notion of *(G-)lux*. The (strong) bisimilarity induced by the synthesised LTS for open terms is a congruence only under very restrictive conditions, hindering the applicability of the framework. It indeed seems that such conditions are satisfied by simple process calculi with a trivial structural congruence, but that this is not always so for richer calculi.

This paper, along the lines of [4], proposes a suitable notion of barbed saturated bisimilarity for open GRSs: this is by definition guaranteed to be a congruence and, under certain conditions, it can alternatively be described in a more efficient way via the G-lux LTS and the semi-saturated game. We present the study for both the strong and the weak equivalence and, as it is usual for any newly proposed technique, we test their adequacy against suitable case studies. We thus consider the asynchronous CCS [1], whose simplicity allows us to use it as running example, and MAs, whose more complex behaviour highlights the usefulness of open semantics. In both cases, we show that the closed version of our semantics exactly captures the standard ones of the calculus.

The paper is organised as follows. §2 recalls the notion of 2-category on which the one of GRS is based, and shows how it can be used to model the syntax of a calculus. §3 introduces open GRSs and the definition of G-lux LTS. The technical core of the paper is presented in §4, where both the strong and weak barbed saturated semantics for open GRSs are studied, and labelled characterisations of them by means of their semi-saturated counterparts are offered. §5 shows an application of the framework to MAs and §6 concludes the paper.

The appendix recalls the categorical notions of GRPO and GRPB (needed in §3), as well as the proof of our main result.

2   A Categorical Representation of Process Calculi

This section briefly introduces a categorical representation of a process calculus, previously introduced in [16], and based on the notion of 2-category [8,15].

Definition 1 (2-category). *A 2-category* **C** *is a category such that, given any two objects a and b, the hom-set of a and b (the collections of arrows between them) is the class of objects of some category* **C**(a, b) *and, correspondingly, whose composition functions* $* : \mathbf{C}(a, b) \times \mathbf{C}(b, c) \to \mathbf{C}(a, c)$ *are functors.*

Arrows of **C**(a, b) are called 2-cells and are denoted by $\alpha : f \Rightarrow g : a \to b$, the composition in **C**(a, b) is instead denoted by $\bullet$.

Definition 2 (G-category). *A groupoidal category (G-category) is a 2-category whose 2-cells are invertible.*

G-categories can be used to model the syntax of process calculi. We will show an example of this by exploiting a G-category based on the PROP-category [18].

Definition 3 (PROP-category). *A product and permutation category (PROP-category)* **C** *has natural numbers $0, 1, 2, \ldots$ as objects and it is equipped with two further structures, as follows*

- *for each n, the group of permutations of n elements, $S(n)$, is a subgroup of all the invertible elements of the hom-set* **C**(n, n). *The identity permutation is the identity morphism $1_n : n \to n$;*
- *there is a functor $\otimes : \mathbf{C} \times \mathbf{C} \to \mathbf{C}$, called* product *and written between its arguments, which acts as addition on the objects, i.e., $m \otimes n = m + n$, and moreover it is required that*
  1. *it is associative: $(f \otimes f') \otimes f'' = f \otimes (f' \otimes f'')$;*
  2. *given $\sigma \in S(n)$ and $\sigma' \in S(n')$, we have $\sigma \otimes \sigma' = \sigma \times \sigma' : n + n' \to n + n'$, where $\times$ denotes the product of permutations;*
  3. *for any two natural numbers n, n', let $\gamma_{n,n'}$ be that permutation in $S(n + n')$ which interchanges the first block of n and the second block of n', $\gamma_{n,n'} : n + n' \to n + n'$. For any maps $f : m \to n$ and $f' : m' \to n'$ we require that $\gamma_{n,n'}(f \otimes f') = (f' \otimes f)\gamma_{m,m'}$.*

$$P ::= \epsilon, a.P, \bar{a}, P \mid P \qquad (P|Q)|R \equiv P|(Q|R) \qquad P|Q \equiv Q|P \qquad a.P \mid \bar{a} \rightsquigarrow P \qquad \dfrac{P \rightsquigarrow Q}{P \mid R \rightsquigarrow Q \mid R}$$

Fig. 1. Syntax, structural congruence and reduction relation of ACCS

A G-PROP is a PROP where the underlying category is a G-category.

Example 1. We now show how the signature of a calculus (modulo term equations) may induce a G-PROP. Inspired by the characterisation of (a fragment of) CCS in [16], we model here the finite, restriction and summation free fragment of the asynchronous version of the calculus [1]. Later on, we will use the same approach for modelling our main case study, i.e., Mobile Ambients (MAs) [9].

The syntax of ACCS is shown on the left of Fig. 1. We assume a set $\mathcal{N}$ of names ranged over by $a, b, c, \ldots$. Also, we let $P, Q, R \ldots$ range over the set of processes. The processes are considered up to the structural congruence $\equiv$ induced by the two axioms in the middle of Fig. 1. To keep the example simple, there is no structural rule guaranteeing that ϵ is the identity for the parallel composition. The transition relation $\rightsquigarrow$ is defined by the rightmost rules of the same figure: it is going to be used just in the forthcoming sections, even if it is introduced here for the sake of presentation.

The signature for ACCS is $\Sigma = \epsilon : 0, a. : 1, \bar{a} : 0, | : 2$. The arrows $p : m \to n$ of G-PROP PA2CP represent n-tuples of terms over Σ quotiented by associativity (the leftmost axiom in the middle of Fig. 1) that altogether contain m distinct holes. Permutations in (n, n) are tuples built from holes, $\otimes$ acts on arrows as tuple juxtaposition, and arrows composition is the standard term composition.

To define the 2-cells of the category, an explicit representation of the arrows of the category is used. A term can be indeed represented as a finite, ordered tree with nodes of any degree, where an immediate child of a node of degree higher than 1 must have degree at most 1. Leaves of such a tree correspond to occurrences either of outputs or of the constants ϵ, so they are labelled with names belonging to $\mathcal{N}$ or with ϵ; nodes of degree 1 correspond to applications of prefix operators and they are indeed labelled with names belonging to $\mathcal{N}$; and nodes of higher degree correspond to term fragments built solely of the parallel operator. So, arrows can be represented as tuples of these trees.

A 2-cell from p to q intuitively says that they are equivalent with respect to the commutative axiom. So, it is defined as a family, indexed by the nodes of (the explicit representation of) p, of permutations on the sets of their immediate children, such that the application of all these permutations to p yields q.

3   Open G-Reactive Systems

This section presents an extension, previously proposed in [16], of the theory of *G-reactive systems* (GRSs) [24]. In general, the aim of this theory is to derive labelled transition systems (LTSs) and bisimulation congruences for those specification formalisms whose operational semantics is provided by reduction rules.

In [24], the technique of LTS derivation is defined for closed GRSs, that is, by considering closed terms and ground reduction rules, while in [16], open terms and parametric rules are considered. The idea is quite simple: a system specified by an open term p has a labelled transition $p \xrightarrow[x]{c} p'$ if p instantiated with the (open) term x and inserted into the context c may evolve into a state p'.

A G-category $\mathbf{C}$ models the syntax of a formalism. An *(open) system* is an arrow $p : a_1 \to a_2$: it can be plugged into $q : a_2 \to a_3$ via arrows composition. Given arrows $p, q : a_1 \to a_2$, a 2-cell $\alpha : p \Rightarrow q$ represents an isomorphism (i.e., a proof of equivalence) between systems p and q. The semantics is given via *reduction rules*: pairs of systems $\langle l, r \rangle$ with the same interfaces.

Definition 4 (Open GRS). *An* open G-reactive system $\mathbb{C}$ *consists of*

1. *a G-category* $\mathbf{C}$;
2. *a composition-reflecting, 2-cell closed subcategory* $\mathbf{D}$ *of reactive contexts;*
3. *a set* $\mathfrak{R} \subseteq \bigcup_{a_1, a_2 \in |\mathbf{C}|} \mathbf{C}(a_1, a_2) \times \mathbf{C}(a_1, a_2)$ *of reduction rules.*

Intuitively, reactive contexts are those arrows inside which a reduction can occur. By 2-cell closed we mean that $d \in \mathbf{D}$ and $\alpha : d \Rightarrow d'$ in $\mathbf{C}$ implies $d' \in \mathbf{D}$, while by composition-reflecting we mean that $d'; d \in \mathbf{D}$ implies $d, d' \in \mathbf{D}$.

Given an open GRS $\mathbb{C}$, the reduction relation over the terms of $\mathbb{C}$ is generated by closing the reduction rules under all reactive contexts, instantiations and 2-cells. Formally, the *reduction relation* is defined by taking $p \rightsquigarrow p'$ if there exist $\langle l, r \rangle \in \mathfrak{R}$, $d \in \mathbf{D}$, $x \in \mathbf{C}$, $\alpha : p \Rightarrow x; l; d$ and $\alpha' : p' \Rightarrow x; r; d$.

Example 2. Consider the G-PROP PA2CP category shown in §2. An open GRS $\mathbb{C}_{ACCS}$ over it takes as reduction rules the set $\bigcup_{a \in \mathcal{N}} \{\langle a.1 \mid \bar{a}, 1 \rangle\}$, and as the subcategory of reactive contexts the smallest composition-reflecting, 2-cell closed subcategory including arrows of the shape $1 \mid p : 1 \to 1$. Note that it contains no context with a hole after the input operator.

The behaviour of an open GRS is given by an unlabelled transition system. To obtain a labelled one, we instantiate an open system p with a sub-term x, plug the result into a context c and observe if a reduction occurs. Categorically, it means that $x; p; c$ is isomorphic to $y; l; d$ (there exists $\alpha : x; p; c \Rightarrow y; l; d$) for an instantiation x, a rule $\langle l, r \rangle$, and a reactive context d, as depicted by diagram (i) of Fig. 2. We shall refer to this kind of diagrams as hexagons.

Clearly, the resulting LTS is often infinite-branching, since all contexts and all instantiations allowing reductions may occur as labels. Moreover, it has redundant transitions: the ACCS open process $a.1$ would have both transitions $a.1 \xrightarrow[1]{1|\bar{a}} 1$ and $a.1 \xrightarrow[1|q]{1|\bar{a}|p} 1 \mid q \mid p$, yet neither p nor q "concur" to the reduction. In order to remove this kind of redundancy, we thus consider only "minimal contexts allowing a reduction" and the "most general instantiations", both modelled by the categorical notion of G-locally universal hexagons (G-luxes). The explicit definition of G-lux is found in [16], here we use a simpler characterisation of it in terms of the better known notions of groupoidal-idem pushouts (GIPOs) and

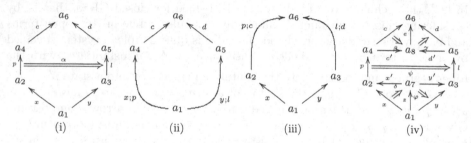

Fig. 2. Hexagon and G-lux

groupoidal-idem pullbacks (GIPBs) in G-categories (see [16, Lemma 10]). We refer to [24] for the notion of GIPO, the one of GIPB being symmetric. For the convenience of the reader we report both definitions in Appendix A.

Definition 5 (G-lux). *The hexagon (i) in Fig. 2 is a G-locally universal hexagon (G-lux) if (ii) is a GIPO and (iii) is a GIPB.*

We say that a GRS has G-luxes if in its underlying category every hexagon as diagram (i) in Fig. 2 has an inner G-lux hexagon as diagram (iv) such that $\alpha = \delta * 1_p * \beta \bullet 1_z * \psi * 1_e \bullet \varphi * 1_l * \gamma$.

Definition 6 (GLUX Transition System). *Let* $\mathbb{C}$ *be an open GRS and* **C** *its underlying G-category. The GLUX LTS (*LLTS$(\mathbb{C})$*) is defined as follows*

- *states:* $p : a_1 \to a_2$ *in* **C**, *for arbitrary* a_1 *and* a_2;
- *transitions:* $p \xrightarrow[x]{c} p'$ *if there exist* $d \in \mathbf{D}$, $y \in \mathbf{C}$, *rule* $\langle l, r \rangle \in \mathfrak{R}$ *and 2-cell* $\alpha : x; p; c \Rightarrow y; l; d$ *such that diagram (i) of Fig. 2 is a G-lux and* $p' = y; r; d$.

Example 3. As for the synchronous version [16], it can be shown that the GRS $\mathbb{C}_{ACCS}$ has G-luxes. The leftmost diagram of Fig. 3 shows the derived transition $a.\bar{a} \xrightarrow{1|\bar{a}} \bar{a}$ (the instantiation 0 is usually omitted). Here the initial state offers a closed process with an input on the channel a at top level, while the environment provides an output action on the same channel, so the communication on it can occur. Note instead that no instantiation is provided. The rightmost diagram of Fig. 3 shows instead the transition $a.\epsilon \mid 1 \xrightarrow[\bar{a}]{1} \epsilon$. Also in this case the initial state offers an input on a at top level but here it is in parallel with a hole, which is going to be replaced with the output action on the same channel provided by the instantiation. So, also in this case, the communication on the same channel can occur. Note however that here the environment does not offer anything.

In [16], (strong) bisimilarity on LLTS is referred to as GLUX-bisimilarity $\sim^L$: despite the flexibility of the G-luxes framework, $\sim^L$ is a congruence (i.e., if $p \sim^L q$, then $p; c \sim^L q; c, \forall c \in \mathbf{C}$) under some (restrictive) assumptions on **C**. In particular, a 2-categorical version of the requirement that all arrows of **C** are mono must be satisfied. These conditions hold for simple process calculi with a

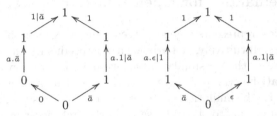

Fig. 3. G-luxes representing the transitions $a.\bar{a} \xrightarrow{1|\bar{a}} \bar{a}$ and $a.\epsilon \mid 1 \xrightarrow[\bar{a}]{1} \epsilon$

trivial structural congruence, such as the choice-free fragment of CCS (as proved in [16]) and thus apparently also by its asynchronous version considered here, even if it is doubtful for richer calculi such as Mobile Ambients.

Remark 1. The standard notion of GRS [24] is very similar to the one presented in Definition 4. One difference is that in the G-category **C** the existence of a chosen object 0 is required, which can never be the codomain of any arrow and which is used to denote the lack of holes. So, arrows having 0 as domain are deemed to represent closed terms. Building on this, the main difference is that the rules are ground, that is, they are pairs of closed terms, and also the states of the reduction relation and of the derived (GIPO) LTS are so. The reduction relation is thus defined by taking $p \rightsquigarrow p'$ if there exist $\langle l, r \rangle \in \mathfrak{R}$, $d \in \mathbf{D}$, $\alpha : p \Rightarrow l; d$ and $\alpha' : q \Rightarrow r; d$. In the LTS, instead, $p \xrightarrow{c} p'$ if for $d \in \mathbf{D}$ and $\langle l, r \rangle \in \mathfrak{R}$, the square p, c, l, d is a GIPO and $p' = r; d$.

It is easy to see that if we specialise the theory of open GRSs to consider closed terms and closed rules, it subsumes the standard theory of GRSs. Indeed, being both terms and rules ground, in the definition of the reduction relation the x will trivially be the identity on 0, so the definition coincides with the one for closed GRSs. Let us now look at the derived LTS. If we consider again closed states and rules, the definition of transition coincides: since both the instantiations x and y will be the identity on 0, the requirement that diagram (i) of Fig. 2 is a G-lux, by definition, means that the square p, c, l, d is a GIPO and the square $0, p; c, 0, l; d$ a GIPB. So, it is easy to see that $p \xrightarrow{c} p'$ is a GLUX transition if and only if it is also a GIPO one.

If instead open rules and closed are considered, then in the definition of the reduction relation the instantiation x of l can be different from 0, and an equivalent, closed GRS is obtained by instantiating all the open rules, i.e., if it contains the ground rule $\langle x; l, x; r \rangle$ for any x. Instead, as far as the definition of LTS is concerned, in diagram (i) of Fig. 2 the x has to be the identity on 0 (since p is ground). By definition, the hexagon is a G-lux if and only if $p, c, y; l, d$ is a GIPO and the square $0, p; c, y, l; d$ is a GIPB, therefore, it is easy to check that $p \xrightarrow{c} p'$ is a GLUX transition if and only if the corresponding closed GRS containing the ground rule $\langle y; l, y; r \rangle$ has the same GIPO transition.

4   Barbed Semantics for Open Reactive Systems

Barbed bisimulation represents a general technique for generating bisimulation-based equivalences. Intuitively, a barb is just a predicate on the states of a system, which detects the possibility of performing some observable action. We write $P \downarrow_o$ if P satisfies the predicate o. Barbed equivalences add the check of such predicates in the bisimulation game: every time that a system shows a barb, the equivalent system has to show the same barb, and vice-versa. For instance, ACCS barbs express the ability of a process to perform an output over a channel. Formally, for a closed process P, $P \downarrow_{\bar{a}}$ if $P \equiv \bar{a} \mid Q$, for some Q. In [1] this notion is used to give the definition of barbed equivalence for the calculus.

In general, the advantage of this kind of semantics is that the flexibility of the definition allows for recasting a wide variety of observational, bisimulation-based equivalences. For example, in [1], the authors prove that the standard bisimilarity for ACCS, (weak) asynchronous bisimilarity, coincides with (weak) barbed equivalence. In [4], instead, a suitable notion of (weak) barbed semantics for closed RSs is introduced, showing that it can be characterised via the LTS labelled with minimal contexts and the semi-saturated game, and that the proposal captures the behavioural semantics for MAs.

Along the lines of [4], we introduce suitable notions of strong and weak barbed saturated semantics for open GRSs and characterise them via the GLUX LTS.

4.1   Strong Open Barbed Saturated Semantics

As for closed systems, barbs for open ones are state predicates. In the following, we fix a 2-cells respected family O of barbs on open systems, and we write $p \downarrow_o$ if p satisfies $o \in O$. By 2-cells respected, we mean that $p \downarrow_o$ and $\alpha : p \Rightarrow q$ implies $q \downarrow_o$. For example, if we consider tuples of open terms, ACCS barbs can observe the presence of an output action at top level of a term.

Definition 7 (Open Barbed Saturated Bisimulation). *A symmetric relation $\mathcal{R}$ is an* open barbed saturated bisimulation *if whenever $p \mathcal{R} q$ then $\forall x, c$*

- *if $x; p; c \downarrow_o$ then $x; q; c \downarrow_o$;*
- *if $x; p; c \rightsquigarrow p'$ then $x; q; c \rightsquigarrow q'$ and $p' \mathcal{R} q'$.*

Open barbed saturated bisimilarity $\sim^{OBS}$ *is the largest such bisimulation.*

Note that $\sim^{OBS}$ is a congruence by definition. It can be efficiently characterised through the GLUX transition system via the semi-saturated game.

Definition 8 (Open Barbed Semi-Saturated Bisimulation). *A symmetric relation $\mathcal{R}$ is an* open barbed semi-saturated bisimulation *if whenever $p \mathcal{R} q$ then*

- *$\forall x, c$, if $x; p; c \downarrow_o$ then $x; q; c \downarrow_o$;*
- *if $p \xrightarrow[x]{c} p'$ then $x; q; c \rightsquigarrow q'$ and $p' \mathcal{R} q'$.*

Open barbed semi-saturated bisimilarity $\sim^{OBSS}$ *is the largest such bisimulation.*

Proposition 1. *In an open GRS having G-luxes,* $\sim^{OBSS}=\sim^{OBS}$.

As an immediate test of the adequacy of our framework, we apply it to ACCS. We know that the standard bisimilarity for ACCS coincides with the barbed equivalence, which in turn is equivalent to the open barbed saturated bisimilarity defined only over closed ACCS processes [11]. Now we can define the open barbed semi-saturated bisimilarity for ACCS by using the GLUX LTS for the calculus, and since the open GRS has G-luxes, thanks to Proposition 1, we can say that it coincides with the open barbed saturated bisimilarity. Therefore, if we consider only closed ACCS processes, we can conclude that barbed semi-saturated bisimilarity is equivalent to asynchronous bisimilarity.

Proof. Clearly $\sim^{OBS}\subseteq\sim^{OBSS}$: it suffices to note that $p \xrightarrow{c}{}_x p'$ implies $x; p; c \rightsquigarrow p'$.

We then focus on $\sim^{OBSS}\subseteq\sim^{OBS}$. We prove it by showing that the closure S of the strong open barbed semi-saturated bisimilarity with respect to instantiations, contexts and 2-cells $\mathcal{S} = \{\langle u, v\rangle | u \Rightarrow x; p; c, v \Rightarrow x; q; c, p \sim^{OBSS} q, x \in \mathbf{C}, c \in \mathbf{C}\}$ is a strong open barbed saturated bisimulation.

Suppose that $x'; u; c' \downarrow_o$. By hypothesis barbs are respected by 2-cells, thus $x'; x; p; c; c' \downarrow_o$. Since $p \sim^{OBSS} q$ also $x'; x; q; c; c' \downarrow_o$ and so $x'; v; c' \downarrow_o$.

Suppose that $x'; u; c' \rightsquigarrow p'$. Then, also $x'; x; p; c; c' \rightsquigarrow p'$. This means that for some $\langle l, r\rangle \in \mathfrak{R}$, $y \in \mathbf{C}$ and $d \in \mathbf{D}$, we have that there exists $\alpha : x'; x; p; c; c' \Rightarrow y; l; d$ (diagram (i)) and $p' = y; r; d$.

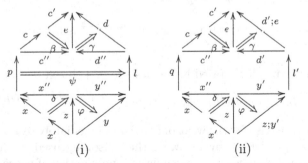

(i)   (ii)

Since the GRS has G-luxes, we can construct a G-LUX as the inner hexagon of diagram (i). This means that $p \xrightarrow{c''}{}_{x''} y''; r; d''$ and since $p \sim^{OBSS} q$, then $x''; q; c'' \rightsquigarrow q'$ and $y''; r; d'' \sim^{OBSS} q'$. Therefore, there exists an hexagon as the inner one of diagram (ii) shown above, with $\langle l', r'\rangle \in \mathfrak{R}$ and $q' = y'; r'; d'$. Thanks to the fact that $\mathbf{D}$ is 2-cell closed and composition reflecting, we know that e is reactive, hence $z; x''; q; c''; e \rightsquigarrow z; q'; e$. Since $x'; x; q; c; c' \Rightarrow z; x''; q; c''; e$ and $x'; v; c' \Rightarrow x'; x; q; c; c'$ also $x'; x; q; c; c' \rightsquigarrow z; q'; e$ (external hexagon of diagram (ii)) and $x'; v; c' \rightsquigarrow z; q'; e$. Moreover, since $y''; r; d'' \sim^{OBSS} q'$ and $p' = y; r; d \Rightarrow z; y''; r; d''; e$, we can conclude that $p' \, \mathcal{S} \, z; q'; e$.

4.2 Weak Open Barbed Saturated Semantics

This section introduces the weak case of open barbed (semi-)saturated bisimilarity. We begin by introducing the notion of weak barb on open states.

Definition 9 (Weak Barb). *Let $o \in O$ be a barb. An open state p satisfies the weak barb o (in symbols $p \Downarrow_o$) if there exists p' such that $p \rightsquigarrow^* p'$ and $p' \downarrow_o$.*

As usual, $\rightsquigarrow^*$ denotes the transitive and reflexive closure of $\rightsquigarrow$.

Definition 10 (Weak Open Barbed Saturated Bisimulation). *A symmetric relation $\mathcal{R}$ is a weak open barbed saturated bisimulation if whenever $p \, \mathcal{R} \, q$ then $\forall x, c$*

- *if $x; p; c \Downarrow_o$ then $x; q; c \Downarrow_o$;*
- *if $x; p; c \rightsquigarrow^* p'$ then $x; q; c \rightsquigarrow^* q'$ and $p' \, \mathcal{R} \, q'$.*

Weak open barbed saturated bisimilarity $\sim^{WOBS}$ is the largest such bisimulation.

As for the strong case, it can be efficiently characterised through the GLUX transition system via the semi-saturated game.

Definition 11 (Weak Open Barbed Semi-Saturated Bisimulation). *A symmetric relation $\mathcal{R}$ is a weak open barbed semi-saturated bisimulation if whenever $p \, \mathcal{R} \, q$ then*

- *$\forall x, c$, if $x; p; c \downarrow_o$ then $x; q; c \Downarrow_o$;*
- *if $p \xrightarrow[x]{c} p'$ then $x; q; c \rightsquigarrow^* q'$ and $p' \, \mathcal{R} \, q'$.*

Weak open barbed semi-saturated bisimilarity $\sim^{WOBSS}$ is the largest such bisimulation.

The correspondence result is stated below: its proof is shown in Appendix B.

Proposition 2. *In an open GRS having G-luxes, $\sim^{WOBSS} = \sim^{WOBS}$.*

As for the strong case, also for the weak one we can immediately test our proposal against ACCS. In particular, by following the reasoning used in the previous section for the strong case, we can analogously conclude that if we consider only closed ACCS processes, weak barbed semi-saturated bisimilarity is equivalent to weak asynchronous bisimilarity.

5   An Open G-Reactive System for Mobile Ambients

In this section we apply the framework of open GRSs to a richer calculus, that is, the calculus of Mobile Ambients (MAs) [9], in order to show the adequacy of the results presented in §4. After a quick introduction of the calculus, we present the corresponding open GRSs and we test our framework on it.

$$P ::= \epsilon, n[P], M.P, P|P \quad M ::= in\, n, out\, n, open\, n \,\Big|\, (P|Q)|R \equiv P|(Q|R) \quad P|Q \equiv Q|P$$

$n[in\ m.P|Q]|m[R] \rightsquigarrow m[n[P|Q]|R]$ if $P \rightsquigarrow Q$ then $n[P] \rightsquigarrow n[Q]$

$m[n[out\ m.P|Q]|R] \rightsquigarrow n[P|Q]|m[R]$ if $P \rightsquigarrow Q$ then $P|R \rightsquigarrow Q|R$

$open\ n.P|n[Q] \rightsquigarrow P|Q$ if $P' \equiv P, P \rightsquigarrow Q, Q \equiv Q'$ then $P' \rightsquigarrow Q'$

Fig. 4. Syntax, structural congruence and reduction relation of MAs

5.1 Mobile Ambients

We consider the finite, restriction free fragment of MAs. The syntax is shown on the left of the upper row of Fig. 4. We assume a set $\mathcal{N}$ of *names* ranged over by $m, n, u, \ldots$ We let $P, Q, R, \ldots$ range over the set $\mathcal{P}$ of closed processes. The semantics of the calculus is given by the combination of an equivalence between processes and a relation among them. The structural congruence $\equiv$ is induced by the two axioms on the right of the first row of Fig. 4. It is used to define the reduction relation $\rightsquigarrow$, which is inductively generated by the set of axioms and inference rules shown in the lower rows of Fig. 4.

As said in the previous section, a *barb* o is a predicate over the states of a system, with $P \downarrow_o$ denoting that P satisfies o. In MAs, $P \downarrow_n$ denotes the presence at top-level of an ambient n. Formally, for a closed process P, $P \downarrow_n$ if $P \equiv n[Q]|R$ for some closed processes Q and R. A closed process P satisfies the *weak barb* n (denoted as $P \Downarrow_n$) if there exists a process P' such that $P \rightsquigarrow^* P'$ and $P' \downarrow_n$. The notions of strong and weak barb above are respectively used to define the strong and weak reduction barbed congruences [19].

Definition 12 (Reduction Barbed Bisimulation). *A symmetric relation $\mathcal{R}$ is a reduction barbed bisimulation if whenever $P \mathcal{R} Q$ then $\forall C[-]$ contexts*

- *if $C[P] \downarrow_o$ then $C[Q] \downarrow_o$;*
- *if $C[P] \rightsquigarrow P'$ then $C[Q] \rightsquigarrow Q'$ and $P' \mathcal{R} Q'$.*

A symmetric relation $\mathcal{R}$ is a weak reduction barbed bisimulation if whenever $P \mathcal{R} Q$ then $\forall C[-]$ contexts

- *if $C[P] \Downarrow_o$ then $C[Q] \Downarrow_o$;*
- *if $C[P] \rightsquigarrow^* P'$ then $C[Q] \rightsquigarrow^* Q'$ and $P' \mathcal{R} Q'$.*

Reduction barbed congruence $\cong$ *and* weak reduction barbed congruence $\cong^W$ *are the largest such bisimulations.*

Labelled characterisations of reduction barbed congruences over MAs processes are presented in [22,4] for the strong case, and in [19,4] for the weak one. In the following, we are going to provide an alternative characterisation of them by using the GLUX LTS and the semi-saturated game.

5.2   An Open GRS for Mobile Ambients

As for ACCS, in order to define the open GRS for MAs, we exploit the G-PROP PA2CP induced from the signature corresponding to the grammar for the calculus modulo the associativity equation. The signature is $\Sigma = \epsilon : 0, n[\ ] : 1, M . : 1, | : 2$, where M stands for $in\,n$, $out\,n$ and $open\,n$.

For the definition of 2-cells, we consider an analogous representation for the terms and consequently for the arrows of the category. The main difference with respect to the representation used for ACCS is that here leaves only correspond to occurrences of the constants ϵ, while a node of degree 1 corresponds to an application of either an ambient operator or a capability. In the former case it is labelled with a name belonging to $\mathcal{N}$, while in the latter case it is labelled with the capability followed by an ambient name.

Now, we use this G-PROP to define the open GRS $\mathbb{C}_{MAs}$ for MAs. We take as reduction rules the set $\bigcup_{n,m\in\mathcal{N}}\{\langle n[in\,m.1 \mid 2] \mid m[3], m[n[1 \mid 2] \mid 3]\rangle, \langle m[n[out\,m.1 \mid 2] \mid 3], n[1 \mid 2] \mid m[3]\rangle, \langle open\,n.1 \mid m[2], 1 \mid 2\rangle\}$, and as the subcategory of reactive contexts the smallest composition-reflecting, 2-cell closed subcategory including arrows of the shapes $1 \mid p : 1 \to 1$ and $n[1] : 1 \to 1$. Note that it does not contain contexts with a hole after a capability.

Fig. 5 shows two interesting labelled transitions derived by applying the synthesis mechanism based on G-luxes. The leftmost G-lux represents $n[1] \xrightarrow[in\,m.1|2]{1|m[Q]} m[n[1 \mid 2] \mid Q]$. The initial open term offers an ambient n with a hole inside it, the instantiation provides the open term $in\,m.1 \mid 2$, which is going to replace the hole inside the ambient n, while the environment provides an ambient $m[Q]$ which is going to be in parallel with n. The rightmost G-lux instead represents $n[1] \xrightarrow[out\,m.1|2]{m[1|R]} n[1 \mid 2] \mid m[R]$. Also in this case the initial open term offers an ambient n with a hole inside it, the instantiation instead provides the open term $out\,m.1 \mid 2$, which is going to replace the hole inside the ambient n, while the environment provides an ambient m with a hole inside it (which will be replaced by n) in parallel with a closed process R.

These transitions show the importance of using instantiation and context at the same time. In both cases, indeed, we can note the interaction of three parts intervening in the reduction: the initial state, the instantiation and the context.

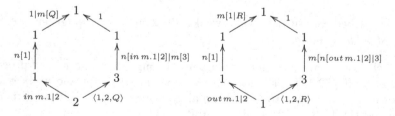

Fig. 5. The transitions $n[1] \xrightarrow[in\,m.1|2]{1|m[Q]} m[n[1 \mid 2] \mid Q]$ and $n[1] \xrightarrow[out\,m.1|2]{m[1|R]} n[1 \mid 2] \mid m[R]$

This never happens in ACCS, where we only have the synchronization between two terms in parallel which thus can be offered by just two of these three parts.

5.3   Open Barbed Semantics for Mobile Ambients

Here we show that the labelled characterisations of $\cong$ and $\cong^W$ via the GIPO LTS and the semi-saturated game presented in [4] are an instance of our results stated in Propositions 1 and 2.

We begin by observing that if we consider only closed terms and the standard notion of strong (respectively weak) barb for them, the notion of strong (respectively weak) open barbed saturated bisimilarity for MAs coincides with the one of strong (respectively weak) reduction barbed congruence, where $C[P]$ stands for $P; C[-]$. Therefore, as shown in § 4, since the open GRS has G-luxes, we can efficiently characterise the strong and weak reduction barbed congruence by instantiating respectively Definitions 8 and 11 for closed MAs processes, with the standard notion of strong and weak MAs barb and the GLUX LTS for the calculus. As shown in [4], the quantification over all contexts can be removed from the first condition of both definitions.

Definition 13 (MAs Closed Barbed Semi-Saturated Bisimulation). *A symmetric relation $\mathcal{R}$ is a MAs closed barbed semi-saturated bisimulation if whenever $P\,\mathcal{R}\,Q$ then*

- *if $P \downarrow_n$ then $Q \downarrow_n$;*
- *if $P \xrightarrow{C[-]} P'$ then $C[Q] \rightsquigarrow Q'$ and $P'\,\mathcal{R}\,Q'$.*

A symmetric relation $\mathcal{R}$ is a MAs weak closed barbed semi-saturated bisimulation if whenever $P\,\mathcal{R}\,Q$ then

- *if $P \downarrow_n$ then $Q \Downarrow_n$;*
- *if $P \xrightarrow{C[-]} P'$ then $C[Q] \rightsquigarrow^* Q'$ and $P'\,\mathcal{R}\,Q'$.*

MAs closed barbed semi-saturated bisimilarity $\sim^{BSS}_{MAs}$ and MAs weak closed barbed semi-saturated bisimilarity $\sim^{WBSS}_{MAs}$ are the largest such bisimulations.

Proposition 3. *Closed barbed semi-saturated bisimilarity for MAs $\sim^{BSS}_{MAs}$ coincides with reduction barbed congruence $\cong$ and weak closed barbed semi-saturated bisimilarity $\sim^{WBSS}_{MAs}$ coincides with weak reduction barbed congruence $\cong^S$.*

According to Remark 1, the results above subsume the ones presented in [4, Theorem 5], thus providing a labelled characterisation for both congruences $\cong$ and $\cong^W$ via the GIPO LTS.

6   Conclusions and Further Works

We introduced the notions of (strong and weak) barbed saturated bisimilarity for open GRSs, and we showed that it can be efficiently characterized via the

GLUX LTS by employing the semi-saturated game. Moreover, in order to show the adequacy of our proposal, we tested it on ACCS and, most importantly, on MAs. In both cases we proved that the framework captures the standard semantics of these calculi. More precisely, we showed that, for both the strong and the weak variant, the open barbed semi-saturated bisimilarity defined on closed ACCS terms coincides with asynchronous bisimilarity, and that on closed MAs terms it coincides with reduction barbed congruence.

However, albeit we showed that the framework is general enough to capture the abstract semantics of well-known formalisms, it is parametric with respect to the choice of the set of barbs: the choice of the "right" barb is left to the ingenuity of the researcher, and often it is not a trivial task [21,14]. So, it would be interesting to extend our framework by considering an automatically derived notion of barb for (open) GRSs. An alternate route would instead be to abstract away from the use of state predicates by providing an efficient labelled characterization of the barbed semantics, as shown for closed GRSs in [3,5].

Our proposal of a semantics for open GRSs builds on [16]. The authors there study only strong bisimilarity, showing that it is a congruence under rather restrictive conditions, hindering the applicability of the framework. Our work aims at partly solving this problem, in order to relax the conditions for proving a congruence property. However, we did not address the issue of the intrinsic redundancy of the G-lux mechanism: the intertwining of context and instantiation may sometimes end up in the offering of components that are not necessary for the reduction. Consider for example the transition represented by the leftmost G-lux in Fig. 5. It is easy to note that Q is not necessary for this reduction, since it appears both in the context on the left and in the instantiation on the right.

Concerning future works, a relatively simpler one should be a comparison with the proposal for parametric rules in GRSs in [10]. A more challenging issue would be to investigate the connections with [2,23]: both works propose a semantic framework for open processes, adopting suitable LTS and alternative behavioural semantics, which is however far removed from the GRSs formalism. In general terms, and stimulated by the latter contributions, we plan to investigate the issue of adequacy for open semantics, and the connection between our proposal and the standard definition of bisimilarity for open terms, considering the standard bisimilarity lifted from closed to open terms by instantiating these ones in all possible ways. It is clear that the latter bisimilarity is included in the former one, and for simpler calculi the two might often coincide. However, the connection between open GRSs and the tile model [12] established in [13] suggests us that the bisimilarity that we obtain is finer, as it is proved already for simple CCS variants and argued for richer calculi in [23,7].

References

1. Amadio, R., Castellani, I., Sangiorgi, D.: On bisimulations for the asynchronous π-calculus. TCS 195(2), 291–324 (1998)
2. Baldan, P., Bracciali, A., Bruni, R.: A semantic framework for open processes. TCS 389(3), 446–483 (2007)

3. Bonchi, F., Gadducci, F., Monreale, G.V.: On barbs and labels in reactive systems. In: Klin, B., Sobocinski, P. (eds.) SOS 2009. EPTCS, vol. 18, pp. 46–61 (2009)
4. Bonchi, F., Gadducci, F., Monreale, G.V.: Reactive systems, barbed semantics, and the mobile ambients. In: de Alfaro, L. (ed.) FOSSACS 2009. LNCS, vol. 5504, pp. 272–287. Springer, Heidelberg (2009)
5. Bonchi, F., Gadducci, F., Monreale, G.V.: Towards a general theory of barbs, contexts and labels. In: Yang, H. (ed.) APLAS 2011. LNCS, vol. 7078, pp. 289–304. Springer, Heidelberg (2011)
6. Bonchi, F., König, B., Montanari, U.: Saturated semantics for reactive systems. In: LICS 2006, pp. 69–80. IEEE Computer Society (2006)
7. Bruni, R., de Frutos-Escrig, D., Martí-Oliet, N., Montanari, U.: Bisimilarity congruences for open terms and term graphs via tile logic. In: Palamidessi, C. (ed.) CONCUR 2000. LNCS, vol. 1877, pp. 259–274. Springer, Heidelberg (2000)
8. Bruni, R., Meseguer, J., Montanari, U.: Symmetric monoidal and cartesian double categories as a semantics framework for tile logic. MSCS 12(1), 53–90 (2002)
9. Cardelli, L., Gordon, A.: Mobile ambients. TCS 240(1), 177–213 (2000)
10. Di Gianantonio, P., Honsell, F., Lenisa, M.: RPO, second-order contexts, and λ-calculus. Logical Methods in Computer Science 5(3) (2009)
11. Fournet, C., Gonthier, G.: A hierarchy of equivalences for asynchronous calculi. JLAP 63(1), 131–173 (2005)
12. Gadducci, F., Montanari, U.: The tile model. In: Plotkin, G.D., Stirling, C., Tofte, M. (eds.) Proof, Language and Interaction: Essays in Honour of Robin Milner, pp. 133–166. MIT Press (2000)
13. Gadducci, F., Monreale, G.V., Montanari, U.: A modular LTS for open reactive systems. In: Baeten, J.C.M., Ball, T., de Boer, F.S. (eds.) TCS 2012. LNCS, vol. 7604, pp. 134–148. Springer, Heidelberg (2012)
14. Honda, K., Yoshida, N.: On reduction-based process semantics. TCS 151(2), 437–486 (1995)
15. Kelly, G., Street, R.: Review of the elements of 2-categories. In: Sydney Category Seminar. LNM, vol. 420, pp. 75–103. Springer, Heidelberg (1974)
16. Klin, B., Sassone, V., Sobociński, P.: Labels from reductions: Towards a general theory. In: Fiadeiro, J.L., Harman, N., Roggenbach, M., Rutten, J. (eds.) CALCO 2005. LNCS, vol. 3629, pp. 30–50. Springer, Heidelberg (2005)
17. Leifer, J.J., Milner, R.: Deriving bisimulation congruences for reactive systems. In: Palamidessi, C. (ed.) CONCUR 2000. LNCS, vol. 1877, pp. 243–258. Springer, Heidelberg (2000)
18. MacLane, S.: Categorical algebra. Bulletin of the AMS 71, 40–106 (1965)
19. Merro, M., Zappa Nardelli, F.: Behavioral theory for mobile ambients. JACM 52(6), 961–1023 (2005)
20. Milner, R., Sangiorgi, D.: Barbed bisimulation. In: Kuich, W. (ed.) ICALP 1992. LNCS, vol. 623, pp. 685–695. Springer, Heidelberg (1992)
21. Rathke, J., Sassone, V., Sobociński, P.: Semantic barbs and biorthogonality. In: Seidl, H. (ed.) FOSSACS 2007. LNCS, vol. 4423, pp. 302–316. Springer, Heidelberg (2007)
22. Rathke, J., Sobociński, P.: Deriving structural labelled transitions for mobile ambients. In: van Breugel, F., Chechik, M. (eds.) CONCUR 2008. LNCS, vol. 5201, pp. 462–476. Springer, Heidelberg (2008)
23. Rensink, A.: Bisimilarity of open terms. IC 156(1-2), 345–385 (2000)
24. Sassone, V., Sobocinski, P.: Deriving bisimulation congruences using 2-categories. Nordic Journal of Computing 10(2), 163–183 (2003)

A   GIPO and GIPB

The first section of the Appendix recalls the notions of GIPO and GIPB.

Definition 14 (GRPO, GIPO). *Let the diagrams in Fig. 6 be in a G-category* **C**. *A candidate for the diagram (i) is a tuple* $\langle a_5, n, o, p, \beta, \gamma, \delta \rangle$ *such that* $1_h *$ $\gamma \bullet \beta * 1_p \bullet 1_g * \delta = \alpha$. *This means that the 2-cells* γ, β, *and* δ, *as illustrated in diagram (ii), paste together to give* α. *A* groupoidal-relative-pushout *(GRPO) is a candidate which satisfies the universal property, i.e., for any other candidate* $\langle a_6, n', o', p', \beta', \gamma', \delta' \rangle$ *there exists a* mediating morphism: *a quadruple* $\langle q : a_5 \rightarrow a_6, \varphi : n' \Rightarrow n; q, \psi : o; q \Rightarrow o', \tau : q; p' \Rightarrow p \rangle$ *illustrated in diagrams (iii) and (iv). The equations to be satisfied are* $1) \gamma' \bullet \varphi * 1_{p'} \bullet 1_n * \tau = \gamma$; $2) 1_o * \tau^{-1} \bullet \psi * 1_{p'} \bullet \delta' = \delta$; $3) 1_h * \varphi \bullet \beta * 1_q \bullet 1_g * \psi = \beta'$. *Such a mediating morphism must be essentially unique, i.e., for any other mediating morphism* $\langle q', \varphi', \psi', \tau' \rangle$ *there exists a unique 2-cell* $\xi : q \Rightarrow q'$ *that makes the two mediating morphisms compatible, i.e.,* $1) \varphi \bullet 1_n * \xi = \varphi'$, $2) 1_o * \xi^{-1} \bullet \psi = \psi'$, *and* $3) \xi * 1_{p'} \bullet \tau' = \tau$.

A square such as diagram (i) of Fig. 6 is called *G-idem pushout (GIPO)* if $\langle a_4, f, m, id_{a_4}, \alpha, 1_f, 1_m \rangle$ is its GRPO.

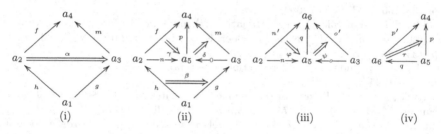

Fig. 6. GRPO

Definition 15 (GRPB, GIPB). *Let the diagrams in Fig. 7 be in a G-category* **C**. *A candidate for the diagram (i) is a tuple* $\langle a_5, n, o, p, \beta, \gamma, \delta \rangle$ *such that* $\gamma * 1_f \bullet 1_p * \beta \bullet \delta * 1_m = \alpha$. *This means that the 2-cells* γ, β, *and* δ, *as illustrated in diagram (ii), paste together to give* α. *A* groupoidal-relative-pullback *(GRPB) is a candidate which satisfies the universal property, i.e., for any other candidate* $\langle a_6, n', o', p', \beta', \gamma', \delta' \rangle$ *there exists a* mediating morphism: *a quadruple* $\langle q : a_6 \rightarrow a_5, \varphi : q; n \Rightarrow n', \psi : q; o \Rightarrow o', \tau : p'; q \Rightarrow p \rangle$ *illustrated in diagrams (iii) and (iv). The equations to be satisfied are* $1) \gamma' \bullet 1_{p'} * \varphi^{-1} \bullet \tau * 1_n = \gamma$; $2) \tau^{-1} * 1_o \bullet 1_{p'} * \psi \bullet \delta' = \delta$; $3) \varphi^{-1} * 1_f \bullet 1_q * \beta \bullet \psi * 1_m = \beta'$. *Such a mediating morphism must be essentially unique, i.e., for any other mediating morphism* $\langle q', \varphi', \psi', \tau' \rangle$ *there exists a unique 2-cell* $\xi : q \Rightarrow q'$ *that makes the two mediating morphisms compatible, i.e.,* $1) \varphi \bullet \xi * 1_n = \varphi'$, $2) \xi^{-1} * 1_o \bullet \psi = \psi'$, *and* $3) 1_{p'} * \xi \bullet \tau' = \tau$.

A square such as diagram (i) of Fig. 7 is called *G-idem pullback (GIPB)* if $\langle a_1, h, g, id_{a_1}, \alpha, 1_h, 1_g \rangle$ is its GRPB.

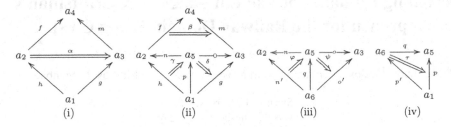

Fig. 7. GRPB

B   Proof of Proposition 2

We first give an alternate definition of weak open barbed saturated bisimulation.

Lemma 1 (Revising Barbed Saturation). *A symmetric relation $\mathcal{R}$ is a weak open barbed saturated bisimulation if whenever $p\,\mathcal{R}\,q$ then $\forall c, x$*

- *if $x; p; c \downarrow_o$ then $x; q; c \Downarrow_o$;*
- *if $x; p; c \rightsquigarrow p'$ then $x; q; c \rightsquigarrow^* q'$ and $p'\,\mathcal{R}\,q'$.*

Let us now move on to the proof of Proposition 2.

We prove that $\sim^{WOBSS} \subseteq \sim^{WOBS}$ by showing that the closure $\mathcal{S}$ of the weak open barbed semi-saturated bisimilarity with respect to instantiations, contexts and 2-cells $\mathcal{S} = \{\langle u, v \rangle | u \Rightarrow x; p, c, v \to x; q; c, p \sim^{WOBSS} q, x \in \mathbf{C}, c \in \mathbf{C}\}$ is a weak open barbed saturated bisimulation according to Lemma 1.

Suppose that $x'; u; c' \downarrow_o$. By hypothesis barbs are respected by 2-cells, therefore $x'; x; p; c; c' \downarrow_o$. Since $p \sim^{WOBSS} q$, then $x'; x; q; c; c' \downarrow_o$ and so $x'; v; c' \Downarrow_o$.

Now, assume that $x'; u; c' \rightsquigarrow p'$. Then, $x'; x; p; c; c' \rightsquigarrow p'$. This means that for some $\langle l, r \rangle \in \mathfrak{R}$, $y \in \mathbf{C}$ and $d \in \mathbf{D}$ there exists $\alpha : x'; x; p; c; c' \Rightarrow y; l; d$ (the exterior hexagon of the diagram below) and $p' = y; r; d$.

Since the GRS has G-luxes, we can construct a G-lux as the inner hexagon of the diagram on the right. Thus $p \xrightarrow[x'']{c''} y''; r; d''$ and, since $p \sim^{WOBSS} q$, then $x''; q; c'' \rightsquigarrow^* q'$ and $y''; r; d'' \sim^{WOBSS} q'$. Since $\mathbf{D}$ is 2-cell closed and composition reflecting, we know that e is reactive, hence $z; x''; q; c''; e \rightsquigarrow^*$ $z; q'; e$. Since $z; x''; q; c''; e \Rightarrow x'; x; q; c; c'$ and $x'; x; q; c; c' \Rightarrow x'; v; c'$ also $x'; x; q; c; c' \rightsquigarrow^*$ $z; q'; e$ and $x'; v; c' \rightsquigarrow^* z; q'; e$. Moreover, since $y''; r; d'' \sim^{WOBSS} q'$ and $p' = y; r; d \Rightarrow z; y''; r; d''; e$, we can conclude that $p' \mathcal{S} z; q'; e$.

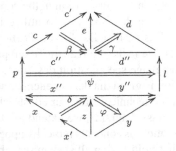

In order to prove that $\sim^{WOBS} \subseteq \sim^{WOBSS}$, it suffices to consider Definition 10 and observe that $x; p; c \downarrow_o$ implies $x; p; c \Downarrow_o$ and $p \xrightarrow{c}_{x} p'$ implies $x; p; c \rightsquigarrow p'$.

Designing Domain Specific Languages – A Craftsman's Approach for the Railway Domain Using CASL

Phillip James[1], Alexander Knapp[2], Till Mossakowski[3], and Markus Roggenbach[1]

[1] Swansea University, UK
[2] Universität Augsburg, Germany
[3] DFKI GmbH Bremen, Germany

Abstract. Domain modelling based on UML Class Diagrams is an established industrial practice. In the context of the Railway industry, we show how to utilize such diagrams for verification. This involves the translation of UML Class Diagrams into the algebraic specification language CASL. To this end, we define new Class Diagram institutions and provide suitable institution comorphisms.

1 Introduction

UML Class Diagrams [24] are industrially accepted for modelling a variety of systems across numerous domains. Often they are used to describe all elements and relationships occurring within a domain. As such, a UML Class Diagram (CD) can be thought of as describing a domain specific language (DSL). A typical example of such an endeavour is given by the Data Model [13] of our research partner Invensys Rail which aims to describe all elements within the railway domain.

Classically, DSLs [19,9] allow even non-experts to create programs or specifications. In the context of programming, additional motivation for DSLs is improved tool support along with ease of use, better readability, and increased productivity. James and Roggenbach [15] have demonstrated an approach where formal DSLs within the railway domain aid verification in the context of algebraic specification using CASL [23].

Here, we describe how to utilize industrial DSLs from the Railway domain, formulated as UML CDs, for verification. As UML CDs only capture the static system aspects, we make the realistic assumption that the CD is accompanied with some natural language specification describing the dynamic system aspects. For Railways this situation is given thanks to generally accepted standard literature, e.g., [16].

On the technical side, our construction is based on institution theory: to capture realistic class diagrams, we extend the UML CD institution by Cengarle and Knapp [6] with numerous concepts typically appearing in applications. UML CDs describe invariants that hold during all system runs, however, do not offer the possibility of capturing a system's dynamics. To this end, we employ MODALCASL [21] as a general framework for specifying Kripke-structures. We give an institution comorphism from UML CD to MODALCASL. Part of this mapping is a general construction, namely that of a Pointed Powerset Institution, which factors out a general construction principle necessary for connecting UML CDs with an arbitrary institution capturing system dynamics. In MODALCASL, we then model system dynamics according to the natural language

N. Martí-Oliet and M. Palomino (Eds.): WADT 2012, LNCS 7841, pp. 178–194, 2013.

specification describing the dynamic system aspects. Finally, for the sake of better proof support, we use an already established comorphism from MODALCASL to CASL [23]. Overall, this approach allows us to directly import DSL specifications from industry into the verification framework proposed by James and Roggenbach [15].

Related work. Our work closely relates to the various approaches that utilize UML as a graphical frontend for formal methods. In this respect, we follow the approach of Cengarle and Knapp [6]: a UML diagram type can be described in its "natural" semantics; its relations to other UML diagram types is expressed by appropriate translations. Overall, this results in a compositional semantics for UML: each diagram type is treated individually. It also separates concerns: first a semantics is given, then a translation is defined. This differs from monolithic constructions, which first select a fixed, usually small number of UML diagrams and then give a semantics by translation into an established formalism. A first such attempt with CASL as the underlying formalism was given in [12]. Yet another example of this second approach is the UML-RSDS method [17]. UML-RSDS translates specifications consisting of CDs, state machines and OCL into B. CD annotations in the form of stereotypes steer the translation. Lano et al. [17] give a railway example, however, not to the extent of defining a DSL. Another example of this kind is the approach taken within the Iness project [7]. The Iness project defines a DSL whose components are mapped into xUML. These are then translated to Promela in order to verify railway systems with the SPIN model checker. Besides the different approach to semantics, our construction allows for theorem proving technology rather than for model checking. A third approach is given by Meng and Aichernig [18], who define various semantics for CDs depending on their use in software development. Their overall approach is of a co-algebraic nature. The most abstract level, defined as the "object type" semantics, is close to the view we take on CDs. The more concrete levels equip objects with a (hidden) state and a transition structure, visibility tags for attributes and methods, or OCL constraints.

Concerning DSL design, Bjørner [4] has studied the endeavour of domain engineering, where he takes the specification languages RSL as the semantic basis. Here, we develop a methodology that designs formal DSLs based upon the outcome of domain engineering undertaken by industry. Therefore, as the domain analysis is undertaken by the domain experts themselves, the resulting DSL captures concepts and their relations "correctly" for people working within the domain. Using the ASF-SDF environment, Andova et al. [1] develop their "Simple Language of Communicating Objects" into a formal DSL by giving it an operational semantics. Our semantics is axiomatic in style, where concepts have a loose semantics.

Organisation. We first review the notion of a DSL, present an established example from academia as a running example, and briefly comment on the Data Model from our industrial research partner Invensys Rail. In Section 3, we recall notions from institution theory and present the CASL and MODALCASL institutions as well as a comorphism between them. Our new institution for UML CDs is given in Section 4. There, we also embed the UML CD institution into the MODALCASL institution via a comorphism. Finally, using the above constructions, we demonstrate that these techniques allow for automated verification in the Railway Domain.

2 Domain Specific Languages

Domain Specific Languages are languages designed and tailored for a specific application area. Examples include Risla [2] for financial products, Hancock [5] for processing large scale customer data concerning telephone calls, and HTML for web-page design. The defining feature of DSLs is exactly the fact that they provide notations and constructs tailored to a particular domain [9,19]. Usually, DSLs are languages designed for programming. Here we consider their use for algebraic specification. Their advantages over general purpose languages concern *readability* (notations as in the application domain), *productivity* (in design, implementation and maintenance of the software life cycle, thanks to ease of notation), and *re-use* (via domain specific components).

The railway community has developed several informal DSLs (usually company specific) for describing elements within the domain. For example, Invensys Rail have created a detailed DSL for the railway domain using UML and natural language [13]. This data model describes a DSL using UML and accompanying explanations in plain English. It is split into various layers, each describing certain aspects such as signalling or geographical information. These layers are mutually dependent. The document gives details of around 150 classes with a textual explanation of each. This paper describes the foundations of how to turn such a rich, informal DSL into a formal one. However, for ease of illustration we consider a simpler DSL from the academic context.

2.1 Bjørner's Railway DSL

The process of identifying, classifying and precisely defining the elements of a domain has been coined as "Domain Engineering" by Dines Bjørner [4]. Bjørner's classification, i.e., DSL [3], for the Railway domain is illustrated using a UML CD in Fig. 1.It contains the following elements: *Classes*, represented by a box, e.g. Net, Unit, Station etc. These represent concepts in the railway domain. *Properties* are listed inside a class, e.g. id: UID in the class Net expresses that all Nets have an identifier of type UID. *Generalisations*, represented by a unfilled arrow head, e.g. Point and Linear are generalisations of Unit. *Associations*, represented by a line/arrow between two classes, e.g. the has link between Unit and Connector. Associations can have direction, and also multiplicities associated with them. The multiplicities on the has association between Unit and Connector can be read as: "One Unit has two or more connectors". *Compositions*, represented by a filled diamond, e.g. the hasLine composition for Net and Line, tell us that one class "is made up of" another class. Compositions can also have multiplicities. Finally, *operations* are also represented inside a class, for example the isOpen operation of type Boolean inside the Route class. Here, we note that the query operations we consider can be seen as parametrised properties. We do not treat operations that can modify the state of an object. Primitive types such as Boolean and structured types such as List, Set, and Pair are considered to be predefined.

To this "pure" UML class diagram a so-called *stereotype*, «dynamic», has been added. In Fig. 1 this appears for the association stateAt and the two operations isClosedAt and isOpen. All other elements are considered to be «rigid». These domain-specific stereotypes are intended to make clear which parts of an object structure complying to the

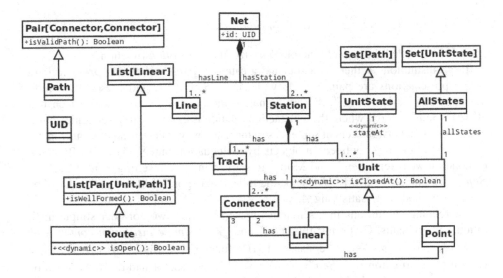

Fig. 1. A UML class diagram for Bjørner's DSL

class diagram can change over time, i.e., are «dynamic» like stateAt, and which parts have to be kept fixed over time, i.e., are «rigid», e.g., the objects of Net, Station, Line, etc.

From a modelling point of view, Bjørner's DSL prescribes that a railway is a Net, built from Stations that are connected via Lines. A station can have a complex, though fixed, structure, including Tracks, switch Points and LinearUnits. Tracks and Lines can contain LinearUnits only. All Units are attached together via Connectors. Such railway Nets gain dynamics by attaching a UnitState to each Unit, see [3]. Each unit can be in one of several states denoted using a Path. Such a Path indicates in which direction a train can pass. However, trains are not an explicit part of the DSL. Instead, Bjørner describes the concept of a Route, as a list of connected units. Finally, Bjørner gives a series of properties using a narrative description. An example of such a narrative is: "*A route is considered to be open if all units within the* Route *are open for the correct connected* Paths *across them.*" Such properties are not expressible purely using CD's. Along with these concepts, Bjørner stipulates further various well-formedness conditions on such a collection of Nets also using natural language.

3 Institutions and Institution Comorphisms

Following [20], we briefly recall the definition of institutions [11] and institution co-morphisms: An institution $\mathscr{I} = (\text{Sig}^{\mathscr{I}}, \textit{Sen}^{\mathscr{I}}, \text{Mod}^{\mathscr{I}}, \models^{\mathscr{I}})$ consists of (i) a category of *signatures* $\text{Sig}^{\mathscr{I}}$; (ii) a *sentence functor* $\textit{Sen}^{\mathscr{I}} : \text{Sig}^{\mathscr{I}} \to \text{Set}$, where Set is the category of sets; (iii) a contra-variant *model functor* $\text{Mod}^{\mathscr{I}} : (\text{Sig}^{\mathscr{I}})^{\text{op}} \to \text{Cat}$, where Cat is the category of categories; and (iv) a family of *satisfaction relations* $\models^{\mathscr{I}}_{\Sigma} \subseteq |\text{Mod}^{\mathscr{I}}(\Sigma)| \times \textit{Sen}^{\mathscr{I}}(\Sigma)$ indexed over $\Sigma \in |\text{Sig}^{\mathscr{I}}|$, such that the following *satisfaction condition* holds for every signature morphism $\sigma : \Sigma \to \Sigma'$ in $\text{Sig}^{\mathscr{I}}$, every sentence $\varphi \in \textit{Sen}^{\mathscr{I}}(\Sigma)$ and for every Σ'-model $M' \in |\text{Mod}^{\mathscr{I}}(\Sigma')|$:

$$\text{Mod}^{\mathscr{I}}(\sigma)(M') \models_{\Sigma}^{\mathscr{I}} \varphi \Leftrightarrow M' \models_{\Sigma'}^{\mathscr{I}} \text{Sen}^{\mathscr{I}}(\sigma)(\varphi) .$$

$\text{Mod}^{\mathscr{I}}(\sigma)$ is called the *reduct* functor, $\text{Sen}^{\mathscr{I}}(\sigma)$ the *translation* function.

To an institution $\mathscr{I}$ there is associated a category $\text{Pres}^{\mathscr{I}}$ of *theory presentations*, where presentations are pairs (Σ, A) with $A \subseteq \text{Sen}^{\mathscr{I}}(\Sigma)$, and presentation morphisms $\theta : (\Sigma, A) \to (\Sigma', A')$ are signature morphisms $\theta : \Sigma \to \Sigma'$ such that $A' \models_{\Sigma'}^{\mathscr{I}} \text{Sen}^{\mathscr{I}}(\theta)(A)$, where the satisfaction relation is extended to a semantic consequence relation between sets of sentences in the usual way. There is an obvious functor $\text{Sig} : \text{Pres}^{\mathscr{I}} \to \text{Sig}^{\mathscr{I}}$ defined on objects by the equation $\text{Sig}(\Sigma, A) = \Sigma$. By abuse of notation we extend $\text{Sen}^{\mathscr{I}}$ and $\text{Mod}^{\mathscr{I}}$ to start from $\text{Pres}^{\mathscr{I}}$ setting $\text{Sen}^{\mathscr{I}}(\Sigma, A) = \text{Sen}^{\mathscr{I}}(\Sigma)$ and letting $\text{Mod}^{\mathscr{I}}(\Sigma, A)$ be the full sub-category of $\text{Mod}^{\mathscr{I}}(\Sigma)$ induced by the class of models M satisfying A.

For relating institutions in a semantics preserving way, we consider simple institution comorphisms. Given institutions $\mathscr{I}$ and $\mathscr{J}$, a *simple institution comorphism* $\mu = (\Phi, \alpha, \beta) : \mathscr{I} \to \mathscr{J}$ consists of (i) a functor $\Phi : \text{Sig}^{\mathscr{I}} \to \text{Pres}^{\mathscr{J}}$; (ii) a natural transformation $\alpha : \text{Sen}^{\mathscr{I}} \dot{\to} \text{Sen}^{\mathscr{J}} \circ \Phi$; and (iii) a natural transformation $\beta : \text{Mod}^{\mathscr{J}} \circ \Phi^{\text{op}} \dot{\to} \text{Mod}^{\mathscr{I}}$, such that the following *representation condition* is satisfied for all $\Sigma \in |\text{Sig}^{\mathscr{I}}|$, $M' \in |\text{Mod}^{\mathscr{J}}(\Phi(\Sigma))|$, and $\varphi \in \text{Sen}^{\mathscr{I}}(\Sigma)$:

$$M' \models_{\text{Sig}(\Phi(\Sigma))}^{\mathscr{J}} \alpha_{\Sigma}(\varphi) \Leftrightarrow \beta_{\Sigma}(M') \models_{\Sigma}^{\mathscr{I}} \varphi .$$

3.1 CASL

The algebraic specification language CASL [23] has many-sorted first order logic with equality as its underlying institution. A CASL *signature* $\Sigma = (S, TF, PF, P)$ consists of a set S of sort symbols, a family of sets TF of total function symbols, a family of sets PF of partial function symbols, and a family of sets P of predicate symbols. Index sets of these families are the possible arities of the function symbols and predicate symbols. CASL *sentences* are the standard first order formulae, including strong equality $t = t'$, existential equality $t \stackrel{e}{=} t'$, and definedness of terms $def(t)$, where t, t' are terms. CASL *models* M interpret sort symbols s with a non empty carrier set s_M, function symbols $f \in PF \cup TF$ with a function f_M of suitable arity, f_M being total for $f \in TF$, and predicate symbols $p \in P$ with a relation p_M of suitable type. CASL *satisfaction* is as is standard for first order logic, where CASL semantics is strict, i.e., an undefined term in a formula yields that the formula does not hold. For details see [23].

On top of the many sorted CASL institution, CASL subsorting is defined: *Subsorted* CASL-*Signatures* $\Sigma = (S, TF, PF, P, \leq)$ are CASL signatures with an additional subsort relation $\leq \subseteq S \times S$. This subsort relation is required to be reflexive and transitive. There is overloading of function symbols and predicate symbols. Each subsorted signature $\Sigma = (S, TF, PF, P, \leq)$ is associated with a many-sorted signature $\Sigma^{\sharp}$ extending (S, TF, PF, P) for each pair of sorts $s \leq s'$ by a total embedding operation $inj_{s,s'}$ (from s into s'), a partial projection $inj_{s',s}$ operation (from s' onto s), and a membership predicate (testing whether values in s' are embeddings of values in s). *Sentences* of subsorted CASL are the ordinary many-sorted sentences over the associated signature.

Models of subsorted CASL are the ordinary many-sorted models over the associated signature that satisfy a list of properties w.r.t. the embeddings, projections, membership, and overloading. *Satisfaction* is as for many sorted CASL. For details see [23].

The datatypes considered to be predefined for class diagrams have been captured in the CASL library of standard datatypes [23].

3.2 MODALCASL

MODALCASL [21] extends CASL by modal logic (see e.g. [10]), providing a multi-modal first-order logic with partiality and subsorting. A MODALCASL *signature* consists of a CASL signature, a predicate on the operation and predicate symbols of the CASL signature, marking some of them as *rigid* (the others are called flexible), a set of *modalities*, including the special modality ϵ, and a subset of the sort set of the CASL signature, called the *set of modality sorts*. A MODALCASL *model M* consists of a set of worlds W, one of which is marked as the initial one[1], and for each world $w \in W$ there is a CASL model M_w, such that carrier sets and the interpretation of rigid operation and predicate symbols are the same for all M_w. A model M has, for each modality, and for each carrier element of each modality sort, a binary accessibility relation on W, which is preserved under the embedding of a carrier element along a subsort injection. MODALCASL *sentences* are built like CASL sentences, with the following modalities as new sentence building (unary prefix) connectives: $\langle m \rangle$, $[m]$ where m is a modality or a term of a modality sort. If $m = \epsilon$, m is omitted, and we obtain the usual (mono-modal) notation $\langle \rangle \varphi$ and $[]\varphi$. *Satisfaction* is based on the definition of satisfaction in CASL, while modalities are interpreted with Kripke semantics in the following sense: $\langle m \rangle \varphi$ means that φ holds in some 1-step m-reachable world, $[m]\varphi$ means that φ holds in all 1-step m-reachable worlds.

A simple institution comorphism from MODALCASL *to* CASL. *Signature translation:* Sorts are unchanged, but a sort w (for "worlds") is added, as well as a constant *init* : w and a binary relation $R : w \times w$. Rigid operation and predicate symbols are kept unchanged. For flexible operation and predicate symbols, w is added as an extra argument. *Sentence translation:* Sentences are translated according to the standard translation. *Model translation:* A CASL model is turned into a MODALCASL by keeping the carrier sets (note that MODALCASL enforces constant domains) as well as the rigid operations and predicates. The set of worlds and the accessibility relation are obtained by the interpretation of w and R respectively. The flexible operations and predicates are obtained by using the current world as the extra argument of the CASL operation respectively predicate.

4 Institutions for (Stereotyped) UML Class Diagrams

We first present an institution for proper UML class diagrams as used in Fig. 1, based on work by Cengarle and Knapp [6]. We then capture the meaning of the stereotype

---

[1] This differs from the original MODALCASL design which refrains from having an initial world.

«dynamic» by a simple general construction on institutions. Finally, we represent this institution in MODALCASL. Technical details and proofs can be found in the accompanying technical report [14].

4.1 An Institution for UML Class Diagrams

The UML class diagram institution is constructed as follows: The signatures capture all classes, relationships, and features of a UML class diagram. The models comprise all objects and links between objects that comply to the UML class diagram. The sentences describe the multiplicities of the associations and compositions. In order to provide generic access to primitive types, like Boolean, and type formers like List, we treat these as built-ins with a standard meaning; all other classes are assumed to be inhabited, i.e., to contain at least one object, like *null*.

Class nets — Signatures. The signatures of the UML class diagram institution are given by *class nets* $\Sigma = ((C, \leq_C), K, P, M, A)$ where $(C, \leq_C)$ describes the *class hierarchy* as a partial order which is closed w.r.t. to the built-in type Boolean (i.e., Boolean $\in C$), and "downwards" closed w.r.t. the unary built-in type formers List and Set, and the binary built-in type former Pair (i.e., if List$[c] \in C$ or Set$[c] \in C$, then $c \in C$; and if Pair$[c_1, c_2] \in C$, then $c_1, c_2 \in C$); K fixes the *instance specifications declarations* of the form $k : c$ with $c \in C$; P contains the *property declarations* of the form $c.p(x_1 : c_1, \ldots, x_n : c_n) : c'$ (where we drop the parentheses if the property has no arguments); M gives the *composition declarations* of the form $c \bullet r : c'$; and A comprises the *association declarations* of the form $a(r_1 : c_1, \ldots, r_n : c_n)$. We require several conditions in order to avoid super-types of built-in types and type formers (e.g., forbidding Boolean $\leq c$ for all $c \neq$ Boolean), overriding of declarations, and also that the composition declarations are cycle-free: if $c_1 \bullet r_1 : c_2, \ldots, c_n \bullet r_n : c_{n+1} \in M$, then $c_{n+1} \neq c_1$.

Example 1. The UML class diagram in Fig. 1 leads to the following class net:

> Classes: {Net, Station, Line, Unit, . . . , Pair[Unit, Path], List[Pair[Unit, Path]]}
>
> Generalisations: Point $\leq$ Unit, Linear $\leq$ Unit, . . . , Route $\leq$ List[Pair[Unit, Path]]
>
> Properties: {Net.ID : UID, . . . , Route.isOpen(r : Route) : Boolean}
>
> Compositions: {Station$\bullet$has : Unit, Station$\bullet$has : Track}
>
> Associations: {stateAt(unit : Unit, state : UnitState), . . . }   □

A class net morphism $\sigma = (\gamma, \kappa, \pi, \mu, \alpha) : \Sigma = ((C, \leq_C), K, P, M, A) \to T = ((D, \leq_D), L, Q, N, B)$ from Σ to T consists of a monotone map γ from $(C, \leq_C)$ to $(D, \leq_D)$ which is homomorphic w.r.t. the type Boolean, and the type formers List, Set, and Pair (i.e., $\gamma($Boolean$) =$ Boolean, $\gamma($List$[c]) =$ List$[\gamma(c)]$, $\gamma($Set$[c]) =$ Set$[\gamma(c)]$, and $\gamma($Pair$[c_1, c_2]) =$ Pair$[\gamma(c_1), \gamma(c_2)]$ for all $c, c_1, c_2 \in C$); maps κ, π, and μ between the instance specification, property, and composition declarations, such that classes are mapped along γ; and a map α from A to B for the association declarations such that n-ary association declarations are mapped to n-ary association declarations along γ, i.e., if $\alpha(a\{r_1 : c_1, \ldots, r_n : c_n\}) = b(s_1 : d_1, \ldots, s_m : d_m) \in B$, then $m = n$ and $d_i = \gamma(c_i)$ for all $1 \leq i \leq n$.

Class nets and class net morphisms form the category of *class nets*, denoted by $\mathcal{C}l$.

Instance nets — Models. A Σ-model of the UML class diagram institution for a class net $\Sigma = ((C, \leq_C), K, P, M, A)$ is given by a Σ-*instance net* $\mathcal{I} = (C^{\mathcal{I}}, K^{\mathcal{I}}, P^{\mathcal{I}}, M^{\mathcal{I}}, A^{\mathcal{I}})$ consisting of interpretations for all declarations where

- $C^{\mathcal{I}}$ maps each $c \in C$ to a non-empty set such that $C^{\mathcal{I}}(c_1) \subseteq C^{\mathcal{I}}(c_2)$ if $c_1 \leq_C c_2$ with $C^{\mathcal{I}}(\text{Boolean}) = \{\mathit{ff}, \mathit{tt}\}$, $C^{\mathcal{I}}(\text{List}[c]) = (C^{\mathcal{I}}(c))^*$ (where V^* denotes the finite sequences over V), $C^{\mathcal{I}}(\text{Set}[c]) = \wp(C^{\mathcal{I}}(c))$ (where $\wp(V)$ denotes the powerset of V), and $C^{\mathcal{I}}(\text{Pair}[c_1, c_2]) = C^{\mathcal{I}}(c_1) \times C^{\mathcal{I}}(c_2)$;
- $K^{\mathcal{I}}$ maps each $k : c \in K$ to an element of $C^{\mathcal{I}}(c)$;
- $P^{\mathcal{I}}$ maps each $c.p(x_1 : c_1, \ldots, x_n : c_n) : c' \in P$ to a partial map $C^{\mathcal{I}}(c) \times C^{\mathcal{I}}(c_1) \times \cdots \times C^{\mathcal{I}}(c_n) \rightharpoonup C^{\mathcal{I}}(c')$;
- $M^{\mathcal{I}}$ maps each $c{\bullet}r : c' \in M$ to a map $C^{\mathcal{I}}(c) \to \wp(C^{\mathcal{I}}(c'))$;
- $A^{\mathcal{I}}$ maps each $a(r_1 : c_1, \ldots, r_n : c_n) \in A$ to a subset of $\prod_{1 \leq i \leq n} C^{\mathcal{I}}(c_i)$,

such that

- instance specifications $k_1 : c$ and $k_2 : c$ with $k_1 \neq k_2$ are interpreted differently: $K^{\mathcal{I}}(k_1 : c) \neq K^{\mathcal{I}}(k_2 : c)$;
- each instance has at most one owner: if $o' \in M^{\mathcal{I}}(c_1{\bullet}r_1 : c_1')(o_1) \cap M^{\mathcal{I}}(c_2{\bullet}r_2 : c_2')(o_2)$, then $o_1 = o_2$.

Example 2. An excerpt of an instance net for the class net for Fig. 1 is:

$C^{\mathcal{I}}(\text{Net}) = \{\text{LondonUnderground}\}$

$C^{\mathcal{I}}(\text{Station}) = \{\text{CharingCross}, \text{OxfordStreet}, \ldots, \text{PicadillyCircus}\}$

$C^{\mathcal{I}}(\text{Connector}) = \{\text{c1}, \text{c2}, \text{c3}, \ldots\}$

$C^{\mathcal{I}}(\text{Pair}[\text{Connector}, \text{Connector}]) = \{(\text{c1}, \text{c1}), (\text{c1}, \text{c2}), (\text{c2}, \text{c1}), (\text{c2}, \text{c2}), \ldots\}$

$P^{\mathcal{I}}(\text{Net.ID} : \text{UID}) : C^{\mathcal{I}}(\text{Net}) \rightharpoonup C^{\mathcal{I}}(\text{UID})$ where
        $\text{LondonUnderground} \mapsto \text{"LUG"}$ etc.

$P^{\mathcal{I}}(\text{Route.isOpen} : \text{Boolean}) : C^{\mathcal{I}}(\text{Route}) \rightharpoonup C^{\mathcal{I}}(\text{Boolean})$ where
        $\text{R1} \mapsto \mathit{tt}$ etc.

$M^{\mathcal{I}}(\text{Net}{\bullet}\text{has} : \text{Station}) : C^{\mathcal{I}}(\text{Net}) \to \wp(C^{\mathcal{I}}(\text{Station}))$ where
        $\text{LondonUnderground} \mapsto \{\text{CharingCross}, \ldots, \text{PicadillyCircus}\}$

$A^{\mathcal{I}}(\text{stateAt}(\text{unit} : \text{Unit}, \text{state} : \text{UnitState})) =$
        $\{(\text{LU1}, \{(\text{c1}, \text{c2})\}), (\text{LU2}, \{(\text{c2}, \text{c3})\}), \ldots\}$ $\square$

A Σ-*instance net morphism* $\zeta : \mathcal{I} = (C^{\mathcal{I}}, K^{\mathcal{I}}, P^{\mathcal{I}}, M^{\mathcal{I}}, A^{\mathcal{I}}) \to \mathcal{J} = (C^{\mathcal{J}}, K^{\mathcal{J}}, P^{\mathcal{J}}, M^{\mathcal{J}}, A^{\mathcal{J}})$ over a class net $\Sigma = ((C, \leq_C), K, P, M, A)$ is given by a map $\zeta \in \prod c \in C . C^{\mathcal{I}}(c) \to C^{\mathcal{J}}(c)$ such that all interpretations of the declarations are mapped homomorphically, e.g., $\zeta(\text{Pair}[c_1, c_2])(o_1, o_2) = (\zeta(c_1)(o_1), \zeta(c_2)(o_2))$ for all $c_1, c_2 \in C$, $o_1 \in C^{\mathcal{I}}(c_1)$, $o_2 \in C^{\mathcal{I}}(c_2)$, and $K^{\mathcal{J}}(k : c) = \zeta(c)(K^{\mathcal{I}}(k : c))$ for each $k : c \in K$. Σ-instance nets and Σ-instance net morphisms as morphisms form the category of Σ-*instance nets*, denoted by $\text{Inst}(\Sigma)$.

For reducts of instance nets and instance net morphisms, let $\sigma = (\gamma, \kappa, \pi, \mu, \alpha) : \Sigma = ((C, \leq_C), K, P, M, A) \to T = ((D, \leq_D), L, Q, N, B)$ be a class net morphism. The *reduct* of a T-instance net $\mathcal{J} = (D^{\mathcal{J}}, L^{\mathcal{J}}, Q^{\mathcal{J}}, N^{\mathcal{J}}, B^{\mathcal{J}})$ along σ is the

Σ-instance net $\mathcal{J}|\sigma = (C^{\mathcal{J}|\sigma}, K^{\mathcal{J}|\sigma}, P^{\mathcal{J}|\sigma}, M^{\mathcal{J}|\sigma}, A^{\mathcal{J}|\sigma})$ formed by component-wise reducts, i.e., $C^{\mathcal{J}|\sigma}(c) = D^{\mathcal{J}}(\gamma(c))$ and similar for the other components. The *reduct* of a T-instance net morphism $\zeta : \mathcal{J}_1 \to \mathcal{J}_2$ along σ is the Σ-instance net morphism $\zeta|\sigma : \mathcal{J}_1|\sigma \to \mathcal{J}_2|\sigma$ with $\zeta|\sigma(c) = \zeta(\gamma(c))$. Setting $\mathrm{Inst}(\sigma)(\mathcal{J}) = \mathcal{J}|\sigma$ and $\mathrm{Inst}(\sigma)(\zeta) = \zeta|\sigma$, $\mathrm{Inst} : \mathrm{Cl}^{\mathrm{op}} \to \mathrm{Cat}$ is a contra-variant functor.

Multiplicity formulae — Sentences. For the sentences of the UML class diagram institution we use *multiplicity formulae* defined by the following grammar:

$$
\begin{aligned}
Frm &::= NumLit \leq FunExpr \mid FunExpr \leq NumLit \mid Composition \,! \\
FunExpr &::= \# \, Composition \mid \# \, Association \, [\, Role(, Role)^* \,] \\
Composition &::= Class\!\bullet\!Role : Class \\
Association &::= Name\{Role : Class(, Role : Class)^*\} \\
Class &::= Name \\
Role &::= Name \\
NumLit &::= 0 \mid 1 \mid \cdots
\end{aligned}
$$

where *Name* is a set of strings. The $\leq$-formulae express constraints on the cardinalities of composition and association declarations, i.e., how many instances are allowed to be in a certain relation with others. The #-expressions return the number of links in an association when some roles are fixed. The !-formulae for compositions express that the owning end must not be empty. The set of sentences or Σ-*multiplicity constraints* $Mult(\Sigma)$ for a class net Σ is given by the multiplicity formulae in *Frm* such that all mentioned elements of *Composition* and *Association* correspond to composition declarations and association declarations of Σ respectively, and the *Role* names mentioned in the last clause of *FunExpr* occur in the mentioned association.

Example 3. Some of the cardinality constraints of the running example in Fig. 1 can be expressed with multiplicity formulae as follows:

Station$\bullet$has : Track !

   — each Track is owned by a Station via has,

$2 \leq$ #has(connector : Connector, unit : Unit)[unit]

   — association has links each Unit to at least two Connectors,

#state(unitState : UnitState, unit : Unit)[unit] $= 1$

   — association state links each Unit to exactly one UnitState.

Note we have used formulae with = instead of two formulae with $\leq$ where the left hand side and the right hand side are switched. $\square$

The *translation* of a formula $\varphi \in Mult(\Sigma)$ along a class net morphism $\sigma = (\gamma, \kappa, \pi, \mu, \alpha) : \Sigma \to T$, written as $\sigma(\varphi)$, is given by applying σ to compositions, associations, and role names.

Satisfaction relation. The Σ-*satisfaction relation* $- \models_\Sigma - \,\subseteq\, |\mathrm{Inst}(\Sigma)| \times Sen(\Sigma)$ of the UML class diagram institution is defined for each Σ-instance net $\mathcal{I} = (C^{\mathcal{I}}, K^{\mathcal{I}}, P^{\mathcal{I}}, M^{\mathcal{I}}, A^{\mathcal{I}})$ as

$$\mathcal{I} \models_{\Sigma} \ell \leq \#c \bullet r : c' \quad \Leftrightarrow \quad \forall o \in C^{\mathcal{I}}(c) . [\![\ell]\!] \leq M^{\mathcal{I}}(c \bullet r : c')(o)$$

$$\mathcal{I} \models_{\Sigma} \#c \bullet r : c' \leq \ell \quad \Leftrightarrow \quad \forall o \in C^{\mathcal{I}}(c) . M^{\mathcal{I}}(c \bullet r : c')(o) \leq [\![\ell]\!]$$

$$\mathcal{I} \models_{\Sigma} c \bullet r : c'! \quad \Leftrightarrow \quad \forall o' \in C^{\mathcal{I}}(c') . \exists o \in C^{\mathcal{I}}(c) . M^{\mathcal{I}}(c \bullet r : c')(o) = o'$$

$$\mathcal{I} \models_{\Sigma} \ell \leq \#a(r_1 : c_1, \ldots, r_n : c_n)[r_{i_1}, \ldots, r_{i_m}] \quad \Leftrightarrow$$
$$\forall o_{i_1} \in C^{\mathcal{I}}(c_{i_1}), \ldots, o_{i_m} \in C^{\mathcal{I}}(c_{i_m}) .$$
$$[\![\ell]\!] \leq |\{t \in A^{\mathcal{I}}(a(r_1 : c_1, \ldots, r_n : c_n)) \mid t_{i_1} = o_{i_1}, \ldots, t_{i_m} = o_{i_m}\}|$$

$$\mathcal{I} \models_{\Sigma} \#a(r_1 : c_1, \ldots, r_n : c_n)[r_{i_1}, \ldots, r_{i_m}] \leq \ell \quad \Leftrightarrow$$
$$\forall o_{i_1} \in C^{\mathcal{I}}(c_{i_1}), \ldots, o_{i_m} \in C^{\mathcal{I}}(c_{i_m}) .$$
$$|\{t \in A^{\mathcal{I}}(a(r_1 : c_1, \ldots, r_n : c_n)) \mid t_{i_1} = o_{i_1}, \ldots, t_{i_m} = o_{i_m}\}| \leq [\![\ell]\!]$$

where $[\![-]\!] : NumLit \to \mathbb{Z}$ maps a numerical literal to an integer.

Theorem 1. $\text{ClDiag} = (\text{Cl}, Mult, \text{Inst}, \models)$ *forms an institution.*

Proof. It remains to prove the satisfaction condition $\mathcal{J} \models_T \sigma(\varphi) \Leftrightarrow \mathcal{J}|\sigma \models_{\Sigma} \varphi$ for each $\sigma : \Sigma \to T$ in Cl, each $\mathcal{J} \in |\text{Inst}(T)|$, and each $\varphi \in Mult(\Sigma)$, which follows by induction on the structure of formulae. $\qquad\qquad\square$

4.2   An Institution for UML Class Diagrams with Rigidity Constraints

Let $\mathscr{I} = (\text{Sig}^{\mathscr{I}}, Sen^{\mathscr{I}}, \text{Mod}^{\mathscr{I}}, \models^{\mathscr{I}})$ be an institution. We define an institution $\wp_*^i \mathscr{I}$ over $\mathscr{I}$ that has as models pointed powersets $(M_0, \mathcal{M})$ of models of $\mathscr{I}$; inherits the sentences from $\mathscr{I}$ which now have to hold in all models of $\mathcal{M}$; and also requires that in each model $(M_0, \mathcal{M})$ over a signature Σ all elements of $\mathcal{M}$ "behave" like M_0 for a certain part Γ of the signature Σ.

Define the signature category $\text{Sig}^{\wp_*^i \mathscr{I}}$ of $\wp_*^i \mathscr{I}$ as follows: The objects of $\text{Sig}^{\wp_*^i \mathscr{I}}$ are the morphisms $\sigma : \Gamma \to \Sigma$ in $\text{Sig}^{\mathscr{I}}$. A morphism $(\gamma, \rho) : (\sigma : \Gamma \to \Sigma) \to (\sigma' : \Gamma' \to \Sigma')$ of $\text{Sig}^{\wp_*^i \mathscr{I}}$ consists of morphisms $\gamma : \Gamma \to \Gamma'$ and $\rho : \Sigma \to \Sigma'$ in $\text{Sig}^{\mathscr{I}}$ such that $\sigma' \circ \gamma = \rho \circ \sigma$.

Define the sentence functor $Sen^{\wp_*^i \mathscr{I}} : \text{Sig}^{\wp_*^i \mathscr{I}} \to \text{Set}$ of $\wp_*^i \mathscr{I}$ by $Sen^{\wp_*^i \mathscr{I}}(\sigma : \Gamma \to \Sigma) = Sen^{\mathscr{I}}(\Sigma)$ and $Sen^{\wp_*^i \mathscr{I}}(\gamma, \rho) = Sen^{\mathscr{I}}(\rho)$.

Define the contra-variant model functor $\text{Mod}^{\wp_*^i \mathscr{I}} : (\text{Sig}^{\wp_*^i \mathscr{I}})^{\text{op}} \to \text{Cat}$ of $\wp_*^i \mathscr{I}$ as follows: Let $(\sigma : \Gamma \to \Sigma) \in |\text{Sig}^{\wp_*^i \mathscr{I}}|$. The objects of $\text{Mod}^{\wp_*^i \mathscr{I}}(\sigma)$ are the pairs $(M_0, \mathcal{M})$ with $M_0 \in |\text{Mod}^{\mathscr{I}}(\Sigma)|$ and $\mathcal{M}$ a sub-*set* of $|\text{Mod}^{\mathscr{I}}(\Sigma)|$ such that $M_0 \in \mathcal{M}$ and $\text{Mod}^{\mathscr{I}}(\sigma)(M) = \text{Mod}^{\mathscr{I}}(\sigma)(M_0)$ for each $M \in \mathcal{M}$. A morphism of $\text{Mod}^{\wp_*^i \mathscr{I}}(\sigma)$ from $(M_0, \mathcal{M})$ to $(M_0', \mathcal{M}')$ is given by a morphism $\chi_0 : M_0 \to M_0'$ in $\text{Mod}^{\mathscr{I}}(\sigma)$ where $\mathcal{M}$ and $\mathcal{M}'$ satisfy the condition that for every $M \in \mathcal{M}$ there is an $M' \in \mathcal{M}'$ and a morphism $\chi : M \to M'$. — Let $(\gamma, \rho) : (\sigma : \Gamma \to \Sigma) \to (\tau : \Delta \to T)$ be a morphism in $\text{Sig}^{\wp_*^i \mathscr{I}}$. Then $\text{Mod}^{\wp_*^i \mathscr{I}}(\gamma, \rho)(N_0, \mathcal{N}) = (\text{Mod}^{\mathscr{I}}(\rho)(N_0), \{\text{Mod}^{\mathscr{I}}(\rho)(N) \mid N \in \mathcal{N}\})$ and $\text{Mod}^{\wp_*^i \mathscr{I}}(\gamma, \rho)(\psi_0) = \text{Mod}^{\mathscr{I}}(\rho)(\psi_0)$.

For each $(\sigma : \Gamma \to \Sigma) \in |\text{Sig}^{\wp_*^i \mathscr{I}}|$, define the satisfaction relation $- \models_{\sigma : \Gamma \to \Sigma}^{\wp_*^i \mathscr{I}} - \subseteq |\text{Mod}^{\wp_*^i \mathscr{I}}(\sigma : \Gamma \to \Sigma)| \times Sen^{\wp_*^i \mathscr{I}}(\sigma : \Gamma \to \Sigma)$ of $\wp_*^i \mathscr{I}$ by $(M_0, \mathcal{M}) \models_{\sigma : \Gamma \to \Sigma}^{\wp_*^i \mathscr{I}} \varphi \Leftrightarrow \forall M \in \mathcal{M} . M \models_{\Sigma}^{\mathscr{I}} \varphi$.

Theorem 2. $(\mathrm{Sig}^{\wp_*^i \mathscr{I}}, Sen^{\wp_*^i \mathscr{I}}, \mathrm{Mod}^{\wp_*^i \mathscr{I}}, \models^{\wp_*^i \mathscr{I}})$ *forms an institution.*   □

We now apply this construction to the UML class diagram institution ClDiag to represent rigidity. In the following, we assume all signature elements of the class diagram to be consistently and completely annotated with stereotypes «rigid» and «dynamic». Consistency rules include: Boolean is «rigid»; if a classifier c is «dynamic», then List$[c]$ is «dynamic» as well, etc. Classifiers c that are added by the signature closure are «rigid» unless forced to be «dynamic» by consistency rules. We form a class net consisting only of those elements that are stereotyped with «rigid»; additionally we form the full class net. Since the signatures of ClDiag are set-based, there is an inclusion morphism which we take as the signature in $\wp_*^i$ClDiag. Thus, the meaning of all elements stereotyped with «rigid» is fixed by their meanings in the distinguished state. In fact, for representing UML class diagrams with rigidity constraints it is enough to work with those signatures in $\wp_*^i$ClDiag which are inclusions in the category of signatures Cl of ClDiag. We denote this institution by $\wp_*^{\hookrightarrow}$ClDiag.

There is a straightforward simple institution comorphism from $\wp_*^i \mathscr{I}$ to $\mathscr{I}$ (which also applies to $\wp_*^{\hookrightarrow}$ClDiag): Define the functor $\Phi^{\wp_*^i \mathscr{I}, \mathscr{I}} : \mathrm{Sig}^{\wp_*^i \mathscr{I}} \to \mathrm{Pres}^{\mathscr{I}}$ by $\Phi^{\wp_*^i \mathscr{I}, \mathscr{I}}(\sigma : \Gamma \to \Sigma) = (\Sigma, \emptyset)$ and $\Phi^{\wp_*^i \mathscr{I}, \mathscr{I}}(\gamma, \rho) = \rho$. Define the natural transformation $\alpha^{\wp_*^i \mathscr{I}, \mathscr{I}} : Sen^{\wp_*^i \mathscr{I}} \xrightarrow{\cdot} Sen^{\mathscr{I}} \circ \Phi^{\wp_*^i \mathscr{I}, \mathscr{I}}$ by $\alpha^{\wp_*^i \mathscr{I}, \mathscr{I}}_{\sigma:\Gamma \to \Sigma}(\varphi) = \varphi$ for $\varphi \in Sen^{\mathscr{I}}(\Sigma)$. Define the natural transformation $\beta^{\wp_*^i \mathscr{I}, \mathscr{I}} : \mathrm{Mod}^{\mathscr{I}} \circ (\Phi^{\wp_*^i \mathscr{I}, \mathscr{I}})^{\mathrm{op}} \xrightarrow{\cdot} \mathrm{Mod}^{\wp_*^i \mathscr{I}}$ by $\beta^{\wp_*^i \mathscr{I}, \mathscr{I}}_{\sigma:\Gamma \to \Sigma}(M) = (M, \{M\})$ and $\beta^{\wp_*^i \mathscr{I}, \mathscr{I}}_{\sigma:\Gamma \to \Sigma}(\chi) = \chi$. Then we obtain:

Theorem 3. $\mu^{\wp_*^i \mathscr{I}, \mathscr{I}} = (\Phi^{\wp_*^i \mathscr{I}, \mathscr{I}}, \alpha^{\wp_*^i \mathscr{I}, \mathscr{I}}, \beta^{\wp_*^i \mathscr{I}, \mathscr{I}})$ *forms a simple institution comorphism from* $\wp_*^i \mathscr{I}$ *to* $\mathscr{I}$.   □

4.3   From UML Class Diagrams with Rigidity Constraints to MODALCASL

In the following, we give a sketch of a simple institution comorphism from UML class diagrams with rigidity constraints to MODALCASL: Let $(\Gamma \subseteq \Sigma) \in |\mathrm{Sig}^{\wp_*^{\hookrightarrow}\mathrm{ClDiag}}|$ with classifier net $\Sigma = ((C, \leq_C), K, P, M, A)$. We map $(\Gamma \subseteq \Sigma)$ to the following MODALCASL-presentation $\Phi(\Gamma \subseteq \Sigma) = (((S, TF, PF, P, \leq), rigid, modalities, modalitySorts), Ax)$:

Let us denote the connected components of $\leq_C$ by $cc(\leq_C)$ and the connected component of a class name[2] $c \in C$ by $[c]_{\leq_C}$, i.e., $cc(\leq_C) = \{[c]_{\leq_C} \mid c \in C\}$. Then,

$$S = C \cup cc(\leq_C), \quad \leq \, = \, \leq_C \cup \{(c, [c]_{\leq_C}) \mid c \in C\} \cup \{([c]_{\leq_C}, [c]_{\leq_C}) \mid c \in C\}$$

For the classifier Boolean and classifiers involving type formers such as Pair$[c_1, c_2]$ we include and (possibly) instantiate specifications from the CASL Basic Datatypes [23]. For example, the sort $PairConnectorConnector$ is specified by instantiating the CASL PAIR specification and renaming the sort name:

PAIR[**sort** *Connector*][**sort** *Connector*] **with**
    Pair[**sort** *Connector*][**sort** *Connector*] $\mapsto$ *PairConectorConector*

---

[2] Here we assume $[c]_{\leq_C} = t$, if t is the top sort of the component of c.

Here, we assume that the semantics of each predefined type and type former is an element of the model class of the respective monomorphic CASL datatype.

Considering the running example, we obtain:

%% Classes:
sorts *Net, Station, Unit, . . . , UID*
%% Hierarchy:
sorts *Point, Linear < Unit; . . . ; Route < ListPairUnitPath*

The total function symbols *TF* just comprise the instance specification declarations, the partial function symbols *PF* the property declarations:

$$TF = \{k : \; \to c \mid k : c \in K\}$$
$$PF = \{p : c \times c_1 \cdots \times c_n \to? \; c' \mid c.p(x_1 : c_1, \ldots, x_n : c_n) : c' \in P\}$$

Considering the running example, we obtain:

%% Properties:
rigid op *id* : *Net* →? *UID*
flexible ops *isClosedAt* : *Unit* →? *Boolean*; . . .

The classification into "rigid" and "flexible" results directly from ($\Gamma \subseteq \Sigma$).

The predicate symbols *P* comprise the composition declarations *M* and the association declarations *A*, as well as a predicate *isAlive* that is used to model "flexible" sort interpretations in MODALCASL:

$$P = \{r : c \times c' \mid c{\bullet}r : c' \in M\} \cup$$
$$\{a : c_1 \times \cdots \times c_n \mid a(r_1 : c_1, \ldots, r_n : c_n) \in A\} \cup$$
$$\{isAlive : c \mid c \in C\}$$

Considering the running example, we obtain:

%% Compositions:
rigid preds *__has__* : *Station* × *Unit*; *__has__* : *Station* × *Track*;
%% Associations:
rigid preds *__has__* : *Unit* × *Connector*; *__has__* : *Linear* × *Connector*; . . .
%% Is Alive preds:
rigid preds *isAlive* : *Net*; *isAlive* : *Station*; *isAlive* : *Unit*; . . . ; *isAlive* : *UID*;

Once again, Γ determines which of the predicates in *P* are rigid. The predicate *isAlive* is rigid if the corresponding class is rigid. In our institution comorphism, the *modalities* set is just $\{\epsilon\}$, and the *modalitySorts* predicate is false for all sorts. That is the models have exactly one accessibility relation defined between worlds. Naturally, representing CDs in MODALCASL imposes no constraints on this relation.

Finally, we need axioms for our comorphism. The first group of axioms stipulates that arguments of operations and predicates need to be "alive". The second group of axioms corresponds to the condition on Σ-instance nets that each instance has at most one owner.

$$Ax = \{isAlive(k : c) \mid k : c \in K\}$$
$$\cup \{\forall x : c, x_1 : c_1, \ldots, x_n : c_n.\mathit{def}\ p(x, x_1, \ldots, x_n) \implies$$
$$isAlive(x) \wedge isAlive(x_1) \wedge \ldots isAlive(x_n) \mid$$
$$c.p(x_1 : c_1, \ldots, x_n : c_n) : c' \in P\}$$
$$\cup \{\forall x : c, x' : c'.r(x, x') \implies isAlive(x) \wedge isAlive(x') \mid c\bullet r : c' \in M\}$$
$$\cup \{\forall x_1 : c_1, \ldots, x_n : c_n.a(x_1, \ldots, x_n) \implies$$
$$isAlive(x_1) \wedge \ldots isAlive(x_n) \mid a(r_1 : c_1, \ldots, r_n : c_n) \in A\}$$
$$\cup \{\forall x_1 : [c_1]_{\leq_C}, x_2 : [c_2]_{\leq_C}, x : [c_1']_{\leq_C} . r(x_1, x) \wedge r(x_2, x) \Rightarrow x_1 = x_2 \mid$$
$$c_1 \bullet r_1 : c_1', c_2 \bullet r_2 : c_2' \in M, [c_1]_{\leq_C} = [c_2]_{\leq_C}, [c_1']_{\leq_C} = [c_2']_{\leq_C}\}$$

The sentences of $\wp_*^{\hookrightarrow}$ClDiag are the sentences of ClDiag. These can be systematically encoded in first order logic using "poor man's counting", e.g. [8], by providing the necessary number of variables, e.g. $\#(c\bullet r : c') \geq n$ translates to

$$\forall x : c\ \exists y_1, \ldots, y_n : c' : pwDifferent(y_1, \ldots, y_n) \wedge r(x, y_1) \wedge \cdots \wedge r(x, y_n)$$

i.e., for every value x in c we find at least n different values in c' of which x comprises of. Here, $pwDifferent(y_1, \ldots, y_n)$ encodes $y_i \neq y_j$ for $i \neq j$, $1 \leq i, j \leq n$. A typical example is:

$$2 \leq \#\mathsf{has}(connector : Connector, unit : Unit)[unit] \mapsto$$
$$\forall u : Unit . \exists c1, c2 : Connector . c1 \neq c2 \wedge has(c1, u) \wedge has(c2, u)$$

Let $\mathcal{M}$ be a MODALCASL model in $|\mathrm{Mod}^{\mathrm{MODALCASL}}(\Phi(\Gamma \subseteq \Sigma))|$, let W be the sets of worlds in $\mathcal{M}$, $i \in W$ the initial world, and M_w the CASL models associated with $w \in W$. In the following, we use the abbreviations $\lceil x \rceil = (inj_{c,[c]})_{M_w}(x)$ for $x \in c_{M_w}$; and $\lfloor x \rfloor = (pr_{[c],c})_{M_w}(x)$ for $x \in [c]_{M_w}$. We first define for each w, how to turn its associated CASL model M_w into a Σ-instance net $\beta(M_w) = (C^{M_w}, K^{M_w}, P^{M_w}, M^{M_w}, A^{M_w})$:

- If c does not involve a type former: $C^{M_w}(c) = \{\lceil x \rceil \mid x \in c_{M_w}, isAlive(x)\}$.
  If c involves a type former, we take the representation of the corresponding type in the instance net.
- $K^{M_w}(k : c) = \lceil (k_{\langle\rangle,c})_{M_w} \rceil$.
- $P^{M_w}(c.p(x_1 : c_1, \ldots, x_n : c_n) : c')(y, y_1, \ldots, y_n) = \lceil (p_{\langle c,c_1,\ldots,c_n\rangle,c'})_{M_w}(\lfloor y \rfloor, \lfloor y_1 \rfloor, \ldots, \lfloor y_n \rfloor) \rceil$ for all $y \in C^{M_w}(c), y_1 \in C^{M_w}(c_1), \ldots, y_n \in C^{M_w}(c_n)$.
- $M^{M_w}(c\bullet r : c')(x) = \{\lceil y \rceil \mid (r_{c,c'})_{M_w}(\lfloor x \rfloor, y)\}$ for all $x \in C^{M_w}(c)$.
- $A^{M_w}(a(r_1 : c_1, \ldots, r_n : c_n)) = \{(\lceil x_1 \rceil, \ldots, \lceil x_n \rceil) \mid (x_1, \ldots, x_n) \in (a_{\langle c_1, \ldots c_n\rangle})_{M_w}\}$.

By construction, $\beta(\mathcal{M}) = (\beta(M_i), \{\beta(M_w) \mid w \in W\})$ is a model in $|\mathrm{Mod}^{\wp_*^{\hookrightarrow}\mathrm{ClDiag}}(\Gamma \subseteq \Sigma)|$.

5   Crafting a Formal DSL for the Railway Domain

Based on the informal DSL as presented in Section 2, we now demonstrate how to create a formal DSL with the techniques developed so far. This process fits with the "Designing DSLs for verification" methodology presented in [15].

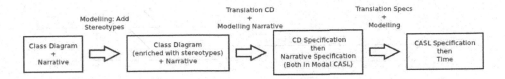

Fig. 2. Process for capturing a DSL

Step 1. Start by creating a UML CD giving concepts and relationships between concepts within the desired domain. Add an accompanying narrative as illustrated in Section 2. Next, in a modelling step, add stereotypes (as discussed in Section 2). These stereotypes are used to describe elements which remain static or are dynamic within the system. *Result:* CD and Narrative – as one can find in industry, see, e.g., [13].

Step 2. Using the comorphism defined in Section 4.3, translate the CD into MODAL-CASL. Next, extend the resulting MODALCASL specification by a modelling of any narrative elements not captured within the CD. *Result:* MODALCASL specification capturing CD and Narrative.

For example, considering the narrative of what it means for the *Route* to be "open" from Bjørner's DSL – see Section 2 – we can produce the following MODALCASL specification:

    flexible op *isOpen* : *Route* →? *Boolean*
    ∀ *r* : *Route*; *upp* : *UnitPathPair*; *u* : *Unit*; *us* : *UnitState*; *p* : *Path*
    • *isOpen(r)* = *tt*
      ⇔ *r has upp* ∧ *upp getUnit u* ∧ *upp getPath p* ∧ *u stateAt us* ∧ *us has p*

Also, given the methodology presented in [15], at this point any verification conditions required can be modelled and added to the specification. In the case of Bjørner's DSL, one would like to check that two *Route*s which share any *Unit*s (i.e. are "conflicting"):

    rigid pred __*inConflict*__ : *Route* × *Route*
    ∀ *r1, r2* : *Route*
    • *r1 inConflict r2*
      ⇔ ¬ *r1* = *r2* ∧ ∃ *upp1, upp2* : *UnitPathPair*; *u* : *Unit*
        • *r1 has upp1* ∧ *r2 has upp2* ∧ *upp1 getUnit u* ∧ *upp2 getUnit u*

are not "open" at the same time:

    %% Verification condition:
    ∀ *r1, r2* : *Route* • *r1 inConflict r2* ⇒ ¬ (*isOpen(r1)* = *tt* ∧ *isOpen(r2)* = *tt*)

Step 3. The next step of the process is to apply the existing MODALCASL to CASL comorphism [21] to gain a CASL Specification of the DSL. This will allow connection to multiple theorem provers for later use in verification. If appropriate, one can specify in CASL how worlds evolve, e.g., by adding a loose specification of discrete time. *Result:* CASL specification of the formal DSL.

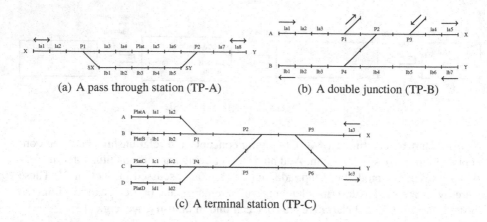

(a) A pass through station (TP-A)   (b) A double junction (TP-B)

(c) A terminal station (TP-C)

Fig. 3. Verified track plans

Using the illustrative examples from Section 4.3, the translation results in the following CASL specification:

> **sorts** *AllStates, Boolean, Connector, Line, Linear,...*
> **sorts** *Line, Track < ListLinear; UnitState < SetPath;*
> **op**   *isClosed : g_World × Unit → Boolean; ...*
> **pred** *__hasStation__ : Net × Station; __hasLine__ : Net × Line;...*
> **pred** *isAlive : Net; isAlive : Path; ...*
> ∀ *g_w1 : g_World; r : Route; upp : UnitPathPair; u : Unit; us : UnitState; p : Path*
> • *isOpen(g_w1, r) = tt*
>   ⇔ *r has upp ∧ upp getUnit u ∧ upp getPath p*
> ...

In this specification, we can see that all elements which were labelled as flexible have now been associated with a *g_World* representing a world in the Kripke structure of MODALCASL. Also, all elements which were labelled as rigid have been translated to elements ignoring *g_World*. The final result of the overall process is a formal DSL that can be used for verification.

5.1   Verification

Once a DSL has been captured in CASL, the techniques described in [15] can be applied. That is, the DSL can be enriched with a series of domain specific lemmas that are proven as a consequence of the DSL. Such lemmas are often specific to the property to be proven. Next, the DSL can be used to describe a particular railway network. For each railway network described, the verification condition can, thanks to the added domain specific lemmas, be automatically discharged using Hets as a broker to various automatic theorem provers [15].

With respect to Bjørner's DSL, this approach has been successfully applied, that is, Bjørner's DSL has been successfully extended for verification. The result of this work is that we have successfully automatically verified a number of track plans of real world complexity against the verification condition that "Conflicting open routes can not be open at the same time". Several of these track plans are illustrated in Figure 3.

Verification times for all track plans are within the region of seconds[3], with TP-A requiring 9.79s, TP-B requiring 2.94s and TP-C requiring 26.35s. The increase in time for track plan TP-C is due to the large number (48 in total) of conflicting routes.

With respect to the Invensys Rail Data Model [13], translation of the CDs required for verification and the modelling of the narrative has resulted in a formal DSL in CASL. The development of domain knowledge for verification is ongoing work. First results on small scale verification problems demonstrate that the full approach is successful.

6   Conclusion

We have suggested a methodology for formal DSL design. Starting with an industrial document in the from a UML class diagram with accompanying natural language descriptions, we have defined a translation based process to obtain a formal specification in CASL that can be used for verification. To this end, we have developed the theoretical concepts required, including an institution for UML CDs and a institution comorphisms to MODALCASL. Finally, we have illustrated the approach works by applying the process to Bjørner's DSL for the railway domain.

Future work. The Heterogeneous Toolset [22] makes extensive use of institutions and and their relations for tool support for various logics. We intend to implement the presented institutions and translations within Hets. Further to this, we hope to complete the development of suitable domain specific lemmas for the Invensys Rail Data Model, and to produce, as outlined in [15], a graphical front end with automatic proof support for use by railway engineers. On the more theoretical side, we want to further investigate the relations of the institutions involved.

Acknowledgements. The authors would like to thank S. Chadwick and D. Taylor from Invensys Rail for their continued support and encouraging feedback. We also thank Erwin R. Catesbeiana for sharing with us his stereotypical views on objects, flies, and the dynamics of the world.

References

1. Andova, S., van den Brand, M., Engelen, L.: Prototyping the Semantics of a DSL using ASF+SDF: Link to Formal Verification of DSL Models. In: AMMSE 2011. Electr. Proc. Theo. Comp. Sci., vol. 56, pp. 65–79 (2011)
2. Arnold, B., van Deursen, A., Res, M.: An Algebraic Specification of a Language for Describing Financial Products. In: Wirsing, M. (ed.) Wsh. Formal Methods Applications in Software Engineering Practice, Seattle, pp. 6–13 (1995)
3. Bjørner, D.: Dynamics of Railway Nets: On an Interface between Automatic Control and Software Engineering. In: CTS 2003. Elsevier (2003)

---

[3] PC with Intel 3.0GHz Quad Core processor with 8GB RAM running Ubuntu 12.10.

4. Bjørner, D.: Domain Engineering – Technology Management, Research and Engineering. JAIST Press (2009)
5. Bonachea, D., Fisher, K., Rogers, A., Smith, F.: Hancock: A Language for Processing Very Large-scale Data. SIGPLAN Notices 35(1) (2000)
6. Cengarle, M.V., Knapp, A., Tarlecki, A., Wirsing, M.: A Heterogeneous Approach to UML Semantics. In: Degano, P., De Nicola, R., Meseguer, J. (eds.) Concurrency, Graphs and Models. LNCS, vol. 5065, pp. 383–402. Springer, Heidelberg (2008)
7. dos Santos, O.M., Woodcock, J., Paige, R.: Using Model Transformation to Generate Graphical Counter-Examples for the Formal Analysis of xUML Models. In: 16th Int. Conf. Engineering of Complex Computer Systems (ICECCS 2011), pp. 117–126. IEEE (2011)
8. Ebbinghaus, H., Flum, J., Thomas, W.: Mathematical logic. Springer (1994)
9. Fowler, M.: Domain Specific Languages. Addison-Wesley (2010)
10. Hughes, G.E., Cresswell, M.J.: A new introduction to modal logic. Routledge (1996)
11. Goguen, J., Burstall, R.: Institutions: Abstract model theory for specification and programming. Journal of the ACM 39, 95–146 (1992)
12. Hussmann, H., Cerioli, M., Baumeister, H.: From uml to casl (static part). Technical Report DISI-TR-00-06, DISI-Universit di Genova (2000)
13. Invensys Rail. Data Model – Version 1A (2010)
14. James, P., Knapp, A., Mossakowski, T., Roggenbach, M.: From UML Class Diagrams to Modal CASL. Technical report, Universität Augsburg (to appear, 2013)
15. James, P., Roggenbach, M.: Designing domain specific languages for verification: First steps. In: Höfner, P., McIver, A., Struth, G. (eds.) 1st Wsh. Automated Theory Engineering (ATE 2011). CEUR Wsh. Proc., vol. 760, pp. 40–45. CEUR-WS.org (2011)
16. Kerr, D., Rowbotham, T.: Introduction to Railway Signalling. Institution of Railway Signal Engineers (2001)
17. Lano, K., Clark, D., Androutsopoulos, K.: UML to B: Formal Verification of Object-Oriented Models. In: Boiten, E.A., Derrick, J., Smith, G.P. (eds.) IFM 2004. LNCS, vol. 2999, pp. 187–206. Springer, Heidelberg (2004)
18. Meng, S., Aichernig, B.: Towards a Coalgebraic Semantics of UML: Class Diagrams and Use Cases. Technical Report 272, UNU/IIST (2003)
19. Mernik, M., Heering, J., Sloane, A.M.: When and How to Develop Domain-specific Languages. ACM Computing Surveys 37(4) (2005)
20. Mossakowski, T.: Relating CASL with Other Specification Languages: The Institution Level. Theo. Comp. Sci. 286, 367–475 (2002)
21. Mossakowski, T.: ModalCASL — Specification with Multi-Modal Logics. Language Summary (2004)
22. Mossakowski, T.: HeTS — the Heterogeneous Tool Set, home page (June 2011), http://www.informatik.uni-bremen.de/agbkb/forschung/formal_methods/CoFI/hets/
23. Mosses, P.D. (ed.): CASL Reference Manual. Springer (2004)
24. Object Managment Group. Unified Modeling Language (UML), v2.4.1 (2011), http://www.omg.org/spec/UML/2.4.1

Satisfiability Calculus: The Semantic Counterpart of a Proof Calculus in General Logics

Carlos Gustavo López Pombo[1,3], Pablo F. Castro[2,3], Nazareno M. Aguirre[2,3],
and Thomas S.E. Maibaum[4]

[1] Departmento de Computación, FCEyN, Universidad de Buenos Aires, Argentina
clpombo@dc.uba.ar
[2] Departmento de Computación, FCEFQyN, Universidad Nacional de Río Cuarto,
Argentina
{pcastro,naguirre}@dc.exa.unrc.edu.ar
[3] Consejo Nacional de Investigaciones Científicas y Técnicas (CONICET), Argentina
[4] Department of Computing & Software, McMaster University, Canada
tom@maibaum.org

Abstract. Since its introduction by Goguen and Burstall in 1984, the theory of institutions has been one of the most widely accepted formalizations of abstract model theory. This work was extended by a number of researchers, José Meseguer among them, who presented *General Logics*, an abstract framework that complements the model theoretical view of institutions by defining the categorical structures that provide a proof theory for any given logic. In this paper we intend to complete this picture by providing the notion of *Satisfiability Calculus*, which might be thought of as the semantical counterpart of the notion of proof calculus, that provides the formal foundations for those proof systems that use model construction techniques to prove or disprove a given formula, thus "implementing" the satisfiability relation of an institution.

1 Introduction

The theory of institutions, presented by Goguen and Burstall in [1], provides a formal and generic definition of what a logical system is, from a model theoretical point of view. This work evolved in many directions: in [2], Meseguer complemented the theory of institutions by providing a categorical characterization for the notions of entailment system (also called π-institutions by other authors in [3]) and the corresponding notion of proof calculi; in [4,5] Goguen and Burstall, and Tarlecki, respectively, extensively investigated the ways in which institutions can be related; in [6], Sannella and Tarlecki studied how specifications in an arbitrary logical system can be structured; in [7], Tarlecki presented an abstract theory of software specification and development; in [8,9] and [10,11], Mossakowski and Tarlecki, and Diaconescu, respectively, proposed the use of institutions as a foundation for heterogeneous environments for software specification. Institutions have also been used as a very general version of abstract

N. Martí-Oliet and M. Palomino (Eds.): WADT 2012, LNCS 7841, pp. 195–211, 2013.

model theory [12], offering a suitable formal framework for addressing heterogeneity in specifications [13,14], including applications to UML [15] and other languages related to computer science and software engineering.

Extensions of institutions to capture proof theoretical concepts have been extensively studied, most notably by Meseguer [2]. Essentially, Meseguer proposes the extension of entailment systems with a categorical concept expressive enough to capture the notion of *proof* in an abstract way. In Meseguer's words:

> *A reasonable objection to the above definition of logic[1] is that it abstracts away the structure of proofs, since we know only that a set Γ of sentences entails another sentence φ, but no information is given about the internal structure of such a $\Gamma \vdash \varphi$ entailment. This observation, while entirely correct, may be a virtue rather than a defect, because the entailment relation is precisely what remains <u>invariant</u> under many equivalent proof calculi that can be used for a logic.*

Before Meseguer's work, there was an imbalance in the definition of a logic in the context of institution theory, since the deductive aspects of a logic were not taken into account. Meseguer concentrates on the proof theoretical aspects of a logic, providing not only the definition of entailment system, but also complementing it with the notion of proof calculus, obtaining what he calls a *logical system*. As introduced by Meseguer, the notion of proof calculus provides, intuitively, an implementation of the entailment relation of a logic. Indeed, Meseguer corrected the inherent imbalance in favour of models in institutions, enhancing syntactic aspects in the definition of logical systems.

However, the same lack of an operational view observed in the definition of entailment systems still appears with respect to the notion of satisfiability, i.e., the satisfaction relation of an institution. In the same way that an entailment system may be "implemented" in terms of different proof calculi, a satisfaction relation may be "implemented" in terms of different satisfiability procedures. Making these satisfiability procedures explicit in the characterization of logical systems is highly relevant, since many successful software analysis tools are based on particular characteristics of these satisfiability procedures. For instance, many automated analysis tools rely on model construction, either for proving properties, as with model-checkers, or for finding counterexamples, as with tableaux techniques or SAT-solving based tools. These techniques constitute an important stream of research in logic, in particular in relation to (semi-)automated software validation and verification.

These kinds of logical systems can be traced back to the works of Beth [16,17], Herbrand [18] and Gentzen [19]. Beth's ideas were used by Smullyan to formulate the tableaux method for first-order predicate logic [20]. Herbrand's and Gentzen's works inspired the formulation of resolution systems presented by Robinson [21]. Methods like those based on resolution and tableaux are strongly related to the semantics of a logic; one can often use them to guide

---

[1] **Authors' note**: Meseguer refers to a logic as a structure that is composed of an entailment system together with an institution, see Def. 6.

the construction of models. This is not possible in *pure* deductive methods, such as natural deduction or Hilbert systems, as formalized by Meseguer. In this paper, our goal is to provide an abstract characterization of this class of semantics based tools for logical systems. This is accomplished by introducing a categorical characterization of the notion of satisfiability calculus which embraces logical tools such as tableaux, resolution, Gentzen style sequents, etc. As we mentioned above, this can be thought of as a formalization of a semantic counterpart of Meseguer's proof calculus. We also explore the concept of mappings between satisfiability calculi and the relation between proof calculi and satisfiability calculi.

The paper is organized as follows. In Section 2 we present the definitions and results we will use throughout this paper. In Section 3 we present a categorical formalization of satisfiability calculus, and prove relevant results underpinning the definitions. We also present examples to illustrate the main ideas. Finally in Section 4 we draw some conclusions and describe further lines of research.

2   Preliminaries

From now on, we assume the reader has a nodding acquaintance with basic concepts from category theory [22,23]. Below we present the basic definitions and results we use throughout the rest of the paper. In the following, we follow the notation introduced in [2].

An *Institution* is an abstract formalization of the model theory of a logic by making use of the relationships existing between signatures, sentences and models. These aspects are reflected by introducing the category of signatures, and by defining functors going from this category to the categories Set and Cat, to capture sets of sentences and categories of models, respectively, for a given signature. The original definition of institutions is the following:

Definition 1. ([1]) *An* institution *is a structure of the form* $\langle$Sign, **Sen**, **Mod**, $\{\models^{\Sigma}\}_{\Sigma \in |\text{Sign}|}\rangle$ *satisfying the following conditions:*

- Sign *is a category of signatures,*
- **Sen** : Sign $\rightarrow$ Set *is a functor. Let* $\Sigma \in |\text{Sign}|$, *then* **Sen**(Σ) *returns the set of* Σ-sentences,
- **Mod** : Sign$^{\text{op}} \rightarrow$ Cat *is a functor. Let* $\Sigma \in |\text{Sign}|$, *then* **Mod**(Σ) *returns the category of* Σ-models,
- $\{\models^{\Sigma}\}_{\Sigma \in |\text{Sign}|}$, *where* $\models^{\Sigma} \subseteq |\textbf{Mod}(\Sigma)| \times \textbf{Sen}(\Sigma)$, *is a family of binary relations,*

and for any signature morphism $\sigma : \Sigma \rightarrow \Sigma'$, Σ-sentence $\phi \in$ **Sen**(Σ) *and* Σ'-model $\mathcal{M}' \in |\textbf{Mod}(\Sigma)|$, *the following* $\models$-invariance condition holds:*

$$\mathcal{M}' \models^{\Sigma'} \textbf{Sen}(\sigma)(\phi) \quad iff \quad \textbf{Mod}(\sigma^{\text{op}})(\mathcal{M}') \models^{\Sigma} \phi .$$

Roughly speaking, the last condition above says that *the notion of truth is invariant with respect to notation change.* Given $\Sigma \in |\text{Sign}|$ and $\Gamma \subseteq$ **Sen**(Σ),

$\mathbf{Mod}(\Sigma, \Gamma)$ denotes the full subcategory of $\mathbf{Mod}(\Sigma)$ determined by those models $\mathcal{M} \in |\mathbf{Mod}(\Sigma)|$ such that $\mathcal{M} \models^{\Sigma} \gamma$, for all $\gamma \in \Gamma$. The relation $\models^{\Sigma}$ between sets of formulae and formulae is defined in the following way: given $\Sigma \in |\mathsf{Sign}|$, $\Gamma \subseteq \mathbf{Sen}(\Sigma)$ and $\alpha \in \mathbf{Sen}(\Sigma)$, $\Gamma \models^{\Sigma} \alpha$ if and only if $\mathcal{M} \models^{\Sigma} \alpha$, for all $\mathcal{M} \in |\mathbf{Mod}(\Sigma, \Gamma)|$.

An *entailment system* is defined in a similar way, by identifying a family of *syntactic* consequence relations, instead of a family of semantic consequence relations. Each of the elements in this family is associated with a signature. These relations are required to satisfy reflexivity, monotonicity and transitivity. In addition, a notion of translation between signatures is considered.

Definition 2. ([2]) *An* entailment system *is a structure of the form* $\langle \mathsf{Sign}, \mathbf{Sen}, \{\vdash^{\Sigma}\}_{\Sigma \in |\mathsf{Sign}|} \rangle$ *satisfying the following conditions:*

- Sign *is a category of signatures,*
- $\mathbf{Sen} : \mathsf{Sign} \to \mathsf{Set}$ *is a functor. Let* $\Sigma \in |\mathsf{Sign}|$*; then* $\mathbf{Sen}(\Sigma)$ *returns the set of* Σ-*sentences, and*
- $\{\vdash^{\Sigma}\}_{\Sigma \in |\mathsf{Sign}|}$*, where* $\vdash^{\Sigma} \subseteq 2^{\mathbf{Sen}(\Sigma)} \times \mathbf{Sen}(\Sigma)$*, is a family of binary relations such that for any* $\Sigma, \Sigma' \in |\mathsf{Sign}|$*,* $\{\phi\} \cup \{\phi_i\}_{i \in \mathcal{I}} \subseteq \mathbf{Sen}(\Sigma)$*,* $\Gamma, \Gamma' \subseteq \mathbf{Sen}(\Sigma)$*, the following conditions are satisfied:*
  1. *reflexivity:* $\{\phi\} \vdash^{\Sigma} \phi$*,*
  2. *monotonicity: if* $\Gamma \vdash^{\Sigma} \phi$ *and* $\Gamma \subseteq \Gamma'$*, then* $\Gamma' \vdash^{\Sigma} \phi$*,*
  3. *transitivity: if* $\Gamma \vdash^{\Sigma} \phi_i$ *for all* $i \in \mathcal{I}$ *and* $\{\phi_i\}_{i \in \mathcal{I}} \vdash^{\Sigma} \phi$*, then* $\Gamma \vdash^{\Sigma} \phi$*, and*
  4. $\vdash$-*translation: if* $\Gamma \vdash^{\Sigma} \phi$*, then for any morphism* $\sigma : \Sigma \to \Sigma'$ *in* Sign*,* $\mathbf{Sen}(\sigma)(\Gamma) \vdash^{\Sigma'} \mathbf{Sen}(\sigma)(\phi)$*.*

Definition 3. ([2]) *Let* $\langle \mathsf{Sign}, \mathbf{Sen}, \{\vdash^{\Sigma}\}_{\Sigma \in |\mathsf{Sign}|} \rangle$ *be an entailment system. Its category* Th *of theories is a pair* $\langle \mathcal{O}, \mathcal{A} \rangle$ *such that:*

- $\mathcal{O} = \{ \langle \Sigma, \Gamma \rangle \mid \Sigma \in |\mathsf{Sign}| \text{ and } \Gamma \subseteq \mathbf{Sen}(\Sigma) \}$*, and*
- $\mathcal{A} = \left\{ \sigma : \langle \Sigma, \Gamma \rangle \to \langle \Sigma', \Gamma' \rangle \;\middle|\; \begin{array}{l} \langle \Sigma, \Gamma \rangle, \langle \Sigma', \Gamma' \rangle \in \mathcal{O}, \\ \sigma : \Sigma \to \Sigma' \text{ is a morphism in } \mathsf{Sign} \text{ and} \\ \text{for all } \gamma \in \Gamma, \Gamma' \vdash^{\Sigma'} \mathbf{Sen}(\sigma)(\gamma) \end{array} \right\}.$

In addition, if a morphism $\sigma : \langle \Sigma, \Gamma \rangle \to \langle \Sigma', \Gamma' \rangle$ satisfies $\mathbf{Sen}(\sigma)(\Gamma) \subseteq \Gamma'$, it is called *axiom preserving*. By retaining those morphisms of Th that are axiom preserving, we obtain the subcategory Th_0. If we now consider the definition of $\mathbf{Mod}$ extended to signatures and sets of sentences, we get a functor $\mathbf{Mod} : \mathsf{Th}^{\mathsf{op}} \to \mathsf{Cat}$ defined as follows: let $T = \langle \Sigma, \Gamma \rangle \in |\mathsf{Th}|$, then $\mathbf{Mod}(T) = \mathbf{Mod}(\Sigma, \Gamma)$.

Definition 4. ([2]) *Let* $\langle \mathsf{Sign}, \mathbf{Sen}, \{\vdash^{\Sigma}\}_{\Sigma \in |\mathsf{Sign}|} \rangle$ *be an entailment system and* $\langle \Sigma, \Gamma \rangle \in |\mathsf{Th}_0|$*. We define* $\bullet : 2^{\mathbf{Sen}(\Sigma)} \to 2^{\mathbf{Sen}(\Sigma)}$ *as follows:* $\Gamma^{\bullet} = \{ \gamma \mid \Gamma \vdash^{\Sigma} \gamma \}$*. This function is extended to elements of* Th_0*, by defining it as follows:* $\langle \Sigma, \Gamma \rangle^{\bullet} = \langle \Sigma, \Gamma^{\bullet} \rangle$*.* $\Gamma^{\bullet}$ *is called the theory generated by* Γ*.*

Definition 5. ([2]) *Let* $\langle \mathsf{Sign}, \mathbf{Sen}, \{\vdash^{\Sigma}\}_{\Sigma \in |\mathsf{Sign}|} \rangle$ *and* $\langle \mathsf{Sign}', \mathbf{Sen}', \{\vdash'^{\Sigma}\}_{\Sigma \in |\mathsf{Sign}'|} \rangle$ *be entailment systems,* $\Phi : \mathsf{Th}_0 \to \mathsf{Th}'_0$ *be a functor and* $\alpha : \mathbf{Sen} \to \mathbf{Sen}' \circ \Phi$ *a natural transformation.* Φ *is said to be* α-sensible *if and only if the following conditions are satisfied:*

1. *there is a functor $\Phi^\diamond$: Sign $\to$ Sign$'$ such that $\mathbf{sign}' \circ \Phi = \Phi^\diamond \circ \mathbf{sign}$, where $\mathbf{sign}$ and $\mathbf{sign}'$ are the forgetful functors from the corresponding categories of theories to the corresponding categories of signatures, that when applied to a given theory project its signature, and*
2. *if $\langle \Sigma, \Gamma \rangle \in |\mathsf{Th}_0|$ and $\langle \Sigma', \Gamma' \rangle \in |\mathsf{Th}'_0|$ such that $\Phi(\langle \Sigma, \Gamma \rangle) = \langle \Sigma', \Gamma' \rangle$, then $(\Gamma')^\bullet = (\emptyset' \cup \alpha_\Sigma(\Gamma))^\bullet$, where $\emptyset' = \alpha_\Sigma(\emptyset)$.*[2]

Φ is said to be α-simple if and only if $\Gamma' = \emptyset' \cup \alpha_\Sigma(\Gamma)$ is satisfied in Condition 2, instead of $(\Gamma')^\bullet = (\emptyset' \cup \alpha_\Sigma(\Gamma))^\bullet$.

It is straightforward to see, based on the monotonicity of $^\bullet$, that α-simplicity implies α-sensibility. An α-sensible functor has the property that the associated natural transformation α depends only on signatures. Now, from Definitions 1 and 2, it is possible to give a definition of *logic* by relating both its model-theoretic and proof-theoretic characterizations; a coherence between the semantic and syntactic relations is required, reflecting the soundness and completeness of standard deductive relations of logical systems.

Definition 6. *([2]) A logic is a structure of the form $\langle \mathsf{Sign}, \mathbf{Sen}, \mathbf{Mod}, \{\vdash^\Sigma\}_{\Sigma \in |\mathsf{Sign}|}, \{\models^\Sigma\}_{\Sigma \in |\mathsf{Sign}|} \rangle$ satisfying the following conditions:*

- *$\langle \mathsf{Sign}, \mathbf{Sen}, \{\vdash^\Sigma\}_{\Sigma \in |\mathsf{Sign}|} \rangle$ is an entailment system,*
- *$\langle \mathsf{Sign}, \mathbf{Sen}, \mathbf{Mod}, \{\models^\Sigma\}_{\Sigma \in |\mathsf{Sign}|} \rangle$ is an institution, and*
- *the following soundness condition is satisfied: for any $\Sigma \in |\mathsf{Sign}|$, $\phi \in \mathbf{Sen}(\Sigma)$, $\Gamma \subseteq \mathbf{Sen}(\Sigma)$: $\Gamma \vdash^\Sigma \phi$ implies $\Gamma \models^\Sigma \phi$.*

A logic is complete *if, in addition, the following condition is also satisfied: for any $\Sigma \in |\mathsf{Sign}|$, $\phi \in \mathbf{Sen}(\Sigma)$, $\Gamma \subseteq \mathbf{Sen}(\Sigma)$: $\Gamma \models^\Sigma \phi$ implies $\Gamma \vdash^\Sigma \phi$.*

Definition 2 associates deductive relations to signatures. As already discussed, it is important to analyze how these relations are obtained. The next definition introduces the notion of *proof calculus*. It formalizes the possibility of associating a proof-theoretic structure to the deductive relations introduced by the definitions of entailment systems. In [2, Ex. 11, pp. 15], Meseguer presents natural deduction as a proof calculus for first-order predicate logic by resorting to *multicategories* (see [2, Definition 10]).

Definition 7. *([2]) A proof calculus is a structure of the form $\langle \mathsf{Sign}, \mathbf{Sen}, \{\vdash^\Sigma\}_{\Sigma \in |\mathsf{Sign}|}, \mathbf{P}, \mathbf{Pr}, \pi \rangle$ satisfying the following conditions:*

- *$\langle \mathsf{Sign}, \mathbf{Sen}, \{\vdash^\Sigma\}_{\Sigma \in |\mathsf{Sign}|} \rangle$ is an entailment system,*
- *$\mathbf{P} : \mathsf{Th}_0 \to \mathsf{Struct}_{PC}$ is a functor. Let $T \in |\mathsf{Th}_0|$, then $\mathbf{P}(T) \in |\mathsf{Struct}_{PC}|$ is the proof-theoretical structure of T,*
- *$\mathbf{Pr} : \mathsf{Struct}_{PC} \to \mathsf{Set}$ is a functor. Let $T \in |\mathsf{Th}_0|$, then $\mathbf{Pr}(\mathbf{P}(T))$ is the set of proofs of T; the composite functor $\mathbf{Pr} \circ \mathbf{P} : \mathsf{Th}_0 \to \mathsf{Set}$ will be denoted by* **proofs***, and*

---

[2] $\emptyset'$ is not necessarily the empty set of axioms. This fact will be clarified later on.

– π : **proofs** $\overset{\cdot}{\rightarrow}$ **Sen** *is a natural transformation such that for each* $T =$ $\langle \Sigma, \Gamma \rangle \in |\mathsf{Th}_0|$ *the image of* π_T : **proofs**$(T) \rightarrow$ **Sen**(T) *is the set* $\Gamma^\bullet$. *The map* π_T *is called the* projection from proofs to theorems *for the theory* T.

Finally, a *logical system* is defined as a logic plus a proof calculus for its proof theory.

Definition 8. ([2]) *A logical system is a structure of the form*

$$\langle \mathsf{Sign}, \mathbf{Sen}, \mathbf{Mod}, \{\vdash^\Sigma\}_{\Sigma \in |\mathsf{Sign}|}, \{\models^\Sigma\}_{\Sigma \in |\mathsf{Sign}|}, \mathbf{P}, \mathbf{Pr}, \pi \rangle$$

satisfying the following conditions:

– $\langle \mathsf{Sign}, \mathbf{Sen}, \mathbf{Mod}, \{\vdash^\Sigma\}_{\Sigma \in |\mathsf{Sign}|}, \{\models^\Sigma\}_{\Sigma \in |\mathsf{Sign}|} \rangle$ *is a logic, and*
– $\langle \mathsf{Sign}, \mathbf{Sen}, \{\vdash^\Sigma\}_{\Sigma \in |\mathsf{Sign}|}, \mathbf{P}, \mathbf{Pr}, \pi \rangle$ *is a proof calculus.*

3   Satisfiability Calculus

In Section 2, we presented the definitions of institutions and entailment systems. Additionally, we presented Meseguer's categorical formulation of proof that provides operational structure for the abstract notion of entailment. In this section, we provide a categorical definition of a satisfiability calculus, providing a corresponding operational formulation of satisfiability. A satisfiability calculus is the formal characterization of a method for constructing models of a given theory, thus providing the semantic counterpart of a proof calculus. Roughly speaking, the semantic relation of satisfaction between a model and a formula can also be *implemented* by means of some kind of structure that depends on the model theory of the logic. The definition of a satisfiability calculus is as follows:

Definition 9. [**Satisfiability Calculus**] *A satisfiability calculus is a structure of the form* $\langle \mathsf{Sign}, \mathbf{Sen}, \mathbf{Mod}, \{\models^\Sigma\}_{\Sigma \in |\mathsf{Sign}|}, \mathbf{M}, \mathbf{Mods}, \mu \rangle$ *satisfying the following conditions:*

– $\langle \mathsf{Sign}, \mathbf{Sen}, \mathbf{Mod}, \{\models^\Sigma\}_{\Sigma \in |\mathsf{Sign}|} \rangle$ *is an institution,*
– $\mathbf{M} : \mathsf{Th}_0 \rightarrow \mathsf{Struct}_{SC}$ *is a functor. Let* $T \in |\mathsf{Th}_0|$, *then* $\mathbf{M}(T) \in |\mathsf{Struct}_{SC}|$ *is the model structure of* T,
– $\mathbf{Mods} : \mathsf{Struct}_{SC} \rightarrow \mathsf{Cat}$ *is a functor. Let* $T \in |\mathsf{Th}_0|$, *then* $\mathbf{Mods}(\mathbf{M}(T))$ *is the category of canonical models of* T; *the composite functor* $\mathbf{Mods} \circ \mathbf{M} :$ $\mathsf{Th}_0 \rightarrow \mathsf{Cat}$ *will be denoted by* $\mathbf{models}$, *and*
– $\mu : \mathbf{models}^{\mathsf{op}} \overset{\cdot}{\rightarrow} \mathbf{Mod}$ *is a natural transformation such that, for each* $T =$ $\langle \Sigma, \Gamma \rangle \in |\mathsf{Th}_0|$, *the image of* $\mu_T : \mathbf{models}^{\mathsf{op}}(T) \rightarrow \mathbf{Mod}(T)$ *is the category of models* $\mathbf{Mod}(T)$. *The map* μ_T *is called the* projection of the category of models *of the theory* T.

The intuition behind the previous definition is that, for any theory T, the functor $\mathbf{M}$ assigns a model structure for T in the category Struct_{SC}[3]. For instance,

---

[3] Notice that the target of functor $\mathbf{M}$, when applied to a theory T, is not necessarily a model, but a structure which, under certain conditions, can be considered a representation of the category of models of T.

in propositional tableaux, a good choice for $Struct_{SC}$ is the collection of legal tableaux, where the functor M maps a theory to the collection of tableaux obtained for that theory. The functor **Mods** projects those particular structures that represent sets of conditions that can produce canonical models of a theory $T = \langle \Sigma, \Gamma \rangle$ (i.e., the structures that represent canonical models of Γ). For example, in the case of propositional tableaux, this functor selects the open branches of tableaux, that represent satisfiable sets of formulae, and returns the collections of formulae obtained by closuring these sets. Finally, for any theory T, the functor μ_T relates each of these sets of conditions to the corresponding canonical model. Again, in propositional tableaux, this functor is obtained by relating a closured set of formulae with the models that can be defined from these sets of formulae in the usual ways [20].

Example 1. [Tableaux Method for First-Order Predicate Logic] Let us start by presenting the tableaux method for first-order logic. Let us denote by $\mathbb{I}_{FOL} = \langle \textbf{Sign}, \textbf{Sen}, \textbf{Mod}, \{\models^{\Sigma}\}_{\Sigma \in |\text{Sign}|} \rangle$ the institution of first-order predicate logic. Let $\Sigma \in |\textbf{Sign}|$ and $S \subseteq \textbf{Sen}(\Sigma)$; then a *tableau* for S is a tree such that:

1. the nodes are labeled with sets of formulae (over Σ) and the root node is labeled with S,
2. if u and v are two connected nodes in the tree (u being an ancestor of v), then the label of v is obtained from the label of u by applying one of the following rules:

$$\frac{X \cup \{A \wedge B\}}{X \cup \{A \wedge B, A, B\}} \; [\wedge] \qquad \frac{X \cup \{A \vee B\}}{X \cup \{A \vee B, A\} \quad X \cup \{A \vee B, B\}} \; [\vee]$$

$$\frac{X \cup \{\neg\neg A\}}{X \cup \{\neg\neg A, A\}} \; [\neg_1] \qquad \frac{X \cup \{A\}}{X \cup \{A, \neg\neg A\}} \; [\neg_2] \qquad \frac{X \cup \{A, \neg A\}}{\textbf{Sen}(\Sigma)} \; [false]$$

$$\frac{X \cup \{\neg(A \wedge B)\}}{X \cup \{\neg(A \wedge B), \neg A \vee \neg B\}} \; [DM_1] \qquad \frac{X \cup \{\neg(A \vee B)\}}{X \cup \{\neg(A \vee B), \neg A \wedge \neg B\}} \; [DM_2]$$

$$\frac{X \cup \{(\forall x)P(x)\}}{X \cup \{(\forall x)P(x), P(t)\}} \; [\forall] \qquad \frac{X \cup \{(\exists x)P(x)\}}{X \cup \{(\exists x)P(x), P(c)\}} \; [\exists]$$

where, in the last rules, c is a new constant and t is a ground term. A sequence of nodes $s_0 \xrightarrow{\tau_0^{\alpha_0}} s_1 \xrightarrow{\tau_1^{\alpha_1}} s_2 \xrightarrow{\tau_2^{\alpha_2}} \ldots$ is a *branch* if: *a)* s_0 is the root node of the tree, and *b)* for all $i \leq \omega$, $s_i \to s_{i+1}$ occurs in the tree, $\tau_i^{\alpha_i}$ is an instance of one of the rules presented above, and α_i are the formulae of s_i to which the rule was applied. A branch $s_0 \xrightarrow{\tau_0^{\alpha_0}} s_1 \xrightarrow{\tau_1^{\alpha_1}} s_2 \xrightarrow{\tau_2^{\alpha_2}} \ldots$ in a tableau is *saturated* if there exists $i \leq \omega$ such that $s_i = s_{i+1}$. A branch $s_0 \xrightarrow{\tau_0^{\alpha_0}} s_1 \xrightarrow{\tau_1^{\alpha_1}} s_2 \xrightarrow{\tau_2^{\alpha_2}} \ldots$ in a tableau is *closed* if there exists $i \leq \omega$ and $\alpha \in \textbf{Sen}(\Sigma)$ such that $\{\alpha, \neg\alpha\} \subseteq s_i$.

Let $s_0 \xrightarrow{\tau_0^{\alpha_0}} s_1 \xrightarrow{\tau_1^{\alpha_1}} s_2 \xrightarrow{\tau_2^{\alpha_2}} \ldots$ be a branch in a tableau. Examining the rules presented above, it is straightforward to see that every s_i with $i < \omega$ is a set of formulae. In each step, we have either the application of a rule decomposing one formula of the set into its constituent parts with respect to its major connective, while preserving satisfiability, or the application of the rule [*false*] denoting the fact that the corresponding set of formulae is unsatisfiable. Thus, the limit set of the branch is a set of formulae containing sub-formulae (and *"instances"* in the case of quantifiers) of the original set of formulae for which the tableau was built. As a result of this, every open branch expresses, by means of a set of formulae, the class of models satisfying them.

In order to define the tableau method as a satisfiability calculus, we provide formal definitions for **M**, **Mods** and μ. The proofs of the lemmas and properties shown below are straightforward using the introduced definitions. The interested reader can find these proofs in [24]. First, we introduce the category $Str^{\Sigma,\Gamma}$ of tableaux for sets of formulae over signature Σ and assuming the set of axioms Γ. In $Str^{\Sigma,\Gamma}$, objects are sets of formulae over signature Σ, and morphisms represent tableaux for the set occurring in their target and having subsets of the set of formulae occurring at the end of open branches, as their source.

Definition 10. *Let* $\Sigma \in |\text{Sign}|$ *and* $\Gamma \subseteq \mathbf{Sen}(\Sigma)$, *then we define* $Str^{\Sigma,\Gamma} = \langle \mathcal{O}, \mathcal{A} \rangle$ *such that* $\mathcal{O} = 2^{\mathbf{Sen}(\Sigma)}$ *and* $\mathcal{A} = \{\alpha : \{A_i\}_{i \in \mathcal{I}} \rightarrow \{B_j\}_{j \in \mathcal{J}} \mid \alpha = \{\alpha_j\}_{j \in \mathcal{J}}\}$, *where for all* $j \in \mathcal{J}$, α_j *is a branch in a tableau for* $\Gamma \cup \{B_j\}$ *with leaves* $\Delta \subseteq \{A_i\}_{i \in \mathcal{I}}$. *It should be noted that* $\Delta \models_\Sigma \Gamma \cup \{B_j\}$.

Lemma 1. *Let* $\Sigma \in |\text{Sign}|$ *and* $\Gamma \subseteq \mathbf{Sen}(\Sigma)$; *then* $\langle Str^{\Sigma,\Gamma}, \cup, \emptyset \rangle$, *where* $\cup : Str^{\Sigma,\Gamma} \times Str^{\Sigma,\Gamma} \rightarrow Str^{\Sigma,\Gamma}$ *is the typical bi-functor on sets and functions, and* $\emptyset$ *is the neutral element for* $\cup$, *is a strict monoidal category.*

Using this definition we can introduce the category of legal tableaux, denoted by $Struct_{SC}$.

Definition 11. $Struct_{SC}$ *is defined as* $\langle \mathcal{O}, \mathcal{A} \rangle$ *where* $\mathcal{O} = \{Str^{\Sigma,\Gamma} \mid \Sigma \in |\text{Sign}| \wedge \Gamma \subseteq \mathbf{Sen}(\Sigma)\}$, *and* $\mathcal{A} = \{\widehat{\sigma} : Str^{\Sigma,\Gamma} \rightarrow Str^{\Sigma',\Gamma'} \mid \sigma : \langle \Sigma, \Gamma \rangle \rightarrow \langle \Sigma', \Gamma' \rangle \in \|\mathsf{Th}_0\|\}$, *the homomorphic extension of the morphisms in* $\|\mathsf{Th}_0\|$.

Lemma 2. $Struct_{SC}$ *is a category.*

The functor **M** must be understood as the relation between a theory in $|\mathsf{Th}_0|$ and its category of structures representing legal tableaux. So, for every theory T, **M** associates the strict monoidal category [22] $\langle Str^{\Sigma,\Gamma}, \cup, \emptyset \rangle$, and for every theory morphism $\sigma : \langle \Sigma, \Gamma \rangle \rightarrow \langle \Sigma', \Gamma' \rangle$, **M** associates a morphism $\widehat{\sigma} : Str^{\Sigma,\Gamma} \rightarrow Str^{\Sigma',\Gamma'}$ which is the homomorphic extension of σ to the structure of the tableaux.

Definition 12. $\mathbf{M} : \mathsf{Th}_0 \rightarrow Struct_{SC}$ *is defined as* $\mathbf{M}(\langle \Sigma, \Gamma \rangle) = \langle Str^{\Sigma,\Gamma}, \cup, \emptyset \rangle$ *and* $\mathbf{M}(\sigma : \langle \Sigma, \Gamma \rangle \rightarrow \langle \Sigma', \Gamma' \rangle) = \widehat{\sigma} : \langle Str^{\Sigma,\Gamma}, \cup, \emptyset \rangle \rightarrow \langle Str^{\Sigma',\Gamma'}, \cup, \emptyset \rangle$, *the homomorphic extension of* σ *to the structures in* $\langle Str^{\Sigma,\Gamma}, \cup, \emptyset \rangle$.

Lemma 3. **M** *is a functor.*

In order to define *Mods*, we need the following auxiliary definition, which resembles the usual construction of maximal consistent sets of formulae.

Definition 13. *Let* $\Sigma \in |\mathsf{Sign}|$, $\Delta \subseteq \mathsf{Sen}(\Sigma)$, *and consider* $\{F_i\}_{i<\omega}$ *an enumeration of* $\mathsf{Sen}(\Sigma)$ *such that for every formula* α, *its sub-formulae are enumerated before* α. *Then* $Cn(\Delta)$ *is defined as follows:*

- $Cn(\Delta) = \bigcup_{i<\omega} Cn^i(\Delta)$
- $Cn^0(\Delta) = \Delta$, $Cn^{i+1}(\Delta) = \begin{cases} Cn^i(\Delta) \cup \{F_i\} & , \textit{if } Cn^i(\Delta) \cup \{F_i\} \textit{ is consistent.} \\ Cn^i(\Delta) \cup \{\neg F_i\} & , \textit{otherwise.} \end{cases}$

Given $\langle \Sigma, \Gamma \rangle \in |\mathsf{Th}_0|$, the functor **Mods** provide the means for obtaining the category containing the closure of those structures in $Str^{\Sigma,\Gamma}$ that represent the closure of the branches in saturated tableaux.

Definition 14. **Mods** : $\mathsf{Struct}_{SC} \to \mathsf{Cat}$ *is defined as:*

$$\mathbf{Mods}(\langle Str^{\Sigma,\Gamma}, \cup, \emptyset \rangle) = \{\langle \Sigma, Cn(\widetilde{\Delta}) \rangle \mid (\exists \alpha : \Delta \to \emptyset \in ||Str^{\Sigma,\Gamma}||)$$
$$(\widetilde{\Delta} \to \emptyset \in \alpha \wedge (\forall \alpha' : \Delta' \to \Delta \in ||Str^{\Sigma,\Gamma}||)(\Delta' = \Delta))\}$$

and for all $\sigma : \Sigma \to \Sigma' \in |\mathsf{Sign}|$ *(and* $\widehat{\sigma} : \langle Str^{\Sigma,\Gamma}, \cup, \emptyset \rangle \to \langle Str^{\Sigma',\Gamma'}, \cup, \emptyset \rangle \in$
$||\mathsf{Struct}_{SC}||$), *the following holds:*

$$\mathbf{Mods}(\widehat{\sigma})(\langle \Sigma, Cn(\widetilde{\Delta}) \rangle) = \langle \Sigma', Cn(\mathsf{Sen}(\sigma)(Cn(\widetilde{\Delta}))) \rangle.$$

Lemma 4. **Mods** *is a functor.*

Finally, the natural transformation μ relates the structures representing saturated tableaux with the model satisfying the set of formulae denoted by the source of the morphism.

Definition 15. *Let* $\langle \Sigma, \Gamma \rangle \in |\mathsf{Th}_0|$, *then we define* $\mu_\Sigma : \mathbf{models}^{\mathrm{op}}(\langle \Sigma, \Gamma \rangle) \to$
$\mathbf{Mod}_{FOL}(\langle \Sigma, \Gamma \rangle)$ *as* $\mu_\Sigma(\langle \Sigma, \Delta \rangle) = \mathbf{Mod}(\langle \Sigma, \Delta \rangle)$.

Fact 1 *Let* $\Sigma \in |\mathsf{Sign}_{FOL}|$ *and* $\Gamma \subseteq \mathsf{Sen}_{FOL}(\Sigma)$. *Then* $\mu_{\langle \Sigma, \Gamma \rangle}$ *is a functor.*

Lemma 5. μ *is a natural transformation.*

Now, from Lemmas 3, 4, and 5, and considering the hypothesis that $\mathbb{I}_{FOL}$ is an institution, the following corollary follows.

Corollary 1. $\langle \mathsf{Sign}_{FOL}, \mathsf{Sen}_{FOL}, \mathbf{Mod}_{FOL}, \{\models^\Sigma_{FOL}\}_{\Sigma \in |\mathsf{Sign}_{FOL}|}, \mathbf{M}, \mathbf{Mods}, \mu \rangle$
is a satisfiability calculus.

Another important kind of system used by automatic theorem provers are the so-called resolution methods. Below, we show how any resolution system conforms to the definition of satisfiability calculus.

Example 2. **[Resolution Method for First-Order Predicate Logic]** Let us describe resolution for first-order logic as introduced in [25]. We use the following notation: [] denotes the empty list; [A] denotes the unitary list containing the formula A; $\ell_0, \ell_1, \ldots$ are variables ranging over lists; and $\ell_i + \ell_j$ denotes the concatenation of lists ℓ_i and ℓ_j. Resolution builds a list of lists representing a disjunction of conjunctions. The rules for resolution are the following:

$$\frac{\ell_0 + [\neg\neg A] + \ell_1}{\ell_0 + [A] + \ell_1} \ [\neg\neg] \qquad \frac{\ell_0 + [\neg A] + \ell_1 \quad \ell_0' + [A] + \ell_1'}{\ell_0 + \ell_1 + \ell_0' + \ell_1'} \ [\neg] \qquad \frac{\ell_0 + [A \wedge A'] + \ell_1}{\ell_0 + [A, A'] + \ell_1} \ [\wedge]$$

$$\frac{\ell_0 + [\neg(A \vee A')] + \ell_1}{\ell_0 + [\neg A, \neg A'] + \ell_1} \ [\neg\wedge] \qquad \frac{\ell_0 + [A \vee A'] + \ell_1}{\ell_0 + [A] + \ell_1 \atop \ell_0 + [A'] + \ell_1} \ [\vee] \qquad \frac{\ell_0 + [\neg(A \wedge A')] + \ell_1}{\ell_0 + [\neg A] + \ell_1 \atop \ell_0 + [\neg A'] + \ell_1} \ [\neg\wedge]$$

$$\text{for any closed term } t \ \frac{\ell_0 + [\forall x : A(x)] + \ell_1}{\ell_0 + [A[x/t]] + \ell_1} \ [\forall]$$

$$\text{for a new constant } c \ \frac{\ell_0 + [\exists x : A(x)] + \ell_1}{\ell_0 + [A[x/c]] + \ell_1} \ [\exists]$$

where $A(x)$ denotes a formula with free variable x, and $A[x/t]$ denotes the formula resulting from replacing variable x by term t everywhere in A. For the sake of simplicity, we assume that lists of formulae do not have repeated elements. A resolution is a sequence of lists of formulae. If a resolution contains an empty list (i.e., $[]$), we say that the resolution is closed; otherwise it is an *open* resolution. For every signature $\Sigma \in |\mathsf{Sign}|$ and each $\Gamma \subset \mathbf{Sen}(\Sigma)$, we denote by $Str^{\Sigma,\Gamma}$ the category whose objects are lists of formulae, and where every morphism $\sigma : [A_0, \ldots, A_n] \to [A_0', \ldots, A_m']$ represents a sequence of application of resolution rules for $[A_0', \ldots, A_m']$. Then, Struct_{SC} is a category whose objects are $Str^{\Sigma,\Gamma}$, for each signature $\Sigma \in |\mathsf{Sign}|$ and set of formulae $\Gamma \in \mathbf{Sen}(\Sigma)$, and whose morphisms are of the form $\hat{\sigma} : Str^{\Sigma,\Gamma} \to Str^{\Sigma',\Gamma'}$, obtained by homomorphically extending $\sigma : \langle \Sigma, \Gamma \rangle \to \langle \Sigma', \Gamma' \rangle$ in $\|\mathsf{Th}_0\|$.

As for the case of Example 1, the functor $\mathbf{M} : \mathbf{Th}_0 \to \mathsf{Struct}_{SC}$ is defined as $\mathbf{M}(\langle \Sigma, \Gamma \rangle) = \langle Str^{\Sigma,\Gamma}, \cup, \emptyset \rangle$, and $\mathbf{Mods} : \mathsf{Struct}_{SC} \to \mathbf{Set}$ is defined as in the previous example.

A typical use for the methods involved in the above described examples is the search for counterexamples of a given logical property. For instance, to search for counterexamples of an intended property in the context of the tableaux method, one starts by applying rules to the negation of the property, and once a saturated tableau is obtained, if all the branches are closed, then there is no model of the axioms and the negation of the property, indicating that the latter is a theorem. On the other hand, if there exists an open branch, the limit set of that branch characterizes a class of counterexamples for the formula. Notice the contrast with Hilbert systems, where one starts from the axioms, and then applies deduction rules until the desired formula is obtained.

3.1   Mapping Satisfiability Calculi

In [4] the original notion of morphism between Institutions was introduced. Meseguer defines the notion of plain map in [2], and in [5] Tarlecki extensively discussed the ways in which different institutions can be related, and how they should be interpreted. More recently, in [26] all these notions of morphism were investigated in detail. In this work we will concentrate only on institution representations (or comorphisms in the terminology introduced by Goguen and Rosu), since this is the notion that we have employed to formalize several concepts arising from software engineering, such as *data refinement* and *dynamic reconfiguration* [27,28]. The study of other important kinds of functorial relations between satisfiability calculi are left as future work. The following definition is taken from [5], and formalizes the notion of institution representation.

Definition 16. ([5]) *Let* $\mathbb{I} = \langle \mathbf{Sign}, \mathbf{Sen}, \mathbf{Mod}, \{\models_\Sigma\}_{\Sigma \in |\mathsf{Sign}|} \rangle$ *and* $\mathbb{I}' = \langle \mathbf{Sign}', \mathbf{Sen}', \mathbf{Mod}', \{\models'_\Sigma\}_{\Sigma \in |\mathsf{Sign}'|} \rangle$ *be institutions. Then,* $\langle \gamma^{Sign}, \gamma^{Sen}, \gamma^{Mod} \rangle : I \to I'$ *is an* institution representation *if and only if:*

- $\gamma^{Sign} : \mathsf{Sign} \to \mathsf{Sign}'$ *is a functor,*
- $\gamma^{Sen} : \mathbf{Sen} \overset{\cdot}{\to} \gamma^{Sign} \circ \mathbf{Sen}'$, *is a natural transformation,*
- $\gamma^{Mod} : (\gamma^{Sign})^{\mathsf{op}} \circ \mathbf{Mod}' \overset{\cdot}{\to} \mathbf{Mod}$, *is a natural transformation,*

such that for any $\Sigma \in |\mathsf{Sign}|$, *the function* $\gamma^{Sen}_\Sigma : \mathbf{Sen}(\Sigma) \to \mathbf{Sen}'(\gamma^{Sign}(\Sigma))$ *and the functor* $\gamma^{Mod}_\Sigma : \mathbf{Mod}'(\gamma^{Sign}(\Sigma)) \to \mathbf{Mod}(\Sigma)$ *preserve the following satisfaction condition: for any* $\alpha \in \mathbf{Sen}(\Sigma)$ *and* $\mathcal{M}' \in |\mathbf{Mod}(\gamma^{Sign}(\Sigma))|$,

$$\mathcal{M}' \models^{\gamma^{Sign}(\Sigma)} \gamma^{Sen}_\Sigma(\alpha) \quad \textit{iff} \quad \gamma^{Mod}_\Sigma(\mathcal{M}') \models^\Sigma \alpha .$$

An institution representation $\gamma : I \to I'$ expresses how the "poorer" set of sentences (respectively, category of models) associated with I is encoded in the "richer" one associated with I'. This is done by:

- constructing, for a given I-signature Σ, an I'-signature into which Σ can be interpreted,
- translating, for a given I-signature Σ, the set of Σ-sentences into the corresponding I'-sentences,
- obtaining, for a given I-signature Σ, the category of Σ-models from the corresponding category of Σ'-models.

The direction of the arrows shows how the whole of I is represented by some parts of I'. Institution representations enjoy some interesting properties. For instance, logical consequence is preserved, and, under some conditions, logical consequence is preserved in a conservative way. The interested reader is referred to [5] for further details.

In many cases, in particular those in which the class of models of a signature in the source institution is completely axiomatizable in the language of the target one, Definition 16 can easily be extended to map signatures of one institution to theories of another. This is done so that the class of models of the richer one can be restricted, by means of the addition of axioms (thus the need for theories in the image of the functor γ^{Sign}), in order to be exactly the class of models obtained by translating to it the class of models of the corresponding signature of the poorer one. In the same way, when the previously described extension is possible, we can obtain what Meseguer calls a *map of institutions* [2, definition 27] by reformulating the definition so that the functor between signatures of one institution and theories of the other is $\gamma^{Th} : \mathsf{Th}_0 \to \mathsf{Th}'_0$. This has to be γ^{Sen}-sensible (see definition 5) with respect to the entailment systems induced by the institutions I and I'. Now, if $\langle \Sigma, \Gamma \rangle \in |\mathsf{Th}_0|$, then γ^{Th_0} can be defined as follows: $\gamma^{Th_0}(\langle \Sigma, \Gamma \rangle) = \langle \gamma^{Sign}(\Sigma), \Delta \cup \gamma^{Sen}_\Sigma(\Gamma) \rangle$, where $\Delta \subseteq \mathbf{Sen}(\rho^{Sign}(\Sigma))$. Then, it is easy to prove that γ^{Th_0} is γ^{Sen}-simple because it is the γ^{Sen}-extension of γ^{Th_0} to theories, thus being γ^{Sen}-sensible.

The notion of a *map of satisfiability calculi* is the natural extension of a map of institutions in order to consider the more material version of the satisfiability relation. In some sense, if a map of institutions provides a means for representing one satisfiability relation in terms of another in a semantics preserving way, the map of satisfiability calculi provides a means for representing a model construction technique in terms of another. This is done by showing how model construction techniques for richer logics express techniques associated with poorer ones.

Definition 17. *Let* $\mathbb{S} = \langle \mathbf{Sign}, \mathbf{Sen}, \mathbf{Mod}, \{\models^\Sigma\}_{\Sigma \in |\mathsf{Sign}|}, \mathbf{M}, \mathbf{Mods}, \mu \rangle$ *and* $\mathbb{S}' = \langle \mathbf{Sign}', \mathbf{Sen}', \mathbf{Mod}', \{\models'^\Sigma\}_{\Sigma \in |\mathsf{Sign}'|}, \mathbf{M}', \mathbf{Mods}', \mu' \rangle$ *be satisfiability calculi. Then,* $\langle \rho^{Sign}, \rho^{Sen}, \rho^{Mod}, \gamma \rangle : \mathbb{S} \to \mathbb{S}'$ *is a map of satisfiability calculi if and only if:*

1. $\langle \rho^{Sign}, \rho^{Sen}, \rho^{Mod} \rangle : \mathbb{I} \to \mathbb{I}'$ *is a map of institutions, and*
2. $\gamma : \mathbf{models}'^{op} \circ \rho^{Th_0} \dot\to \mathbf{models}^{op}$ *is a natural transformation such that the following equality holds:*

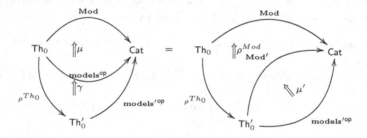

Roughly speaking, the 2-cell equality in the definition says that the translation of saturated tableaux is coherent with respect to the mapping of institutions.

Example 3. [***Mapping Modal Logic to First-Order Logic***] A simple example of a mapping between satisfiability calculi is the mapping between the tableau method for propositional logic, and the one for first-order logic. It is straightforward since the tableau method for first-order logic is an extension of that of propositional logic.

Let us introduce a more interesting example. We will map the tableau method for modal logic (as presented by Fitting [25]) to the first-order predicate logic tableau method. The mapping between the institutions is given by the standard translation from modal logic to first-order logic. Let us recast here the tableau method for the system K of modal logic. Recall that formulae of standard modal logic are built from boolean operators and the "diamond operator" $\Diamond$. Intuitively, formula $\Diamond\varphi$ says that φ is possibly true in some alternative state of affairs.

The semantics for modal logic is given by means of Kripke structures. A Kripke structure is a tuple $\langle W, R, L \rangle$, where W is a set of states, $R \subseteq W \times W$ is a relation between states, and $L : W \to 2^{AP}$ is a labeling function (AP is a set of atomic propositions). Note that a signature in modal logic is given by a set of propositional letters: $\langle \{p_i\}_{i \in \mathcal{I}} \rangle$. The interested reader can consult [29].

In [25] modal formulae are prefixed by labels denoting semantic states. Labeled formulae are then terms of the form $\ell : \varphi$, where φ is a modal formula and ℓ is a sequence of natural numbers $n_0, \ldots, n_k$. The relation R between these labels is then defined in the following way: $\ell R \ell' \equiv \exists n : \ell, n = \ell'$. The new rules are the following:

$$\text{For all } \ell' \text{ such that } \ell R \ell' \text{ and such that } \ell' \text{ appears in } X \quad \frac{X \cup \{\ell : \Box\varphi\}}{X \cup \{\ell : \Box\varphi, \ell' : \varphi\}} \;\; [\Box]$$

$$\text{For } \ell' \text{ such that } \ell R \ell' \quad \frac{X \cup \{\ell : \Diamond\varphi\}}{X \cup \{\ell : \Diamond\varphi, \ell' : \varphi\}} \;\; [\Diamond]$$

The rules for the propositional connectives are the usual ones, obtained by labeling the formulae with a given label. Notice that labels denote states of a Kripke structure. This is related in some way to the tableau method used for first-order predicate logic. Branches, saturated branches and closed branches are defined in the same way as in Example 1, but considering the relations between sets to be also indexed by the relation used at that point. Thus, $s_i \xrightarrow[R_i]{\tau_{\alpha_i}} s_{i+1}$ must be understood as follows: the set s_{i+1} is obtained from s_i by applying rule τ_i to formula $\alpha_i \in s_i$ under the accessibility relation R_i.

Assume $\langle \text{Sign}_{FOL}, \textbf{Sen}_{FOL}, \textbf{M}_{FOL}, \textbf{Mods}_{FOL}, \{\models^\Sigma_{FOL}\}_{\Sigma \in |\text{Sign}_{FOL}|}, \mu_{FOL} \rangle$ is the satisfiability calculus for first-order predicate logic, denoted by $\mathbb{SC}_{FOL}$, and $\langle \text{Sign}_K, \textbf{Sen}_K, \textbf{M}_K, \textbf{Mods}_K, \{\models^\Sigma_K\}_{\Sigma \in |\text{Sign}_K|}, \mu_K \rangle$ is the satisfiability calculus for modal logic, denoted by $\mathbb{SC}_K$. Consider now the standard translation from modal logic to first-order logic. Therefore, the tuple $\langle \rho^{Sign}, \rho^{Sen}, \rho^{Mod} \rangle$ is defined as follows [29]:

Definition 18. $\rho^{Sign} : \text{Sign}_K \to \text{Sign}_{FOL}$ *is defined as* $\rho^{Sign}(\langle \{p_i\}_{i \in \mathcal{I}} \rangle) = \langle R, \{p_i\}_{i \in \mathcal{I}} \rangle$ *by mapping each propositional variable* p_i, *for all* $i \in \mathcal{I}$, *to a first-order unary logic predicate* p_i, *and adding a binary predicate* R, *and* $\rho^{Sign}(\sigma : \langle \{p_i\}_{i \in \mathcal{I}} \rangle \to \langle \{p'_{i'}\}_{i' \in \mathcal{I}'} \rangle) = \sigma' : \langle R, \{p_i\}_{i \in \mathcal{I}} \rangle \to \langle R', \{p'_{i'}\}_{i' \in \mathcal{I}'} \rangle$ *mapping* R *to* R', *and* p_i *to* p'_i *for all* $i \in \mathcal{I}$.

Lemma 6. ρ^{Sign} *is a functor.*

Definition 19. *Let* $\langle \{p_i\}_{i \in \mathcal{I}} \rangle \in |\text{Sign}_K|$. *Then* $\rho^{Sen}_{\langle \{p_i\}_{i \in \mathcal{I}} \rangle} : \textbf{Sen}_K(\langle \{p_i\}_{i \in \mathcal{I}} \rangle) \to \rho^{Sign} \circ \textbf{Sen}_{FOL}(\langle \{p_i\}_{i \in \mathcal{I}} \rangle)$ *is defined recursively as* $\rho^{Sen}_{\langle \{p_i\}_{i \in \mathcal{I}} \rangle}(\alpha) = T_{\langle \{p_i\}_{i \in \mathcal{I}} \rangle, x}(\alpha)$ *where:*

$$T_{\langle \{p_i\}_{i \in \mathcal{I}} \rangle, x}(p_i) = p'_i(x), \text{ for all } i \in \mathcal{I}.$$
$$T_{\langle \{p_i\}_{i \in \mathcal{I}} \rangle, x}(\neg\alpha) = \neg T_{\langle \{p_i\}_{i \in \mathcal{I}} \rangle, x}(\alpha)$$
$$T_{\langle \{p_i\}_{i \in \mathcal{I}} \rangle, x}(\alpha \vee \beta) = T_{\langle \{p_i\}_{i \in \mathcal{I}} \rangle, x}(\alpha) \vee T_{\langle \{p_i\}_{i \in \mathcal{I}} \rangle, x}(\beta)$$
$$T_{\langle \{p_i\}_{i \in \mathcal{I}} \rangle, x}(\Diamond\alpha) = (\exists y)(R(x, y) \wedge T_{\langle \{p_i\}_{i \in \mathcal{I}} \rangle, y}(\alpha))$$

Lemma 7. ρ^{Sen} is a natural transformation.

Definition 20. Let $\langle\{p_i\}_{i\in\mathcal{I}}\rangle \in |\mathsf{Sign}_K|$. Then we define $\rho^{Mod}_{\langle\{p_i\}_{i\in\mathcal{I}}\rangle} : \rho^{Sign} \circ \mathbf{Mod}_{FOL}(\langle\{p_i\}_{i\in\mathcal{I}}\rangle) \to \mathbf{Mod}_K(\langle\{p_i\}_{i\in\mathcal{I}}\rangle)$ as follows:

- for all $\mathcal{M} = \langle S, \overline{R}, \{\overline{p_i}\}_{i\in\mathcal{I}}\rangle \in |\mathbf{Mod}_{FOL}(\langle R, \{p_i\}_{i\in\mathcal{I}}\rangle)|$, $\rho^{Mod}_{\langle\{p_i\}_{i\in\mathcal{I}}\rangle}(\mathcal{M}) = \langle S, \overline{R}, L\rangle$, with $L(p_i) = \{s \in S | \overline{p_i}(s)\}$.[4]
- let $\langle\{p_i\}_{i\in\mathcal{I}}\rangle \in |\mathsf{Sign}_K|$; then for all homomorphism $h : \langle S_1, \overline{R_1}, \{\overline{p_{1_i}}\}_{i\in\mathcal{I}}\rangle \to \langle S_2, \overline{R_2}, \{\overline{p_{2_i}}\}_{i\in\mathcal{I}}\rangle \in ||\mathbf{Mod}_{FOL}(\langle R, \{p_i\}_{i\in\mathcal{I}}\rangle)||$, we define $\rho^{Mod}_{\langle\{p_i\}_{i\in\mathcal{I}}\rangle}(h)$ to be $\widehat{h}$, where $\widehat{h}(s_1) = s_2$ if and only if $h(s_1) = s_2$ for all $s_1 \in S_1$.

Lemma 8. Let $\langle\{p_i\}_{i\in\mathcal{I}}\rangle \in |\mathsf{Sign}_K|$. Then $\rho^{Mod}_{\langle\{p_i\}_{i\in\mathcal{I}}\rangle}$ is a functor.

The proof that this is a mapping between institutions relies on the correctness of the translation presented in [29]. Using this map we can define a mapping between the corresponding satisfiability calculi. The natural transformation: $\gamma : \rho^{Tho} \circ \mathbf{models}'^{\mathrm{op}} \overset{.}{\to} \mathbf{models}^{\mathrm{op}}$ is defined as follows.

Definition 21. Let $\langle\langle\{p_i\}_{i\in\mathcal{I}}\rangle, \Gamma\rangle \in |\mathsf{Th}_0^K|$; then

$$\gamma_{\langle\{p_i\}_{i\in\mathcal{I}}\rangle} : \rho^{Tho} \circ \mathbf{models}^{\mathrm{op}}_{FOL}(\langle\langle\{p_i\}_{i\in\mathcal{I}}\rangle, \Gamma\rangle) \to \mathbf{models}^{\mathrm{op}}_K(\langle\langle\{p_i\}_{i\in\mathcal{I}}\rangle, \Gamma\rangle)$$

is defined as:

$$\gamma_{\langle\{p_i\}_{i\in\mathcal{I}}\rangle}(\langle\langle R, \{p_i\}_{i\in\mathcal{I}}\rangle, \Delta\rangle) = \langle\langle\{p_i\}_{i\in\mathcal{I}}\rangle, \{\varphi \in |\mathbf{Sen}_K(\langle\{p_i\}_{i\in\mathcal{I}}\rangle)| \mid \rho^{Sen}_{\langle\{p_i\}_{i\in\mathcal{I}}\rangle}(\varphi) \in \Delta\})$$

Lemma 9. Let $\langle\{p_i\}_{i\in\mathcal{I}}\rangle \in |\mathsf{Sign}_K|$; then $\gamma_{\langle\{p_i\}_{i\in\mathcal{I}}\rangle}$ is a functor.

Lemma 10. $\gamma : \rho^{Tho} \circ \mathbf{models}^{\mathrm{op}}_{FOL} \to \mathbf{models}^{\mathrm{op}}_K$ is a natural transformation.

Finally, the following lemma prove the equivalence of the two cells shown in Definition 17.

Lemma 11. Let $\langle\{p_i\}_{i\in\mathcal{I}}\rangle \in |\mathsf{Sign}_K|$, then

$$\mu_K \langle\{p_i\}_{i\in\mathcal{I}}\rangle \circ \gamma_{\langle\{p_i\}_{i\in\mathcal{I}}\rangle} = \rho^{Mod}_{\rho^{Sign}(\langle\{p_i\}_{i\in\mathcal{I}}\rangle)} \circ \mu_{FOL} \rho^{Sign}(\langle\{p_i\}_{i\in\mathcal{I}}\rangle) \ .$$

This means that building a tableau using the first-order rules for the translation of a modal theory, then obtaining the corresponding canonical model in modal logic using γ, and therefore obtaining the class of models by using μ, is exactly the same as obtaining the first-order models by μ' and then the corresponding modal models by using ρ^{Mod}.

4 Conclusions and Further Work

Methods like resolution and tableaux are strongly related to the semantics of a logic. They are often employed to construct models, a characteristic that is missing in purely deductive methods, such as natural deduction or Hilbert systems, as

---

[4] Notice that $\rho^{Sign}(\langle\{p_i\}_{i\in\mathcal{I}}\rangle) = \langle R, \{p_i\}_{i\in\mathcal{I}}\rangle$, where $\langle\{p_i\}_{i\in\mathcal{I}}\rangle \in |\mathsf{Sign}_K|$.

formalized by Meseguer. In this paper, we provided an abstract characterization of this class of semantics-based tecniques for logical systems. This was accomplished by introducing a categorical characterization of the notion of satisfiability calculus, which covers logical tools such as tableaux, resolution, Gentzen style sequents, etc. Our new characterization of a logical system, that includes the notion of satisfiability calculus, provides both a proof calculus and a satisfiability calculus, which essentially implement the entailment and satisfaction relations, respectively. There clearly exist connections between these calculi that are worth exploring, especially when the underlying structure used in both definitions is the same (see Example 1).

A close analysis of the definitions of proof calculus and satisfiability calculus takes us to observe that the constraints imposed over some elements (e.g., the natural family of functors $\pi_{\langle \Sigma, \Gamma \rangle} : \mathbf{proofs}(\langle \Sigma, \Gamma \rangle) \to \mathbf{Sen}(\langle \Sigma, \Gamma \rangle)$ and $\mu_{\langle \Sigma, \Gamma \rangle} : \mathbf{models}^{op}(\langle \Sigma, \Gamma \rangle) \to \mathbf{Mod}(\langle \Sigma, \Gamma \rangle))$ may be too restrictive, and working on generalizations of these concepts is part of our further work. In particular, it is worth noticing that partial implementations of both the entailment relation and the satisfiability relation are gaining visibility in the software engineering community. Examples on the syntactic side are the implementation of less expressive calculi with respect to an entailment, as in the case of the finitary definition of the reflexive and transitive closure in the Kleene algebras with tests [30], the case of the implementation of rewriting tools like Maude [31] as a partial implementation of equational logic, etc. Examples on the semantic side are the bounded model checkers and model finders for undecidable languages, such as Alloy [32] for relational logic, the growing family of SMT-solvers [33] for languages including arithmetic, etc. Clearly, allowing for partial implementations of entailment/satisfiability relations would enable us to capture the behaviors of some of the above mentioned logical tools. In addition, functorial relations between partial proof calculi (resp., satisfiability calculi) may provide a measure for how good the method is as an approximation of the ideal entailment relation (resp., satisfaction relation). We plan to explore this possibility, as future work.

Acknowledgements. The authors would like to thank the anonymous referees for their helpful comments. This work was partially supported by the Argentinian Agency for Scientific and Technological Promotion (ANPCyT), through grants PICT PAE 2007 No. 2772, PICT 2010 No. 1690, PICT 2010 No. 2611 and PICT 2010 No. 1745, and by the MEALS project (EU FP7 programme, grant agreement No. 295261). The fourth author gratefully acknowledges the support of the National Science and Engineering Research Council of Canada and McMaster University.

References

1. Goguen, J.A., Burstall, R.M.: Introducing institutions. In: Clarke, E., Kozen, D. (eds.) Logic of Programs 1983. LNCS, vol. 164, pp. 221–256. Springer, Heidelberg (1984)

2. Meseguer, J.: General logics. In: Ebbinghaus, H.D., Fernandez-Prida, J., Garrido, M., Lascar, D., Artalejo, M.R. (eds.) Proceedings of the Logic Colloquium 1987, Granada, Spain, vol. 129, pp. 275–329. North Holland (1989)

3. Fiadeiro, J.L., Maibaum, T.S.E.: Generalising interpretations between theories in the context of π-institutions. In: Burn, G., Gay, D., Ryan, M. (eds.) Proceedings of the First Imperial College Department of Computing Workshop on Theory and Formal Methods, London, UK, pp. 126–147. Springer (1993)

4. Goguen, J.A., Burstall, R.M.: Institutions: abstract model theory for specification and programming. Journal of the ACM 39(1), 95–146 (1992)

5. Tarlecki, A.: Moving between logical systems. In: Haveraaen, M., Owe, O., Dahl, O.J. (eds.) Abstract Data Types 1995 and COMPASS 1995. LNCS, vol. 1130, pp. 478–502. Springer, Heidelberg (1996)

6. Sannella, D., Tarlecki, A.: Specifications in an arbitrary institution. Information and Computation 76(2-3), 165–210 (1988)

7. Tarlecki, A.: Abstract specification theory: an overview. In: Broy, M., Pizka, M. (eds.) Proceedings of the NATO Advanced Study Institute on Models, Algebras and Logic of Engineering Software, Marktoberdorf, Germany. NATO Science Series, pp. 43–79. IOS Press (2003)

8. Mossakowski, T.: Comorphism-based Grothendieck logics. In: Diks, K., Rytter, W. (eds.) MFCS 2002. LNCS, vol. 2420, pp. 593–604. Springer, Heidelberg (2002)

9. Mossakowski, T., Tarlecki, A.: Heterogeneous logical environments for distributed specifications. In: Corradini, A., Montanari, U. (eds.) WADT 2008. LNCS, vol. 5486, pp. 266–289. Springer, Heidelberg (2009)

10. Diaconescu, R., Futatsugi, K.: Logical foundations of CafeOBJ. Theoretical Computer Science 285(2), 289–318 (2002)

11. Diaconescu, R.: Grothendieck institutions. Applied Categorical Structures 10(4), 383–402 (2002)

12. Diaconescu, R. (ed.): Institution-independent Model Theory. Studies in Universal Logic, vol. 2. Birkhäuser (2008)

13. Mossakowski, T., Maeder, C., Lüttich, K.: The heterogeneous tool set, HETS. In: Grumberg, O., Huth, M. (eds.) TACAS 2007. LNCS, vol. 4424, pp. 519–522. Springer, Heidelberg (2007)

14. Tarlecki, A.: Towards heterogeneous specifications. In: Gabbay, D., de Rijke, M. (eds.) Frontiers of Combining Systems. Studies in Logic and Computation, vol. 2, pp. 337–360. Research Studies Press (2000)

15. Cengarle, M.V., Knapp, A., Tarlecki, A., Wirsing, M.: A heterogeneous approach to UML semantics. In: Degano, P., De Nicola, R., Meseguer, J. (eds.) Concurrency, Graphs and Models. LNCS, vol. 5065, pp. 383–402. Springer, Heidelberg (2008)

16. Beth, E.W.: The Foundations of Mathematics. North Holland (1959)

17. Beth, E.W.: Semantic entailment and formal derivability. In: Hintikka, J. (ed.) The Philosophy of Mathematics, pp. 9–41. Oxford University Press (1969) (reprinted from [34])

18. Herbrand, J.: Investigation in proof theory. In: Goldfarb, W.D. (ed.) Logical Writings, pp. 44–202. Harvard University Press (1969) (translated to English from [35])

19. Gentzen, G.: Investigation into logical deduction. In: Szabo, M.E. (ed.) The Collected Papers of Gerhard Gentzen, pp. 68–131. North Holland (1969) (translated to English from [36])

20. Smullyan, R.M.: First-order Logic. Dover Publishing (1995)

21. Robinson, J.A.: A machine-oriented logic based on the resolution principle. Journal of the ACM 12(1), 23–41 (1965)

22. McLane, S.: Categories for working mathematician. Graduate Texts in Mathematics. Springer, Berlin (1971)
23. Fiadeiro, J.L.: Categories for software engineering. Springer (2005)
24. Lopez Pombo, C.G., Castro, P., Aguirre, N.M., Maibaum, T.S.E.: Satisfiability calculus: the semantic counterpart of a proof calculus in general logics. Technical report, McMaster University, Centre for Software Certification (2011)
25. Fitting, M.: Tableau methods of proof for modal logics. Notre Dame Journal of Formal Logic 13(2), 237–247 (1972) (Lehman College)
26. Goguen, J.A., Rosu, G.: Institution morphisms. Formal Asp. Comput. 13(3-5), 274–307 (2002)
27. Castro, P.F., Aguirre, N.M., López Pombo, C.G., Maibaum, T.S.E.: Towards managing dynamic reconfiguration of software systems in a categorical setting. In: Cavalcanti, A., Deharbe, D., Gaudel, M.-C., Woodcock, J. (eds.) ICTAC 2010. LNCS, vol. 6255, pp. 306–321. Springer, Heidelberg (2010)
28. Castro, P.F., Aguirre, N., López Pombo, C.G., Maibaum, T.: A categorical approach to structuring and promoting Z specifications. In: Păsăreanu, C.S., Salaün, G. (eds.) FACS 2012. LNCS, vol. 7684, pp. 73–91. Springer, Heidelberg (2013)
29. Blackburn, P., de Rijke, M., Venema, Y.: Modal logic. Cambridge Tracts in Theoretical Computer Science, vol. 53. Cambridge University Press (2001)
30. Kozen, D.: Kleene algebra with tests. ACM Transactions on Programming Languages and Systems 19(3), 427–443 (1997)
31. Clavel, M., Durán, F., Eker, S., Lincoln, P., Martí-Oliet, N., Meseguer, J., Talcott, C.: All About Maude. LNCS, vol. 4350. Springer, Heidelberg (2007)
32. Jackson, D.: Alloy: a lightweight object modelling notation. ACM Transactions on Software Engineering and Methodology 11(2), 256–290 (2002)
33. Moura, L.D., Djørner, N.: Satisfiability modulo theories: introduction and applications. Communications of the ACM 54(9), 69–77 (2011)
34. Beth, E.W.: Semantic entailment and formal derivability. Mededlingen van de Koninklijke Nederlandse Akademie van Wetenschappen, Afdeling Letterkunde 18(13), 309–342 (1955) (reprinted in [17])
35. Herbrand, J.: Recherches sur la theorie de la demonstration. PhD thesis, Université de Paris (1930) (English translation in [18])
36. Gentzen, G.: Untersuchungen tiber das logische schliessen. Mathematische Zeitschrijt 39, 176–210, 405–431 (1935) (English translation in [19])

Semantics of the Distributed Ontology Language: Institutes and Institutions

Till Mossakowski[1,3], Oliver Kutz[1], and Christoph Lange[1,2]

[1] Research Center on Spatial Cognition, University of Bremen
[2] School of Computer Science, University of Birmingham
[3] DFKI GmbH Bremen

Abstract. The Distributed Ontology Language (DOL) is a recent development within the ISO standardisation initiative 17347 Ontology Integration and Interoperability (OntoIOp). In DOL, heterogeneous and distributed ontologies can be expressed, i.e. ontologies that are made up of parts written in ontology languages based on various logics. In order to make the DOL meta-language and its semantics more easily accessible to the wider ontology community, we have developed a notion of institute which are like institutions but with signature partial orders and based on standard set-theoretic semantics rather than category theory. We give an institute-based semantics for the kernel of DOL and show that this is compatible with institutional semantics. Moreover, as it turns out, beyond their greater simplicity, institutes have some further surprising advantages over institutions.

1 Introduction

OWL is a popular language for ontologies. Yet, the restriction to a decidable description logic often hinders ontology designers from expressing knowledge that cannot (or can only in quite complicated ways) be expressed in a description logic. A current practice to deal with this problem is to intersperse OWL ontologies with first-order axioms in the comments or annotate them as having temporal behaviour [35,3], e.g. in the case of bio-ontologies where mereological relations such as parthood are of great importance, though not definable in OWL. However, these remain informal annotations to inform the human designer, rather than first-class citizens of the ontology with formal semantics, and will therefore unfortunately be ignored by tools with no impact on reasoning. Moreover, foundational ontologies such as DOLCE, BFO or SUMO use full first-order logic or even first-order modal logic.

A variety of languages is used for formalising ontologies.[1] Some of these, such as RDF, OBO and UML, can be seen more or less as fragments and notational variants of OWL, while others, such as F-logic and Common Logic (CL), clearly go beyond the expressiveness of OWL.

This situation has motivated the Distributed Ontology Language (DOL), a language currently under active development within the ISO standard 17347 Ontology Integration and Interoperability (OntoIOp). In DOL, heterogeneous and distributed ontologies can be expressed. At the heart of this approach is a graph of ontology languages and translations [27], shown in Fig. 1.

---

[1] For the purposes of this paper, "ontology" can be equated with "logical theory".

N. Martí-Oliet and M. Palomino (Eds.): WADT 2012, LNCS 7841, pp. 212–230, 2013.

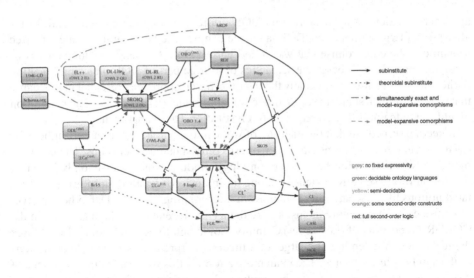

Fig. 1. An initial logic graph for the Distributed Ontology Language DOL

This graph enables users to

- relate ontologies that are written in different formalisms (e.g. prove that the OWL version of the foundational ontology DOLCE is logically entailed by the first-order version);
- re-use ontology modules even if they have been formulated in a different formalism;
- re-use ontology tools such as theorem provers and module extractors along translations between formalisms.

What is the semantics of DOL? Previous presentations of the semantics of heterogeneous logical theories [36,7,30,16,27] relied heavily on the theory of institutions [10]. The central insight of the theory of institutions is that logical notions such as model, sentence, satisfaction and derivability should be indexed over signatures (vocabularies). In order to abstract from any specific form of signature, category theory is used: nothing more is assumed about signatures other than that (together with suitable signature morphisms) they form a category.

However, the use of category theory diminishes the set of potential readers:

> "Mathematicians, and even logicians, have not shown much interest in the theory of institutions, perhaps because their tendency toward Platonism inclines them to believe that there is just one true logic and model theory; it also *doesn't much help that institutions use category theory extensively.*"
>
> (J. Goguen and G. Roşu in [9], our emphasis)

Indeed, during the extensive discussions within the ISO standardisation committee in TC37/SC3 to find an agreement concerning the right semantics for the DOL language, we (a) encountered strong reservations to base the semantics entirely on the institutional approach in order not to severely limit DOL's potential adoption by users, and (b) realised that a large kernel of the DOL language can be based on a simpler, category-free semantics. The compromise that was found within OntoIOp therefore adopted a two-layered approach: (i) it bases the semantics of a large part of DOL on a simplification of the notion of institutions, namely the institute-based approach presented in this paper that relies purely on standard set-theoretic semantics, and (ii) allows an elegant addition of additional features that do require a full institution-based approach.

Indeed, it turned out that the majority of work in the ontology community either disregards signature morphisms altogether, or uses only signature inclusions. The latter are particularly important for the notion of ontology module, which is essentially based on the notion of conservative extension along an inclusion signature morphisms, and related notions like inseparability and uniform interpolation (see also Def. 5 below). Another use case for signature inclusions are theory interpretations, which are used in the COLORE repository of (first-order) Common Logic ontologies. Indeed, COLORE uses the technique of extending the target of a theory interpretation by suitable definitions of the symbols in the source. The main motivation for this is probably the avoidance of derived signature morphisms; as a by-product, also renamings of symbols are avoided.

There are only rare cases where signature morphisms are needed in their full generality: the renaming of ontologies, which so far has only been used for combinations of ontologies by colimits. Only here, the full institution-based approach is needed. However, only relatively few papers are explicitly concerned with colimits of ontologies.[2]

Another motivation for our work is the line of *signature-free* thinking in logic and ontology research; for example, the ISO/IEC standard 24707:2007 Common Logic [5] names its signature-free approach to sentence formation a chief novel feature:

> "Common Logic has some novel features, chief among them being a syntax which is *signature-free* ..." [5]

Likewise, many abstract studies of consequence and satisfaction systems [8,34,2,4] disregard signatures. Hence, we base our semantics on the newly introduced notion of *institutes*. These start with the signature-free approach, and then introduce signatures a posteriori, assuming that they form a partial order. While this approach covers only signature inclusions, not renamings, it is much simpler than the category-based approach of institutions. Of course, for features like colimits, full institution theory is needed. We therefore show that institutes and institutions can be integrated smoothly.

2   Institutes: Semantics for a DOL Kernel

The notion of institute follows the insight that central to a model-theoretic view on logic is the notion of satisfaction of sentences in models. We also follow the insight of institution theory that signatures are essential to control the vocabulary of symbols used in sentences and models. However, in many logic textbooks as well as in the Common Logic standard [5], sentences are defined independently of a specific signature,

---

[2] To make this more explicit, as of January 2013, Google Scholar returns about 1 million papers for the keyword 'ontology', around 10.000 for the keyword 'colimits', but only around 200 for the conjunctive query.

while models always interpret a given signature. The notion of institute reflects this common practice. Note that the satisfaction relation can only meaningfully be defined between models and sentences where the model interprets all the symbols occurring in the sentence; this is reflected in the fact that we define satisfaction per signature. We also require a partial order on models; this is needed for minimisation in the sense of circumscription.

Moreover, we realise the goal of avoiding the use of category theory by relying on partial orders of signatures as the best possible approximation of signature categories. This also corresponds to common practice in logic, where signature extensions and (reducts against these) are considered much more often than signature morphisms.

Definition 1 (Institutes). *An* **institute** $\mathcal{I} = (\mathbf{Sen}, \mathbf{Sign}, \leq, sig, \mathbf{Mod}, \models, .|.)$ *consists of*

- *a class* **Sen** *of sentences;*
- *a partially ordered class* $(\mathbf{Sign}, \leq)$ *of signatures (which are arbitrary sets);*
- *a function sig* : **Sen** $\to$ **Sign***, giving the (minimal) signature of a sentence (then for each signature Σ, let* $\mathbf{Sen}(\Sigma) = \{\varphi \in \mathbf{Sen} \mid sig(\varphi) \leq \Sigma\}$*);*
- *for each signature Σ, a partially ordered class* $\mathbf{Mod}(\Sigma)$ *of Σ-models;*
- *for each signature Σ, a satisfaction relation* $\models_\Sigma \subseteq \mathbf{Mod}(\Sigma) \times \mathbf{Sen}(\Sigma)$*;*
- *for any Σ_2-model M, a Σ_1-model $M|_{\Sigma_1}$ (called the* **reduct**)*, provided that $\Sigma_1 \leq \Sigma_2$,*

such that the following properties hold:

- *given $\Sigma_1 \leq \Sigma_2$, for any Σ_2-model M and any Σ_1-sentence φ*

$$M \models \varphi \text{ iff } M|_{\Sigma_1} \models \varphi$$

(satisfaction is **invariant under reduct***),*
- *for any Σ-model M, given $\Sigma_1 \leq \Sigma_2 \leq \Sigma$,*

$$(M|_{\Sigma_2})|_{\Sigma_1} = M|_{\Sigma_1}$$

(reducts are compositional), and
- *for any Σ-models $M_1 \leq M_2$, if $\Sigma' \leq \Sigma$, then $M_1|_{\Sigma'} \leq M_2|_{\Sigma'}$ (reducts preserve the model ordering).* $\qquad\qquad\Box$

We give two examples illustrating these definitions, by phrasing the description logic $\mathcal{ALC}$ and Common Logic CL in institute style:

Example 2 (Description Logics $\mathcal{ALC}$). An institute for $\mathcal{ALC}$ is defined as follows: sentences are subsumption relations $C_1 \sqsubseteq C_2$ between concepts, where concepts follow the grammar

$$C ::= A \mid \top \mid \bot \mid C_1 \sqcup C_2 \mid C_1 \sqcap C_2 \mid \neg C \mid \forall R.C \mid \exists R.C$$

Here, A stands for atomic concepts. Such sentences are also called TBox sentences. Sentences can also be ABox sentences, which are membership assertions of individuals in concepts (written $a : C$, where a is an individual constant) or pairs of individuals in roles (written $R(a, b)$, where R is a role, and a, b are individual constants).

Signatures consist of a set $\mathcal{A}$ of atomic concepts, a set $\mathcal{R}$ of roles and a set $\mathcal{I}$ of individual constants. The ordering on signatures is component-wise inclusion. For a sentence φ, $sig(\varphi)$ contains all symbols occurring in φ.

Σ-models consist of a non-empty set Δ, the universe, and an element of Δ for each individual constant in Σ, a unary relation over Δ for each concept in Σ, and a binary relation over Δ for each role in Σ. The partial order on models is defined as coincidence of the universe and the interpretation of individual constants plus subset inclusion for the interpretation of concepts and roles. Reducts just forget the respective components of models. Satisfaction is the standard satisfaction of description logics.

An extension of $\mathcal{ALC}$ named $\mathcal{SROIQ}$ [13] is the logical core of the Web Ontology Language OWL 2 DL[3].

Example 3 (Common Logic - CL). Common Logic (CL) has first been formalised as an institution in [16]. We here formalise it as an institute.

A CL-sentence is a first-order sentence, where predications and function applications are written in a higher-order like syntax as $t(s)$. Here, t is an arbitrary term, and s is a sequence term, which can be a sequence of terms $t_1 \ldots t_n$, or a sequence marker. However, a predication $t(s)$ is interpreted like the first-order formula $holds(t, s)$, and a function application $t(s)$ like the first-order term $app(t, s)$, where $holds$ and app are fictitious symbols (denoting the semantic objects *rel* and *fun* defined in models below). In this way, CL provides a first-order simulation of a higher-order language. Quantification variables are partitioned into those for individuals and those for sequences.

A CL signature Σ (called vocabulary in CL terminology) consists of a set of names, with a subset called the set of discourse names, and a set of sequence markers. The partial order on signatures is componentwise inclusion with the requirement that the a name is a discourse name in the smaller signature if and only if is in the larger signature. *sig* obviously collects the names and sequence markers present in a sentence.

A Σ-model consists of a set *UR*, the universe of reference, with a non-empty subset $UD \subseteq UR$, the universe of discourse, and four mappings:

- *rel* from *UR* to subsets of $UD^* = \{< x_1, \ldots, x_n > | x_1, \ldots, x_n \in UD\}$ (i.e., the set of finite sequences of elements of *UD*);
- *fun* from *UR* to total functions from UD^* into *UD*;
- *int* from names in Σ to *UR*, such that $int(v)$ is in *UD* if and only if v is a discourse name;
- *seq* from sequence markers in Σ to UD^*.

The partial order on models is defined as $M_1 \leq M_2$ iff M_1 and M_2 agree on all components except perhaps *rel*, where we require $rel_1(x) \subseteq rel_2(x)$ for all $x \in UR_1 = UR_2$. Model reducts leave *UR*, *UD*, *rel* and *fun* untouched, while *int* and *seq* are restricted to the smaller signature.

Interpretation of terms and formulae is as in first-order logic, with the difference that the terms at predicate resp. function symbol positions are interpreted with *rel* resp. *fun* in order to obtain the predicate resp. function, as discussed above. A further difference is the presence of sequence terms (namely sequence markers and juxtapositions of terms),

---

[3] See also http://www.w3.org/TR/owl2-overview/

which denote sequences in UD^*, with term juxtaposition interpreted by sequence concatenation. Note that sequences are essentially a second-order feature. For details, see [5].

Working within an Arbitrary but Fixed Institute

Like with institutions, many logical notions can be formulated in an arbitrary but fixed institute. However, institutes are more natural for certain notions used in the ontology community.

The notions of 'theory' and 'model class' in an institute are defined as follows:

Definition 4 (Theories and Model Classes). *A theory* $T = (\Sigma, \Gamma)$ *in an institute* $\mathcal{I}$ *consists of a signature* Σ *and a set of sentences* $\Gamma \subseteq \mathbf{Sen}(\Sigma)$. *Theories can be partially ordered by letting* $(\Sigma_1, \Gamma_1) \leq (\Sigma_2, \Gamma_2)$ *iff* $\Sigma_1 \leq \Sigma_2$ *and* $\Gamma_1 \subseteq \Gamma_2$. *The* **class of models** $\mathbf{Mod}(\Sigma, \Gamma)$ *is defined as the class of those* Σ-*models satisfying* Γ. *This data is easily seen to form an institute* $\mathcal{I}^{th}$ *of theories in* $\mathcal{I}$ *(with theories as "signatures").*   □

The following definition is taken directly from [21],[4] showing that central notions from the ontology modules community can be seamlessly formulated in an arbitrary institute:

Definition 5 (Entailment, inseparability, conservative extension)

- *A theory* T_1 Σ-*entails* T_2, *written* $T_1 \sqsubseteq T_2$, *if* $T_2 \models \varphi$ *implies* $T_1 \models \varphi$ *for all sentences* φ *with* $sig(\varphi) \leq \Sigma$;
- T_1 *and* T_2 *are* Σ-*inseparable if* T_1 Σ-*entails* T_2 *and* T_2 Σ-*entails* T_1;
- T_2 *is a* Σ-*conservative extension of* T_1 *if* $T_2 \geq T_1$ *and* T_1 *and* T_2 *are* Σ-*inseparable*;
- T_2 *is a conservative extension of* T_1 *if* T_2 *is a* Σ-*conservative extension of* T_2 *with* $\Sigma = sig(T_1)$.

Note the use of sig here directly conforms to institute parlance. In contrast, since there is no global set of sentences in institutions, one would need to completely reformulate the definition for the institution representation and fiddle with explicit sentence translations.

From time to time, we will need the notion of 'unions of signatures':

Definition 6 (Signature unions). *A* **Signature union** *is a supremum (least upper bound) in the signature partial order. Note that signature unions need not always exist, nor be unique. In either of these cases, the enclosing construct containing the union is undefined.*   □

3   Institute Morphisms and Comorphisms

Institute morphisms and comorphisms relate two given institutes. A typical situation is that an institute morphism expresses the fact that a "larger" institute *is built upon* a

---

[4] There are two modifications: 1. We use $\leq$ where [21] write $\subseteq$. 2. In [21], all these notions are defined relative to a query language. This can also be done in an institute by singling out a subinstitute (see end of Sect. 3 below), which then becomes an additional parameter of the definition.

"smaller" institute by *projecting* the "larger" institute onto the "smaller" one. Somewhat dually to institute morphisms, institute comorphisms allow to express the fact that one institute is *included* in another one. (Co)morphisms play an essential role for DOL: the DOL semantics is parametrised over a graph of institutes and institute morphisms and comorphisms. The formal definitions are as follows:

Definition 7 (Institute morphism). *Given* $\mathcal{I}_1 = (\mathbf{Sen}_1, \mathbf{Sign}_1, \leq_1, sig_1, \mathbf{Mod}_1, \models_1, .|.)$ *and* $\mathcal{I}_2 = (\mathbf{Sen}_2, \mathbf{Sign}_2, \leq_2, sig_2, \mathbf{Mod}_2, \models_2, .|.)$, *an* **institute morphism** $\rho = (\Phi, \alpha, \beta):$ $\mathcal{I}_1 \longrightarrow \mathcal{I}_2$ *consists of*

- *a monotone map* $\Phi : (\mathbf{Sign}^1, \leq^1) \to (\mathbf{Sign}^2, \leq^2),$
- *a sentence translation function* $\alpha: \mathbf{Sen}_2 \longrightarrow \mathbf{Sen}_1,$ *and*
- *for each* $\mathcal{I}_1$-*signature* Σ, *a monotone model translation function* $\beta_\Sigma : \mathbf{Mod}_1(\Sigma) \to$ $\mathbf{Mod}_2(\Phi(\Sigma)),$

such that

- $M_1 \models_1 \alpha(\varphi_2)$ *if and only if* $\beta_\Sigma(M_1) \models_2 \varphi_2$ *holds for each* $\mathcal{I}_1$-*signature* Σ, *each model* $M_1 \in \mathbf{Mod}_1(\Sigma)$ *and each sentence* $\varphi_2 \in \mathbf{Sen}_2(\Sigma)$ *(satisfaction condition)*
- $\Phi(sig^1(\alpha(\varphi_2))) \leq sig^2(\varphi_2)$ *for any sentence* $\varphi_2 \in Sen^2$ *(sentence coherence);*
- **model translation commutes with reduct,** *that is, given* $\Sigma_1 \leq \Sigma_2$ *in* $\mathcal{I}_1$ *and a* Σ_2-*model* M,

$$\beta_{\Sigma_2}(M)|_{\Phi(\Sigma_1)} = \beta_{\Sigma_1}(M|_{\Sigma_1}). \qquad \square$$

The dual notion of institute comorphism is then defined as:

Definition 8 (Institute comorphism). *Given* $\mathcal{I}_1 = (\mathbf{Sen}_1, \mathbf{Sign}_1, \leq_1, sig_1, \mathbf{Mod}_1, \models_1, .|.)$ *and* $\mathcal{I}_2 = (\mathbf{Sen}_2, \mathbf{Sign}_2, \leq_2, sig_2, \mathbf{Mod}_2, \models_2, .|.)$, *an* **institute comorphism** $\rho = (\Phi, \alpha, \beta):$ $\mathcal{I}_1 \longrightarrow \mathcal{I}_2$ *consists of*

- *a monotone map* $\Phi : (\mathbf{Sign}^1, \leq^1) \to (\mathbf{Sign}^2, \leq^2),$
- *a sentence translation function* $\alpha: \mathbf{Sen}_1 \longrightarrow \mathbf{Sen}_2,$ *and*
- *for each* $\mathcal{I}_1$-*signature* Σ, *a monotone model translation function* $\beta_\Sigma : \mathbf{Mod}_2(\Phi(\Sigma)) \to$ $\mathbf{Mod}_1(\Sigma),$

such that

- $M_2 \models_2 \alpha(\varphi_1)$ *if and only if* $\beta_\Sigma(M_2) \models_1 \varphi_1$ *holds for each* $\mathcal{I}_1$-*signature* Σ, *each model* $M_2 \in \mathbf{Mod}_2(\Sigma)$ *and each sentence* $\varphi_1 \in \mathbf{Sen}_1(\Sigma)$ *(satisfaction condition)*
- $sig_2(\alpha(\varphi_1)) \leq \Phi(sig_1(\varphi_1))$ *for any sentence* $\varphi_1 \in \mathbf{Sen}_1$ *(sentence coherence);*
- **model translation commutes with reduct,** *that is, given* $\Sigma_1 \leq \Sigma_2$ *in* $\mathcal{I}_1$ *and a* $\Phi(\Sigma_2)$-*model* M *in* $\mathcal{I}_2$,

$$\beta_{\Sigma_2}(M)|_{\Sigma_1} = \beta_{\Sigma_1}(M|_{\Phi(\Sigma_1)}). \qquad \square$$

Some important properties of institution (co-)morphisms will be needed in the technical development below:

Definition 9 (Model-expansive, (weakly) exact, (weak) amalgamation). *An institute comorphism is **model-expansive**, if each β_Σ is surjective. It is easy to show that model-expansive comorphisms faithfully encode logical consequence, that is, $\Gamma \models \varphi$ iff $\alpha(\Gamma) \models \alpha(\varphi)$.*

*An institute comorphism $\rho = (\Phi, \alpha, \beta) \colon \mathcal{I}_1 \longrightarrow \mathcal{I}_2$ is **(weakly) exact**, if for each signature extension $\Sigma_1 \leq \Sigma_2$ the diagram*

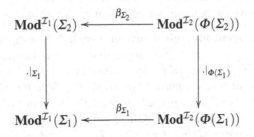

*admits **(weak) amalgamation**, i.e. for any $M_2 \in \mathbf{Mod}^I(\Sigma_2)$ and $M_1' \in \mathbf{Mod}^J(\Phi(\Sigma_1))$ with $M_2|_{\Sigma_1} = \beta_{\Sigma_1}(M_1')$, there is a (not necessarily unique) $M_2' \in \mathbf{Mod}^J(\Phi(\Sigma_2))$ with $\beta_{\Sigma_2}(M_2') = M_2$ and $M_2'|_{\Phi(\Sigma_1)} = M_1'$.* □

Given these definitions, a **simple theoroidal institute comorphism** $\rho \colon \mathcal{I}_1 \longrightarrow \mathcal{I}_2$ is an ordinary institute comorphism $\rho \colon \mathcal{I}_1 \longrightarrow \mathcal{I}_2^{th}$ (for $\mathcal{I}_2^{th}$, see Def. 4). Moreover, an institute comorphism is said to be **model-isomorphic** if β_Σ is an isomorphism. It is a **subinstitute comorphism** (cf. also [25]), if moreover the signature translation is an embedding and sentence translation is injective. The intuition is that theories should be embedded, while models should be represented exactly (such that model-theoretic results carry over).

4   A DOL Kernel and Its Semantics

The Distributed Ontology Language (DOL) shares many features with the language HetCASL [26] which underlies the Heterogeneous Tool Set Hets [29]. However, it also adds a number of new features:

- minimisation of models following the circumscription paradigm [24,19];
- ontology module extraction, i.e. the extraction of a subtheory that contains all relevant logical information w.r.t. some subsignature [14];
- projections of theories to a sublogic;
- ontology alignments, which involve partial or even relational variants of signature morphisms [6];
- combination of theories via colimits, which has been used to formalise certain forms of ontology alignment [37,17];
- referencing of all items by URLs, or, more general, IRIs [18].

Sannella and Tarlecki [32,33] show that the structuring of logical theories (specifications) can be defined independently of the underlying logical system. They define a kernel language for structured specification that can be interpreted over an arbitrary institution.

Similar to [32] and also integrating heterogeneous constructs from [36,30], we now introduce a kernel language for heterogeneous structured specifications for DOL. We will use the term "structured ontology" instead of "structured specification" to stress the intended use for DOL.

Since DOL involves not only one, but possibly several ontology languages, we need to introduce the notion of a 'heterogeneous logical environment'.

Definition 10 (Heterogeneous logical environment). *A **heterogeneous logical environment** is defined to be a graph of institutes and institute morphisms and (possibly simple theoroidal) comorphisms, where we assume that some of the comorphisms (including all obvious identity comorphisms) are marked as default inclusions. The default inclusions are assumed to form a partial order on the institutes of the logic graph. If $\mathcal{I}_1 \leq \mathcal{I}_2$, the default inclusion is denoted by $\iota \colon \mathcal{I}_1 \longrightarrow \mathcal{I}_2$. For any pair of institutes $\mathcal{I}_1$ and $\mathcal{I}_2$, if their supremum exists, we denote it by $\mathcal{I}_1 \cup \mathcal{I}_2$, and the corresponding default inclusions by $\iota_i \colon \mathcal{I}_i \longrightarrow \mathcal{I}_1 \cup \mathcal{I}_2$.*

We are now ready for the definition of heterogeneous structured ontology.

Definition 11 (Heterogeneous structured ontology - DOL kernel language). *Let a heterogeneous logical environment be given. We inductively define the notion of heterogeneous structured ontology (in the sequel: ontology). Simultaneously, we define functions **Ins**, **Sig** and **Mod** yielding the institute, the signature and the model class of such an ontology.*

presentations: *For any institute $\mathcal{I}$, signature $\Sigma \in |\mathbf{Sign}^{\mathcal{I}}|$ and finite set $\Gamma \subseteq \mathbf{Sen}^{\mathcal{I}}(\Sigma)$ of Σ-sentences, the presentation $\langle \mathcal{I}, \Sigma, \Gamma \rangle$ is an ontology with:*

$$\mathbf{Ins}(\langle \mathcal{I}, \Sigma, \Gamma \rangle) := \mathcal{I}$$
$$\mathbf{Sig}(\langle \mathcal{I}, \Sigma, \Gamma \rangle) := \Sigma$$
$$\mathbf{Mod}(\langle \mathcal{I}, \Sigma, \Gamma \rangle) := \{M \in \mathbf{Mod}(\Sigma) \mid M \models \Gamma\}$$

union: *For any signature $\Sigma \in |\mathbf{Sign}|$, given ontologies O_1 and O_2 with the same institute $\mathcal{I}$ and signature Σ, their union O_1 **and** O_2 is an ontology with:*

$$\mathbf{Ins}(O_1 \text{ and } O_2) := \mathcal{I}$$
$$\mathbf{Sig}(O_1 \text{ and } O_2) := \Sigma$$
$$\mathbf{Mod}(O_1 \text{ and } O_2) := \mathbf{Mod}(O_1) \cap \mathbf{Mod}(O_2)$$

extension: *For any ontology O with institute $\mathcal{I}$ and signature Σ and any signature extension $\Sigma \leq \Sigma'$ in $\mathcal{I}$, O **with** Σ' is an ontology with:*

$$\mathbf{Ins}(O \text{ with } \Sigma') := \mathcal{I}$$
$$\mathbf{Sig}(O \text{ with } \Sigma') := \Sigma'$$
$$\mathbf{Mod}(O \text{ with } \Sigma') := \{M' \in \mathbf{Mod}(\Sigma') \mid M'|_{\Sigma} \in \mathbf{Mod}(O)\}$$

hiding: *For any ontology O' with institute $\mathcal{I}$ and signature Σ' and any signature extension $\Sigma \leq \Sigma'$ in $\mathcal{I}$, O' **hide** Σ is an ontology with:*

$$\mathbf{Ins}(O' \text{ hide } \Sigma) := \mathcal{I}$$
$$\mathbf{Sig}(O' \text{ hide } \Sigma) := \Sigma$$
$$\mathbf{Mod}(O' \text{ hide } \Sigma) := \{M'|_{\Sigma} \mid M' \in \mathbf{Mod}(O')\}$$

minimisation: *Let O be an ontology with institute $\mathcal{I}$ and signature Σ and let $\Sigma_{min}, \Sigma_{fixed}$ be subsignatures of Σ such that $\Sigma_{min} \cup \Sigma_{fixed}$ is defined. Intuitively, the interpretation of the symbols in Σ_{min} will be minimised among those models interpreting the symbols in Σ_{fixed} in the same way, while the interpretation of all symbols outside $\Sigma_{min} \cup \Sigma_{fixed}$ may vary arbitrarily. Then O **minimize** $\Sigma_{min}, \Sigma_{fixed}$ is an ontology with:*

Ins$(O$ **minimize** $\Sigma_{min}, \Sigma_{fixed}) := \mathcal{I}$
Sig$(O$ **minimize** $\Sigma_{min}, \Sigma_{fixed}) := \Sigma$
Mod$(O$ **minimize** $\Sigma_{min}, \Sigma_{fixed}) := \{M \in \mathbf{Mod}(O) \mid M|_{\Sigma_{min} \cup \Sigma_{fixed}}$ *is minimal in* $Fix(M)\}$
where $Fix(M) = \{M' \in \mathbf{Mod}(O)|_{\Sigma_{min} \cup \Sigma_{fixed}} \mid M'|_{\Sigma_{fixed}} = M|_{\Sigma_{fixed}}\}$

translation along a comorphism: *For any ontology O with institute $\mathcal{I}$ and signature Σ and any institute comorphism $\rho = (\Phi, \alpha, \beta): \mathcal{I} \to \mathcal{I}'$, O **with** ρ is a ontology with:*

Ins$(O$ **with** $\rho) := \mathcal{I}'$
Sig$(O$ **with** $\rho) := \Phi(\Sigma)$
Mod$(O$ **with** $\rho) := \{M' \in \mathbf{Mod}^{\mathcal{I}'}(\Phi(\Sigma)) \mid \beta_{\Sigma}(M') \in \mathbf{Mod}(O)\}$
If ρ is simple theoroidal, then **Sig**$(O$ **with** $\rho)$ *is the signature component of* $\Phi(\Sigma)$.
hiding along a morphism: *For any ontology O' with institute $\mathcal{I}$ and signature Σ' and any institute morphism $\mu = (\Phi, \alpha, \beta): \mathcal{I} \to \mathcal{I}'$, O' **hide** μ is a ontology with:*

Ins$(O'$ **hide** $\mu) := \mathcal{I}'$
Sig$(O'$ **hide** $\mu) := \Phi(\Sigma)$
Mod$(O'$ **hide** $\mu) := \{\beta_{\Sigma}(M') \mid M' \in \mathbf{Mod}(O')\}$ □

Derived operations. We also define the following derived operation generalising union to arbitrary pairs of ontologies: For any ontologies O_1 and O_2 with institutes $\mathcal{I}_1$ and $\mathcal{I}_2$ and signatures Σ_1 and Σ_2, if the supremum $\mathcal{I}_1 \cup \mathcal{I}_2$ exists and the union $\Sigma = \Phi(\Sigma_1) \cup \Phi(\Sigma_2)$ is defined, the *generalised union* of O_1 and O_2, by abuse of notation also written as O_1 **and** O_2, is defined as

$$(O_1 \text{ with } \iota_1 \text{ with } \Sigma) \text{ and } (O_2 \text{ with } \iota_2 \text{ with } \Sigma)$$

The full DOL language adds further language constructs that can be expressed in terms of this kernel language. Furthermore, DOL allows the omission of translations along default inclusion comorphisms, since these can be reconstructed in a unique way.

Logical consequence. We say that a sentence φ is a logical consequence of a heterogeneous structured ontology O, written $O \models \varphi$, if any model of O satisfies φ.

Monotonicity. Similar to [33], Ex. 5.1.4, we get:

Proposition 12. *All structuring operations of the DOL kernel language except minimisation are monotone in the sense that they preserve model class inclusion:* $\mathbf{Mod}(O_1) \subseteq \mathbf{Mod}(O_2)$ *implies* $\mathbf{Mod}(op(O_1)) \subseteq \mathbf{Mod}(op(O_2))$. *(Union is monotone in both arguments.)*

Indeed, the minimisation is a deliberate exception: its motivation is to capture non-monotonic reasoning.

Proposition 13. *If reducts are surjective,* **minimize** *is anti-monotone in* Σ_{min}.

Proof. Let O be an ontology with institute $\mathcal{I}$ and signature Σ and let $\Sigma_{min}^1, \Sigma_{min}^2, \Sigma_{fixed} \subseteq \Sigma$ be subsignatures such that $\Sigma_{min}^1 \leq \Sigma_{min}^2$, and $\Sigma_{min}^i \cup \Sigma_{fixed}$ is defined for $i = 1, 2$. Let Fix^1 and Fix^2 defined as Fix above, using Σ_{min}^1 and Σ_{min}^2 respectively. Let $M \in$ **Mod**$(O$ **minimize** $\Sigma_{min}^2, \Sigma_{fixed})$. Then M is an O-model such that $M|_{\Sigma_{min}^2 \cup \Sigma_{fixed}}$ is minimal in $Fix^2(M)$. We show that $M|_{\Sigma_{min}^1 \cup \Sigma_{fixed}}$ is minimal in $Fix^1(M)$: Let M' be in $Fix^1(M)$. By surjectivity of reducts, it can be expanded to a $\Sigma_{min}^2 \cup \Sigma_{fixed}$-model M''. Now $M'' \in Fix^2(M)$, because all involved models agree on Σ_{fixed}. Since $M|_{\Sigma_{min}^2 \cup \Sigma_{fixed}}$ is minimal in $Fix^2(M)$, $M|_{\Sigma_{min}^2 \cup \Sigma_{fixed}} \leq M''$. Since reducts preserve the model ordering, $M|_{\Sigma_{min}^1 \cup \Sigma_{fixed}} \leq M'$. Hence, $M \in$ **Mod**$(O$ **minimize** $\Sigma_{min}^1, \Sigma_{fixed})$. □

5   Relations between Ontologies

Besides heterogeneous structured ontologies, DOL features the following statements about relations between heterogeneous structured ontologies:

interpretations Given heterogeneous structured ontologies O_1 and O_2 with institutes $\mathcal{I}_1$ and $\mathcal{I}_2$ and signatures Σ_1 and Σ_2, we write

$$O_1 \rightsquigarrow O_2$$

(read: O_1 can be interpreted in O_2) for the conjunction of
1. $\mathcal{I}_1 \leq \mathcal{I}_2$ with default inclusion $\iota = (\Phi, \alpha, \beta) \colon \mathcal{I}_1 \longrightarrow \mathcal{I}_2$,
2. $\Phi(\Sigma_1) \leq \Sigma_2$, and
3. $\beta(\textbf{Mod}(O_2)|_{\Phi(\Sigma_1)}) \subseteq \textbf{Mod}(O_1)$.

modules Given heterogeneous structured ontologies O_1 and O_2 over the same institute $\mathcal{I}$ with signatures Σ_1 and Σ_2, and given another signature $\Sigma \leq \Sigma_1$ (called the restriction signature), we say that

O_1 is a model-theoretic (consequence-theoretic) module of O_2 w.r.t. Σ

if for any O_1-model M_1, $M_1|_\Sigma$ can be extended to an O_2-model (resp. O_2 is a conservative extension of O_1, see Def. 5). It is easy to see that the model-theoretic module relation implies the consequence-theoretic one. However, the converse is not true in general, compare [20] for an example from description logic, and see [15] for more general conservativity preservation results.

We first considered to integrate a module extraction operator into the kernel language of heterogeneous structured ontologies. However, there are so many different notions of ontology module and techniques of module extraction used in the literature that we would have to define a whole collection of module extraction operators, a collection that moreover would quickly become obsolete and incomplete. We refrained from this, and instead provide a relation between heterogeneous structured ontologies that is independent of the specificities of particular module extraction operators.

Still, it is possible to define all the relevant notions used in the ontology modules community within an arbitrary institute, namely the notions of conservative extension, inseparability, uniform interpolant etc. The reason is that these notions typically are defined in set-theoretic parlance about signatures (see Def. 5).

The full DOL language is based on the DOL kernel and also includes a construct for colimits (which is omitted here, because its semantics requires institutions) and ontology alignments (which are omitted here, because they do not have a model-theoretic semantics). The full DOL language is detailed in the current OntoIOp ISO 17347 working draft, see `ontoiop.org`. There, also an alternative semantics to the above direct set-theoretic semantics is given: a translational semantics. It assumes that all involved institutes can be translated to Common Logic, and gives the semantics of an arbitrary ontology by translation to Common Logic (and then using the above direct semantics). The two semantics are compatible, see [28] for details. However, the translational semantics has some important drawbacks. In particular, the semantics of ontology modules (relying on the notion of conservative extension) is not always preserved when translating to Common Logic. See [28] for details.

6 An Example in DOL

As an example of a heterogeneous ontology in DOL, we formalise some notions of mereology. Propositional logic is not capable of describing mereological *relations*, but of describing the basic categories over which the DOLCE foundational ontology [23] defines mereological relations. The same knowledge can be formalised more conveniently in OWL, which additionally allows for describing (not defining!) basic parthood properties. As our OWL ontology redeclares as classes the same categories that the propositional logic ontology *Taxonomy* had introduced as propositional variables, using different names but satisfying the same disjointness and subsumption axioms, we observe that it *interprets* the former. Mereological relations are frequently *used* in lightweight OWL ontologies, e.g. biomedical ontologies in the EL profile (designed for efficient reasoning with a large number of entities, a frequent case in this domain), but these languages are not fully capable of *defining* these relations. Therefore, we finally provide a full definition of several mereological relations in first order logic, in the Common Logic language, by importing, translating and extending the OWL ontology. We use Common Logic's second-order facility of quantifying over predicates to concisely express the restriction of the variables x, y, and z to the same taxonomic category.

```
%prefix( :      <http://www.example.org/mereology#>      %% prefix for this distributed ontology
      owl:   <http://www.w3.org/2002/07/owl#>            %% OWL basic ontology language
      log:   <http://purl.net/dol/logics/>               %% DOL-conforming logics (Fig. 1)
      trans: <http://purl.net/dol/translations/>         %% translations between these logics
      ser:   <http://purl.net/dol/serializations/> %)    %% serializations, i.e. concrete syntaxes

distributed-ontology Mereology

logic log:Propositional syntax ser:Prop/Hets            %% non-standard serialization built into Hets
ontology Taxonomy =                       %% basic taxonomic information about mereology reused from DOLCE
   props PT, T, S, AR, PD
   . S ∨ T ∨ AR ∨ PD ⟶ PT                                %% PT is the top concept
   . S ∧ T ⟶ ⊥                            %% PD, S, T, AR are pairwise disjoint
   . T ∧ AR ⟶ ⊥                                          %% and so on

logic log:SROIQ syntax ser:OWL2/Manchester              %% OWL Manchester syntax
ontology BasicParthood =                  %% Parthood in SROIQ, as far as expressible
   Class: ParticularCategory SubClassOf: Particular %% omitted similar declarations of the other classes
```

```
DisjointUnionOf: SpaceRegion, TimeInterval, AbstractRegion, Perdurant
                            %% pairwise disjointness more compact thanks to an OWL built-in
ObjectProperty: isPartOf        Characteristics: Transitive
ObjectProperty: isProperPartOf  Characteristics: Asymmetric  SubPropertyOf: isPartOf
Class: Atom EquivalentTo: inverse isProperPartOf only owl:Nothing

interpretation TaxonomyToParthood : Taxonomy to BasicParthood =
  translate with trans:PropositionalToSROIQ,          %% translate the logic, then rename the entities
  PT ↦ Particular, S ↦ SpaceRegion, T ↦ TimeInterval, A ↦ AbstractRegion, %[ and so on ]%

logic log:CommonLogic syntax ser:CommonLogic/CLIF %% syntax: the Lisp-like CLIF dialect of Common Logic
ontology ClassicalExtensionalParthood =
  BasicParthood translate with trans:SROIQtoCL %% import the OWL ontology from above, translate it ...
  then {                                       %% ... to Common Logic, then extend it there:
   (forall (X) (if (or (= X S) (= X T) (= X AR) (= X PD))
                     (forall (x y z) (if (and (X x) (X y) (X z))
                                    (and                               %% now list all the axioms
     (if (and (isPartOf x y) (isPartOf y x)) (= x y))                  %% antisymmetry
     (if (and (isProperPartOf x y) (isProperPartOf y z)) (isProperPartOf x z))
                          %% transitivity; can't be expressed in OWL together with asymmetry
     (iff (overlaps x y) (exists (pt) (and (isPartOf pt x) (isPartOf pt y))))
     (iff (isAtomicPartOf x y) (and (isPartOf x y) (Atom x)))
     (iff (sum z x y) (forall (w) (iff (overlaps w z) (and (overlaps w x) (overlaps w y)))))
     (exists (s) (sum s x y)))))))               %% existence of the sum
  }
```

7   Relating Institutes and Institutions

In this section, we show that institutes are a certain restriction of institutions. We first recall Goguen's and Burstall's notion of *institution* [10], which they have introduced as a formalisation of the intuitive notion of logical system. We assume some acquaintance with the basic notions of category theory and refer to [1] or [22] for an introduction.

Definition 14. *An **institution** is a quadruple* $I = (\mathbf{Sign}, \mathbf{Sen}, \mathbf{Mod}, \models)$ *consisting of the following:*

- *a category* **Sign** *of signatures and signature morphisms,*
- *a functor* **Sen**: **Sign** $\longrightarrow$ **Set**[5] *giving, for each signature* Σ, *the set of* sentences **Sen**(Σ), *and for each signature morphism* $\sigma\colon \Sigma \longrightarrow \Sigma'$, *the sentence translation map* **Sen**$(\sigma)\colon$ **Sen**$(\Sigma) \longrightarrow$ **Sen**(Σ'), *where often* **Sen**$(\sigma)(\varphi)$ *is written as* $\sigma(\varphi)$,
- *a functor* **Mod**: **Sign**^{op} $\longrightarrow \mathcal{CAT}$[6] *giving, for each signature* Σ, *the category of* models **Mod**(Σ), *and for each signature morphism* $\sigma\colon \Sigma \longrightarrow \Sigma'$, *the reduct functor* **Mod**$(\sigma)\colon$ **Mod**$(\Sigma') \longrightarrow$ **Mod**(Σ), *where often* **Mod**$(\sigma)(M')$ *is written as* $M'|_\sigma$, *and* $M'|_\sigma$ *is called the* σ-*reduct of* M', *while* M' *is called a* σ-*expansion of* $M'|_\sigma$,
- *a satisfaction relation* $\models_\Sigma \subseteq |\mathbf{Mod}(\Sigma)| \times \mathbf{Sen}(\Sigma)$ *for each* $\Sigma \in |\mathbf{Sign}|$,

such that for each $\sigma\colon \Sigma \longrightarrow \Sigma'$ *in* **Sign** *the following **satisfaction condition** holds:*

$$(\star) \qquad M' \models_{\Sigma'} \sigma(\varphi) \; iff \; M'|_\sigma \models_\Sigma \varphi$$

---

[5] **Set** is the category having all small sets as objects and functions as arrows.

[6] $\mathcal{CAT}$ is the category of categories and functors. Strictly speaking, $\mathcal{CAT}$ is not a category but only a so-called quasicategory, which is a category that lives in a higher set-theoretic universe.

for each $M' \in |\mathbf{Mod}(\Sigma')|$ *and* $\varphi \in \mathbf{Sen}(\Sigma)$, *expressing that truth is invariant under change of notation and context.*[7]   □

With institutions, a few more features of DOL can be equipped with a semantics:

- renamings along signature morphisms [32],
- combinations (colimits), and
- monomorphic extensions.

Due to the central role of inclusions of signatures for institutes, we also need to recall the notion of inclusive institution.

Definition 15 ([31]). *An **inclusive category** is a category having a broad subcategory which is a partially ordered class.*

*An **inclusive institution** is one with an inclusive signature category such that the sentence functor preserves inclusions. We additionally require that such institutions*

- *have inclusive model categories,*
- *have signature intersections (i.e. binary infima), which are preserved by* $\mathbf{Sen}$,[8] *and*
- *have well-founded sentences, which means that there is no sentence that occurs in all sets of an infinite chain of strict inclusions*

$$\ldots \hookrightarrow \mathbf{Sen}(\Sigma_n) \hookrightarrow \ldots \hookrightarrow \mathbf{Sen}(\Sigma_1) \hookrightarrow \mathbf{Sen}(\Sigma_o)$$

that is the image (under $\mathbf{Sen}$) *of a corresponding chain of signature inclusions.*   □

Definition 16. *Given institutions I and J, an* institution morphism *[10] written* $\mu = (\Phi, \alpha, \beta) \colon I \longrightarrow J$ *consists of*

- *a functor* $\Phi \colon \mathbf{Sign}^I \longrightarrow \mathbf{Sign}^J$,
- *a natural transformation* $\alpha \colon \mathbf{Sen}^J \circ \Phi \longrightarrow \mathbf{Sen}^I$ *and*
- *a natural transformation* $\beta \colon \mathbf{Mod}^I \longrightarrow \mathbf{Mod}^J \circ \Phi^{op}$,

such that the following satisfaction condition holds for all $\Sigma \in \mathbf{Sign}^I$, $M \in \mathbf{Mod}^I(\Sigma)$ *and* $\varphi' \in \mathbf{Sen}^J(\Phi(\Sigma))$:

$$M \models_{\Sigma}^I \alpha_{\Sigma}(\varphi') \text{ iff } \beta_{\Sigma}(M) \models_{\Phi(\Sigma)}^J \varphi'$$

Definition 17. *Given institutions I and J, an* institution comorphism *[9] denoted as* $\rho = (\Phi, \alpha, \beta) \colon I \longrightarrow J$ *consists of*

- *a functor* $\Phi \colon \mathbf{Sign}^I \longrightarrow \mathbf{Sign}^J$,
- *a natural transformation* $\alpha \colon \mathbf{Sen}^I \longrightarrow \mathbf{Sen}^J \circ \Phi$,
- *a natural transformation* $\beta \colon \mathbf{Mod}^J \circ \Phi^{op} \longrightarrow \mathbf{Mod}^I$

---

[7] Note, however, that non-monotonic formalisms can only indirectly be covered this way, but compare, e.g., [12].

[8] This is a quite reasonable assumption met by practically all institutions. Note that by contrast, preservation of unions is quite unrealistic—the union of signatures normally leads to new sentences combining symbols from both signatures.

such that the following satisfaction condition *holds for all* $\Sigma \in \mathbf{Sign}^I$, $M' \in \mathbf{Mod}^J(\Phi(\Sigma))$ *and* $\varphi \in \mathbf{Sen}^I(\Sigma)$:

$$M' \models^J_{\Phi(\Sigma)} \alpha_\Sigma(\varphi) \text{ iff } \beta_\Sigma(M') \models^I_\Sigma \varphi.$$

Let **InclIns** (**CoInclIns**) denote the quasicategory of inclusive institutions and morphisms (comorphisms). Furthermore, let **Class** denote the quasicategory of classes and functions. Note that (class-indexed) colimits of sets in **Class** can be constructed in the same way as in **Set**. Finally, call an institute **locally small**, if each $\mathbf{Sen}(\Sigma)$ is a set. Let **Institute** (**CoInstitute**) be the quasicategory of locally small institutes and morphisms (comorphisms).

Proposition 18. *There are functors* F^{co} : **CoInstitute** $\to$ **CoInclIns** *and* F : **Institute** $\to$ **InclIns**.

Proof. Given an institute $\mathcal{I} = (\mathbf{Sen}^{\mathcal{I}}, \mathbf{Sign}^{\mathcal{I}}, \leq^{\mathcal{I}}, sig^{\mathcal{I}}, \mathbf{Mod}^{\mathcal{I}}, \models^{\mathcal{I}}, .|.)$, we construct an institution $F(\mathcal{I}) = F^{co}(\mathcal{I})$ as follows: $(\mathbf{Sign}^{\mathcal{I}}, \leq^{\mathcal{I}})$ is a partially ordered class, hence a (thin) category. We turn it into an inclusive category by letting all morphisms be inclusions. This will be the signature category of $F(\mathcal{I})$.

For each signature Σ, we let $\mathbf{Sen}^{F(\mathcal{I})}(\Sigma)$ be $\mathbf{Sen}^{\mathcal{I}}(\Sigma)$ (here we need local smallness of $\mathcal{I}$). Then $\mathbf{Sen}^{F(\mathcal{I})}$ easily turns into an inclusion-preserving functor. Also, $\mathbf{Mod}^{F(\mathcal{I})}(\Sigma)$ is $\mathbf{Mod}^{\mathcal{I}}(\Sigma)$ turned into a thin category using the partial order on $\mathbf{Mod}^{\mathcal{I}}$. Since reducts are compositional and preserve the model ordering, we obtain reduct functors for $F(\mathcal{I})$. Satisfaction in $F(\mathcal{I})$ is defined as in $\mathcal{I}$. The satisfaction condition holds because satisfaction is invariant under reduct.

Given an institute comorphism $\rho = (\Phi, \alpha, \beta) : \mathcal{I}_1 \longrightarrow \mathcal{I}_2$, we define an institution comorphism $F^{co}(\rho) : F(\mathcal{I}_1) \longrightarrow F(\mathcal{I}_2)$ as follows. Φ obviously is a functor from $\mathbf{Sign}^{F(\mathcal{I}_1)}$ to $\mathbf{Sign}^{F(\mathcal{I}_2)}$. If $sig(\varphi) \leq \Sigma$, by sentence coherence, $sig(\alpha(\varphi)) \leq \Phi(\Sigma)$. Hence, $\alpha : \mathbf{Sen}_1 \longrightarrow \mathbf{Sen}_2$ can be restricted to $\alpha_\Sigma : \mathbf{Sen}_1(\Sigma) \longrightarrow \mathbf{Sen}_2(\Sigma)$ for any $\mathcal{I}_1$-signature Σ. Naturality of the family $(\alpha_\Sigma)_{\Sigma \in \mathbf{Sign}_1}$ follows from the fact that the α_Σ are restrictions of a global α. Each β_Σ is functorial because it is monotone. Naturality of the family $(\beta_\Sigma)_{\Sigma \in \mathbf{Sign}_1}$ follows from model translation commuting with reduct. The satisfaction condition is easily inherited from the institute comorphism.

The translation of institute morphisms is similar. ☐

Proposition 19. *There are functors* G^{co} : **CoInclIns** $\to$ **CoInstitute** *and* G : **InclIns** $\to$ **Institute**, *such that* $G^{co} \circ F^{co} \cong id$ *and* $G \circ F \cong id$.

Proof. Given an inclusive institution $\mathcal{I} = (\mathbf{Sign}^{\mathcal{I}}, \mathbf{Sen}^{\mathcal{I}}, \mathbf{Mod}^{\mathcal{I}}, \models^{\mathcal{I}})$, we construct an institute $G(\mathcal{I}) = G^{co}(\mathcal{I})$ as follows: $(\mathbf{Sign}^{\mathcal{I}}, \leq^{\mathcal{I}})$ is the partial order given by the inclusions.

$\mathbf{Sen}^{G(\mathcal{I})}$ is the colimit of the diagram of all inclusions $\mathbf{Sen}^{\mathcal{I}}(\Sigma_1) \hookrightarrow \mathbf{Sen}^{\mathcal{I}}(\Sigma_1)$ for $\Sigma_1 \leq \Sigma_2$. This colimit is taken in the quasicategory of classes and functions. It exists because all involved objects are sets (the construction can be given as a quotient of the disjoint union, following the usual construction of colimits as coequalisers of coproducts). Let $\mu_\Sigma : \mathbf{Sen}^{\mathcal{I}}(\Sigma) \longrightarrow \mathbf{Sen}^{G(\mathcal{I})}$ denote the colimit injections. For a sentence φ, let $\mathbf{S}(\varphi)$ be the set of signatures Σ such that φ is in the image of μ_Σ. We show that $\mathbf{S}(\varphi)$ has a least element. For if not, choose some $\Sigma_0 \in \mathbf{S}(\varphi)$. Assume that we have chosen

$\Sigma_n \in \mathbf{S}(\varphi)$. Since Σ_n is not the least element of $\mathbf{S}(\varphi)$, there must be some $\Sigma \in \mathbf{S}(\varphi)$ such that $\Sigma_n \not\subseteq \Sigma$. Then let $\Sigma_{n+1} = \Sigma_n \cap \Sigma$; since $\mathbf{Sen}$ preserves intersections, $\Sigma_{n+1} \in \mathbf{S}(\varphi)$. Moreover, $\Sigma_{n+1} < \Sigma_n$. This gives an infinite descending chain of signature inclusions in $\mathbf{S}(\varphi)$, contradicting $\mathcal{I}$ having well-founded sentences. Hence, $\mathbf{S}(\varphi)$ must have a least element, which we use as $sig(\varphi)$.

$\mathbf{Mod}^{G(\mathcal{I})}(\Sigma)$ is the partial order of inclusions in $\mathbf{Mod}^{\mathcal{I}}(\Sigma)$, and also reduct is inherited. Since $\mathbf{Mod}^{G(\mathcal{I})}$ is functorial, reducts are compositional. Since each $\mathbf{Mod}^{G(\mathcal{I})}(\sigma)$ is functorial, reducts preserve the model ordering. Satisfaction in $G(\mathcal{I})$ is defined as in $\mathcal{I}$. The satisfaction condition implies that satisfaction is invariant under reduct.

Given an institution comorphism $\rho = (\Phi, \alpha, \beta) : \mathcal{I}_1 \longrightarrow \mathcal{I}_2$, we define an institute comorphism $G^{co}(\rho) : G(\mathcal{I}_1) \longrightarrow G(\mathcal{I}_2)$ as follows. Φ obviously is a monotone map from $\mathbf{Sign}^{G(\mathcal{I}_1)}$ to $\mathbf{Sign}^{G(\mathcal{I}_2)}$.

$\alpha : \mathbf{Sen}^{G(\mathcal{I}_1)} \longrightarrow \mathbf{Sen}^{G(\mathcal{I}_2)}$ is defined by exploiting the universal property of the colimit $\mathbf{Sen}^{G(\mathcal{I}_1)}$: it suffices to define a cocone $\mathbf{Sen}^{\mathcal{I}_1}(\Sigma) \to \mathbf{Sen}^{G(\mathcal{I}_2)}$ indexed over signatures Σ in $\mathbf{Sign}^{\mathcal{I}_1}$. The cocone is given by composing α_Σ with the inclusion of $\mathbf{Sen}^{\mathcal{I}_1}(\Phi(\Sigma))$ into $\mathbf{Sen}^{G(\mathcal{I}_2)}$. Commutativity of a cocone triangle follows from that of a cocone triangle for the colimit $\mathbf{Sen}^{G(\mathcal{I}_2)}$ together with naturality of α. This construction also ensures sentence coherence.

Model translation is just given by the β_Σ; the translation of institution morphisms is similar.

Finally, $G \circ F \cong id$ follows because $\mathbf{Sen}$ can be seen to be the colimit of all $\mathbf{Sen}(\Sigma_1) \hookrightarrow \mathbf{Sen}(\Sigma_2)$. This means that we can even obtain $G \circ F = id$. However, since the choice of the colimit in the definition of G is only up to isomorphism, generally we obtain only $G \circ F \cong id$. The argument for $G^{co} \circ F^{co} \cong id$ is similar, since isomorphism institution morphisms are also isomorphism institution comorphisms. $\qquad\square$

It should be noted that $F^{co} : \mathbf{CoInstitute} \to \mathbf{CoInclIns}$ is "almost" left adjoint to $G^{co} : \mathbf{CoInclIns} \to \mathbf{CoInstitute}$: By the above remarks, w.l.o.g., the unit $\eta : Id \longrightarrow G^{co} \circ F^{co}$ can be chosen to be the identity. Hence, we need to show that for each institute comorphism $\rho : \mathcal{I}_1 \longrightarrow G(\mathcal{I}_2)$, there is a unique institution comorphism $\rho^{\#} : F(\mathcal{I}_1) \longrightarrow \mathcal{I}_2$ with $G(\rho^{\#}) = \rho$. The latter condition easily ensures uniqueness. Let $\rho = (\Phi, \alpha, \beta)$. We construct $\rho^{\#}$ as $(\Phi, \alpha^{\#}, \beta)$. Clearly, Φ also is a functor from $\mathbf{Sign}^{F(\mathcal{I}_1)}$ into $\mathbf{Sign}^{\mathcal{I}_2}$ (which is a supercategory of $\mathbf{Sign}^{G(\mathcal{I}_2)}$. A similar remark holds for β, but only if the model categories in $\mathcal{I}_2$ consist of inclusions only. $\alpha^{\#}$ can be constructed from α by passing to the restrictions α_Σ. Altogether we get:

Proposition 20. $F^{co} : \mathbf{CoInstitute} \to \mathbf{CoInclIns}$ *is left adjoint to* $G^{co} : \mathbf{CoInclIns} \to \mathbf{CoInstitute}$ *if institutions are restricted to model categories in consisting of inclusions only.*

Since also $G^{co} \circ F^{co} \cong id$, $\mathbf{CoInstitute}$ comes close to being a coreflective subcategory of $\mathbf{CoInclIns}$.

We also obtain:

Proposition 21. *For the DOL kernel language, the institute-based semantics (over some institute-based heterogeneous logical environment $\mathcal{E}$) and the institution-based semantics (similar to that given in [32,30], over F applied to $\mathcal{E}$) coincide up to application of G to the* **Ins** *component of the semantics.*

8   Conclusion

We have taken concepts from the area of formal methods for software specification and applied them to obtain a kernel language for the Distributed Ontology Language (DOL), including a semantics, and have thus provided the syntax and semantics of a heterogeneous structuring language for ontologies. The standard approach here would be to use institutions to formalise the notion of logical system. However, aiming at a more simple presentation of the heterogeneous semantics, we here develop the notion of *institute* which allows us to obtain a set-based semantics for a large part of DOL. Institutes can be seen as institutions without category theory.

Goguen and Tracz [11] have a related set-theoretic approach to institutions: they require signatures to be tuple sets. Our approach is more abstract, because signatures can be any partial order. Moreover, the results of Sect. 7 show that institutes integrate nicely with institutions. That is, we can have the cake and eat it, too: we can abstractly formalise various logics as institutes, a formalisation which, being based on standard set-theoretic methods, can be easily understood by the broader ontology communities that are not necessarily acquainted with category theoretic methods. Moreover, the possibility to extend the institute-based formalisation to a full-blown institution which is compatible with the institute (technically, this means that the functor G defined in Prop. 19, applied to the institution, should yield the institute), allows a smooth technical integration of further features into the framework which do require institutions, such as colimits.

This work provides the semantic backbone for the Distributed Ontology Language DOL, which is being developed in the ISO Standard 17347 Ontology Integration and Interoperability, see ontoiop.org. An experimental repository for ontologies written in different logics and also in DOL is available at ontohub.org.

Acknowledgements. We would like to thank the OntoIOp working group within ISO/TC 37/SC 3 for providing valuable feedback, in particular Michael Grüninger, Pat Hayes, Maria Keet, Chris Menzel, and John Sowa. We also want to thank Andrzej Tarlecki, with whom we collaborate(d) on the semantics of heterogeneous specification, Thomas Schneider for help with the semantics of modules, and Christian Maeder, Eugen Kuksa and Sören Schulze for implementation work. This work has been supported by the DFG-funded Research Centre on Spatial Cognition (SFB/TR 8), project I1-[OntoSpace], and EPSRC grant "EP/J007498/1".

References

1. Adámek, J., Herrlich, H., Strecker, G.: Abstract and Concrete Categories. Wiley, New York (1990)
2. Avron, A.: Simple consequence relations. Inf. Comput. 92(1), 105–140 (1991)
3. Beisswanger, E., Schulz, S., Stenzhorn, H., Hahn, U.: BioTop: An upper domain ontology for the life sciences - a description of its current structure, contents, and interfaces to OBO ontologies. Applied Ontology 3(4), 205–212 (2008)
4. Carnielli, W.A., Coniglio, M., Gabbay, D.M., Gouveia, P., Sernadas, C.: Analysis and synthesis of logics: how to cut and paste reasoning systems. Applied logic series. Springer (2008)

5. Common Logic Working Group. Common Logic: Abstract syntax and semantics. Technical report (2003), http://iso-commonlogic.org

6. David, J., Euzenat, J., Scharffe, F., dos Santos, C.T.: The alignment API 4.0. Semantic Web 2(1), 3–10 (2011)

7. Diaconescu, R.: Grothendieck institutions. Applied Categorical Structures 10, 383–402 (2002)

8. Gentzen, G.: Investigations into logical deduction. In: Szabo, M.E. (ed.) The Collected Papers of Gerhard Gentzen, pp. 68–213. North-Holland, Amsterdam (1969)

9. Goguen, J., Rosu, G.: Institution morphisms. Formal Aspects of Computing 13, 274–307 (2002)

10. Goguen, J.A., Burstall, R.M.: Institutions: Abstract model theory for specification and programming. Journal of the Association for Computing Machinery 39, 95–146 (1992); Predecessor in: Clarke, E., Kozen, D. (eds.): Logic of Programs 1983. LNCS, vol. 164, pp. 221–256. Springer, Heidelberg (1984)

11. Goguen, J.A., Tracz, W.: An implementation-oriented semantics for module composition. In: Leavens, G.T., Sitaraman, M. (eds.) Foundations of Component-Based Systems, ch. 11, pp. 231–263. Cambridge University Press, New York (2000)

12. Guerra, S.: Composition of Default Specifications. J. Log. Comput. 11(4), 559–578 (2001)

13. Horrocks, I., Kutz, O., Sattler, U.: Even More Irresistible $\mathcal{SROIQ}$. In: Proc. of the 10th Int. Conf. on Principles of Knowledge Representation and Reasoning (KR 2006), pp. 57–67. AAAI Press (June 2006)

14. Konev, B., Lutz, C., Walther, D., Wolter, F.: Formal properties of modularisation. In: Stuckenschmidt, H., Parent, C., Spaccapietra, S. (eds.) Modular Ontologies. LNCS, vol. 5445, pp. 25–66. Springer, Heidelberg (2009)

15. Kutz, O., Mossakowski, T.: Conservativity in Structured Ontologies. In: 18th European Conf. on Artificial Intelligence (ECAI 2008), Patras, Greece. IOS Press (2008)

16. Kutz, O., Mossakowski, T., Lücke, D.: Carnap, Goguen, and the Hyperontologies: Logical Pluralism and Heterogeneous Structuring in Ontology Design. Logica Universalis 4(2), 255–333 (2010); Special Issue on 'Is Logic Universal?'

17. Kutz, O., Normann, I., Mossakowski, T., Walther, D.: Chinese Whispers and Connected Alignments. In: Proc. of the 5th International Workshop on Ontology Matching (OM 2010), 9th International Semantic Web Conference, ISWC 2010, Shanghai, China (November 7, 2010)

18. Lange, C., Mossakowski, T., Kutz, O.: LoLa: A Modular Ontology of Logics, Languages, and Translations. In: Schneider, T., Walther, D. (eds.) Modular Ontologies, Aachen. CEUR Workshop Proceedings, vol. 875 (2012)

19. Lifschitz, V.: Circumscription. In: Handbook of Logic in Artificial Intelligence and Logic Programming, vol. 3, pp. 297–352. Oxford University Press (1994)

20. Lutz, C., Walther, D., Wolter, F.: Conservative Extensions in Expressive Description Logics. In: Proceedings of IJCAI 2007, pp. 453–458. AAAI Press (2007)

21. Lutz, C., Wolter, F.: Deciding inseparability and conservative extensions in the description logic $\mathcal{EL}$. Journal of Symbolic Computation 45(2), 194–228 (2010)

22. Mac Lane, S.: Categories for the Working Mathematician, 2nd edn. Springer, Berlin (1998)

23. Masolo, C., Borgo, S., Gangemi, A., Guarino, N., Oltramari, A.: Ontology library. WonderWeb Deliverable 18. Laboratory for Applied Ontology – ISTC-CNR (December 2003)

24. McCarthy, J.: Circumscription - A Form of Non-Monotonic Reasoning. Artif. Intell. 13(1-2), 27–39 (1980)

25. Meseguer, J.: General logics. In: Logic Colloquium 1987, pp. 275–329. North Holland (1989)

26. Mossakowski, T.: HetCASL - Heterogeneous Specification. Language Summary (2004), http://www.informatik.uni-bremen.de/agbkb/forschung/formal_methods/CoFI/HetCASL/HetCASL-Summary.pdf
27. Mossakowski, T., Kutz, O.: The Onto-Logical Translation Graph. In: Modular Ontologies—Proceedings of the Fifth International Workshop (WoMO 2011). Frontiers in Artificial Intelligence and Applications, vol. 230, pp. 94–109. IOS Press (2011)
28. Mossakowski, T., Lange, C., Kutz, O.: Three Semantics for the Core of the Distributed Ontology Language. In: Donnelly, M., Guizzardi, G. (eds.) FOIS 2012: 7th International Conference on Formal Ontology in Information Systems, pp. 337–352. IOS Press, Amsterdam (2012) (Best paper award), http://www.iospress.nl
29. Mossakowski, T., Maeder, C., Lüttich, K.: The Heterogeneous Tool Set, HETS. In: Grumberg, O., Huth, M. (eds.) TACAS 2007. LNCS, vol. 4424, pp. 519–522. Springer, Heidelberg (2007)
30. Mossakowski, T., Tarlecki, A.: Heterogeneous logical environments for distributed specifications. In: Corradini, A., Montanari, U. (eds.) WADT 2008. LNCS, vol. 5486, pp. 266–289. Springer, Heidelberg (2009)
31. Goguen, J., Roşu, G.: Composing hidden information modules over inclusive institutions. In: Owe, O., Krogdahl, S., Lyche, T. (eds.) From Object-Orientation to Formal Methods. LNCS, vol. 2635, pp. 96–123. Springer, Heidelberg (2004)
32. Sannella, D., Tarlecki, A.: Specifications in an arbitrary institution. Information and Computation 76, 165–210 (1988)
33. Sannella, D., Tarlecki, A.: Foundations of Algebraic Specification and Formal Software Development. EATCS Monographs on theoretical computer science. Springer (2012)
34. Scott, D.: Rules and derived rules. In: Stenlund, S. (ed.) Logical Theory and Semantic Analysis, pp. 147–161. Reidel (1974)
35. Smith, B., Ceusters, W., Klagges, B., Kohler, J., Kumar, A., Lomax, J., Mungall, C.J., Neuhaus, F., Rector, A., Rosse, C.: Relations in biomedical ontologies. Genome Biology 6, R46 (2005)
36. Tarlecki, A.: Towards heterogeneous specifications. In: Gabbay, D., de Rijke, M. (eds.) Frontiers of Combining Systems 2, 1998. Studies in Logic and Computation, pp. 337–360. Research Studies Press (2000)
37. Zimmermann, A., Krötzsch, M., Euzenat, J., Hitzler, P.: Formalizing Ontology Alignment and its Operations with Category Theory. In: Bennett, B., Fellbaum, C. (eds.) Proceedings of the Fourth International Conference on Formal Ontology in Information Systems (FOIS 2006). Frontiers in Artificial Intelligence and Applications, vol. 150, pp. 277–288. IOS Press (November 2006)

Formal Specification of the Kademlia
and the Kad Routing Tables in Maude⋆

Isabel Pita and María-Inés Fernández-Camacho

Dept. Sistemas Informáticos y Computación,
Universidad Complutense de Madrid, Spain
ipandreu@sip.ucm.es, minesfc@sip.ucm.es

Abstract. Kad is the implementation by eMule and aMule of the Kademlia peer-to-peer distributed hash table protocol. Although it agrees with the basic behaviour of the protocol, there are some significant differences. This paper presents the specification of both the Kademlia and the Kad routing tables, using the specification language Maude. As far as we know, this is the first such a formal development. The routing tables present a dynamic behavior in the sense that they should be able to send messages to other peers and they should have a notion of time for raising events and detect no answered messages. Our main contribution is the integration of these dynamic aspects in the protocol specification.

Keywords: Kademlia, Kad, distributed specification, formal analysis, Maude, Real-Time Maude.

1 Introduction

Distributed hash tables (DHTs) are designed to locate objects in distributed environments, like peer-to-peer (P2P) systems, without the need for a centralized server. There are a large number of existing DHTs proposals; the best known are: Chord [12], Pastry [11], CAN [10], and Kademlia [4]. However, only Kademlia has been implemented in P2P networks through the eMule[1] and aMule[2] clients. Kad is the name given to the implementation of Kademlia incorporated to eMule and aMule, and it shows important differences from the original. There are also two BitTorrent overlays that use Kademlia: one by Azureus clients[3] and one by many other clients including Mainline[4] and BitComet[5].

Although the different DHTs have been extensively studied through theoretical simulations and analysis, there is a lack of formal specifications for all of them. Bakhshi and Gurov give in [1] a formal verification of Chord's stabilization algorithm using the π-calculus. Lately Lu, Merz, and Weidenbach [3] have

---

⋆ Research supported by MEC Spanish project *DESAFIOS10* (TIN2009-14599-C03-01) and Comunidad de Madrid program *PROMETIDOS* (S2009/TIC1465).

[1] http://www.emule-project.net
[2] http://www.amule.org
[3] http://azureus.sourceforge.net
[4] http://www.bittorrent.com
[5] http://www.bitcomet.com

N. Martí-Oliet and M. Palomino (Eds.): WADT 2012, LNCS 7841, pp. 231–247, 2013.
© IFIP International Federation for Information Processing 2013

modeled Pastry's core routing algorithms in the specification language TLA$^+$. They consider the complete P2P network and focused their study on the lookup correctness property: *the lookup message for a particular key is answered by at most one 'ready' node covering the key.* The paper provides a detailed model of the network, the routing table and other Pastry structures, while abstract from an explicit notion of time. Periodic actions are performed non-deterministically. The TLA$^+$ model checker is used to check the model and illustrate some open issues related with the algorithm. Finally the TLA$^+$ theorem prover is been used to verify the correctness property.

There is a preliminary study of the Kademlia searching process protocol by the first author in [10], and a distributed specification of the protocol in [11]. Based on them, we have identified the need of a detailed study of the routing tables, where each peer stores contact information about others. This information is used in the network look-up process.

The original version of Kademlia differs from the real implementation made in Kad. A Kademlia routing table is a proper binary tree whose leaves are lists of at most k contacts, called k-buckets. A bucket is kept sorted by the time contacts were last seen, and is identified by the common prefix of the IDs it contains. Each bucket is responsible for a range of the node ID space. In fact, Kademlia routing table is a list of buckets. The bucket in the first level contains contacts whose first bit is the opposite to the first bit of the owner of the routing table, while the ID of the owner is in the range of the last bucket. Kademlia uses, when add the contact to the routing table, a tighter splitting rule where it is only allowed to split the bucket with common prefix equal to the one of the routing table owner. A KAD routing table is a left-balanced binary tree, and may be a complete tree up to level 4. The tree nodes are called routing zones, and the buckets, called now routing bins, are placed in the leaf nodes. Kad divides the table structure in levels and index (the horizontal distance in number of routing zones from the leftmost one on the same level) and uses the bitwise exclusive-or of the n-bit quantities (XOR) distance to the owner ID of the routing table, to store contacts in routing bins. Kad has a looser splitting rule that Kademlia, it allows to split a bin if its level is smaller than 4 or its index is smaller than 5, which results in more possible contacts. Kademlia has no specific actions to update its routing table periodically, checking for offline contacts only when inserting into full buckets, while KAD has actual maintenance tasks that run at set periods. Each peer has a creation and expiration time, and also a type has scale from 0 to 4. Type 4 means that the contact is rarely connected and will probably be eliminated at the next occasion.

In this paper we present our specification of the routing table in the formal language Maude, as well as some preliminary results on its properties. The Maude algebraic specification language, is based on rewriting logic [2]. It supports both equational and rewriting logic computations and offers simple and elegant time simulation resources. Since the specifications are directly executable, Maude can be used to prototype the systems as well as to prove properties of them.

Our specification includes features not previously modeled as sending messages to update the Kademlia routing table or raising events for populate and removing offline nodes in the Kad routing table. We have focused our work on the routing tables and abstract from the rest of the network. In order to model maintenance cycles, and eventually other effects of time on the system, we turn to the real time extension of Maude [6]. Our model should allow us to test some correctness properties such as: *If there is an alive node in the network and there is space in the routing table, the node will be included in a certain period of time.* We are also interested in checking the interleaving of actions that take place at different points in time, and prove consistency properties like: *if two bins are consolidated, they do not split again until a certain period of time.*

As far as we know there is no other formal specification of the Kademlia and Kad routing tables. Right now the best sources to understand both protocol details are the original paper on the Kademlia DHT and the source code of the Kad implementation. Thus our first contribution is the benefits of having a formal specification of a system that is being consulted by many developers. In addition, the formalization of the routing tables allows us to compare the original version of Kademlia with the real implementation made in Kad.

However, our main contribution is the integration of the dynamic aspects of the routing table in the full protocol specification. The specification includes the ability of the routing tables to send and receive messages autonomously from the node by using different levels of configurations. Detection of non-answered messages is done by assigning a timeout when sending the message and triggering the appropriate action when the time expires. The actions that are performed automatically in Kad periodically, like populate almost empty buckets or remove offline contacts from the buckets, are triggered when their time expires. These actions require having a notion of time defined in different parts of the routing table and allow us to study the interleaving of actions that take place at different points in time.

The structure of the paper is as follows: In Section 2 we give some notions about the Maude language and the Kademlia and Kad DHTs. Sections 3 and 4 present the specification of the Kademlia and the Kad networks respectively. Section 5 defines a Kad routing table and proves some properties about it. Finally Section 6 presents some conclusions and future work.

2   Preliminaries

We present in this section the basic notions about Maude, Kademlia, and Kad.

2.1   Maude

Maude [2] is a formal specification language based on rewriting logic. In Maude, the state of a system is formally specified as an algebraic data type by means of a membership equational specification. We define new types (by means of keyword sort(s)); subtype relations between types (subsort); operators (op) for building values of these types; equations (eq) that identify terms built with these operators; and membership axioms (mb) which specify terms as having a given sort.

Both equations and membership axioms may be conditional (ceq and cmb). Conditions, introduced by the keyword if can be either a single equation, a single membership, or a conjunction of equations and memberships using the binary conjunction connective /\. The equations can be ordinary equations (t = t'), matching equations (t := t'), or abbreviate Boolean equations (t).

The sorts connected by a subtype relationship are grouped into equivalence classes called *Kinds*. Kinds are implicitly associated with connected components of sorts and are considered as *error supersorts*. They are written by enclosing the name of one of the connected sorts in square brackets.

The *dynamic* behaviour of the system is specified by rewrite rules of the form $t \longrightarrow t'$ *if* C, that describe the local, concurrent transitions of the system. That is, when a part of a system matches the pattern t and satisfies the condition C, it can be transformed into the corresponding instance of the pattern t'.

In object-oriented specifications, *classes* are declared with the syntax class $C \mid a_1 : S_1, \ldots, a_n : S_n$, where C is the class name, a_i is an attribute identifier, and S_i is the sort of the values this attribute can have. An *object* is represented as a term $< O : C \mid a_1 : v_1, \ldots, a_n : v_n >$ where O is the object's name, belonging to a set Oid of object identifiers, and the v_i's are the current values of its attributes. The order of the attributes in an object is irrelevant. *Messages* are defined by the user for each application (introduced with syntax msg).

In a concurrent object-oriented system the concurrent state, which is called a *configuration*, has the structure of a multiset made up of objects and messages that evolves by concurrent rewriting. The behaviour associated with the messages is specify by rules of the form: $m_1 \ldots m_k o_1 \ldots o_l \longrightarrow m'_1 \ldots m'_s o'_1 \ldots o'_t$ *if* C where the m_i are messages of type Msg, and the o_i are objects, with their class and attributes. Objects that appear only in the lefthand side of the rule are removed from the configuration, and objects that appear only in the righthand side of the rule are added to the configuration. Regarding the attributes, the only object attributes made explicit in a rule are those relevant for that rule. Attributes mentioned only in the lefthand side of a rule are preserved unchanged, the original values of the attributes mentioned only on the righthand side do not matter, and all attributes not explicitly mentioned are left unchanged. We use Full Maude's object-oriented notation and conventions [2, Part II] throughout the whole paper.

The notion of time is specified by using the real time extension of Maude [6]. In the Real-Time Maude language, one can make explicit the way in which change in a system is dependent on time. The language provides two sorts: GlobalSystem and System to represent the whole system and the state of the system. The whole system is obtained from the state by the constructor {_} : System -> GlobalSystem .. The transitions of the system that are assumed to take zero time are specified over the System sort. The elapse of time in a system is modelled by the *tick* rule: $l : \{t\} \longrightarrow \{u\}$ *if* C . Tick rules must only be applied on the system as a whole, i.e., on terms of sort GlobalSystem, to ensure that time advances uniformly in all parts of the system.

The time domain may be discrete or dense. In particular, we use the discrete time definition with natural numbers, that provides us with a sort `Time` such that `subsort Nat < Time .`, a sort extension with the infinite value: `subsort Time < TimeInf .`, and the infinite value: `op INF : -> TimeInf .`

2.2 Kademlia

Kademlia [4] is a P2P distributed hash table used by the peers to access files shared by other peers. In Kademlia both peers and files are identified with n-bit quantities, computed by a hash function. Information of shared files is kept in the peers with an ID *close* to the ID file, where the notion of distance between two IDs is defined as the bitwise exclusive-or of the n-bit quantities. Then, the lookup algorithm which is based on locating successively *closer* nodes to any desired key has $\mathcal{O}(\log n)$ complexity.

Each node stores contact information about others in what is called its *routing table*. In Kademlia, every node keeps a list of: IP address, UDP port, and node ID, for nodes of distance between 2^i and 2^{i+1} from itself, for $i = 0, \ldots, n$ and n the ID length. Note that each list contains IDs that differ in the *ith* bit. These lists, called *k-buckets*, have at most k elements, where k is chosen such that any given k nodes are very unlikely to leave the network within an hour of each other. Buckets are kept sorted by the time contacts were last seen.

The routing table is organized as a binary tree whose leaves are buckets. Each bucket is identified by the common prefix of the IDs it contains. The binary tree in the basic version of Kademlia is a list of buckets, since the only buckets that can be split are those with common prefix equal to the one of the routing table owner. Figure 1 (partially borrowed from [5]) shows a routing table for node 00000000.

Kademlia does not have explicit operations to maintain the information of the routing table. When a node receives any message (request or reply) from another node, it updates the appropriate bucket for the sender's node ID. If the sender node exists, it is moved to the tail of the list. If it does not exist and there is free space, it is inserted at the tail of the list. Otherwise, if the bucket has not free space, the node at the head of the list is contacted and if it fails to respond it is removed from the list and the new contact is added at the tail. In the case the node at the head of the list responds, if the bucket can be divided, it is split and the new contact added to the appropriate bucket; if the bucket cannot be divided the first contact is moved to the tail, and the new node is discarded.

2.3 The Kad Routing Table

Kad [5] is an implementation of the Kademlia distributed hash table. Its routing tables allow for more contacts than the Kademlia ones. They are not lists anymore, but left-balanced binary trees. The routing tree nodes are called *routing zones*. The buckets, now called *routing bins* are located at the leaf nodes and are lists of at most 10 contacts. Routing zones may be identified by their level and their zone index,

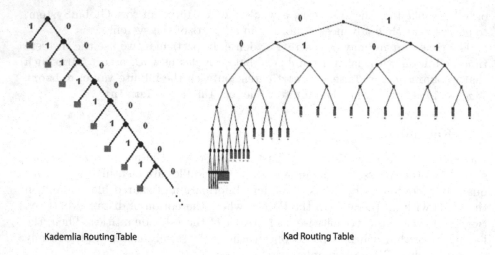

Kademlia Routing Table   Kad Routing Table

Fig. 1. Kademlia and Kad Routing table structures

that is the distance of the node to the leftmost node of its level. Kad has a looser splitting rule that Kademlia, it allows to split a bin if its level is smaller than 4 or its zone index is smaller than 5. Figure 1 shows a Kad routing table.

Kad has explicit operations to maintain the information of the routing table, and does not rely on receiving messages as Kademlia. The process that adds new contacts to a bin is done once an hour. If the bin has less than three contacts or if splitting is allowed, the process selects a random ID from the bin ID space and starts a search for it. It will find new contacts that will be inserted into the bin. The process that deletes dead contacts from the bin is executed once a minute. It checks if the *expire time* of the first contact in the bin is ok. Every contact has an expire time variable which holds the time when the contact has to be checked whether it is still alive. If this time has expired, the routing table sends a HELLO-REQ message to the contact, increases a type variable, with values from 0 to 4, that indicates how long the contact has been online, and sets its expire time to 2 minutes. The process also removes all contacts of type 4 from the bin. If the node receives a HELLO-RES message indicating that the contact is alive, it places the contact at the end of the list, re-adjusts the type of the contact and sets a new expire time. If the contact does not reply, it will be removed from the bin when its type raises level 4.

Finally, there is a process that consolidates almost empty bins. It is done on adjacent bins that have the same parent routing zone if they both together have less than 5 contacts. It is run every 45 minutes on the whole routing tree.

3   The Maude Specification of Kademlia

Kademlia buckets are represented by the sort Bucket{X} as a list of contacts, where the contacts are parameters of the specification, represented by the sort X$Contact. The subsort NeBucket{X} represents non empty buckets.

```
sorts NeBucket{X} Bucket{X} .
subsort NeBucket{X} < Bucket{X} .

op empty-bucket : -> Bucket{X} [ctor] .
op _!_ : Bucket{X} X$Contact -> [Bucket{X}] [ctor] .
```

Non-empty k-buckets are bounded and do not have repeated contacts by means
of the membership axiom, where the bucketDim constant is set to k:

```
var T : X$Contact .    var B : Bucket{X} .
cmb B ! T : NeBucket{X} if length-b(B) < bucketDim /\ not T in B .
```

We have defined operations to perform all the bucket functions, like moving a
contact to the tail, adding contacts, and removing contacts (see [8]).

Routing tables are defined as lists of buckets. We do not consider in this
version of the specification the optimizations proposed in [4] that allow keeping
more contacts in the routing table, therefore there is no need for a tree of buckets.
Notice that routing tables have at least one bucket that cannot be empty.

```
sort RoutingTable{X} .
subsort NeBucket{X} < RoutingTable{X} .

op _!!_ : Bucket{X} [RoutingTable{X}] -> [RoutingTable{X}] [ctor] .
```

The following membership axiom sets when a list with more than one bucket is
a routing table. It checks whether the table does not exceed the maximum size,
that the contacts in each bucket are the appropriate and that at least one of the
two last buckets is not empty. A single non-empty bucket is a routing table by
the previous subsort relationship.

```
var KR : [RoutingTable{X}] .    vars B B1 B2 : Bucket{X} .
cmb KR : RoutingTable{X} if num-buckets(KR) > 1 /\
         num-buckets(KR) <= length(give-contact(KR)) /\
         atLeastOne?(KR) /\ is-RT(KR,1,peer-prefix(KR)) .

op is-RT : [RoutingTable{X}] NzNat BitString -> Bool .
eq [is-RT1] : is-RT(B1 !! B2,Nz,prefix) =
    fix-bucket?(B1,Nz,prefix) and fix-last-bucket?(B2,Nz,prefix)
    and (not empty-bucket?(B1) or not empty-bucket?(B2)) .
eq [is-RT2] : is-RT(B1 !! (B2 !! KR),Nz,prefix) =
    fix-bucket?(B1,Nz,prefix) and is-RT(B2 !! KR,s Nz,prefix) .
eq [is-RT3] : is-RT(B,Nz,prefix) = not empty-bucket?(B) .
```

We have defined operations to divide a bucket and to compute the nearest contact
to a given one among others.

The operation that adds a contact to the routing table checks, when the
bucket is full, if the first contact is still alive by sending it a PING message [8].
To encapsulate the behaviour of the routing table in the peer we define the RT
attribute of the object Peer as a configuration of a routing table, a message, and

a value of sort `Time`. By means of the rules `RT-receive` and `RT-send` the routing table messages are sent out from the peer and the reply messages go inside the routing table. Notice that the reply is captured by the routing table only when a message to that object has been sent before.

```
op Peer : -> Cid .
op RT :_ : Configuration -> Attribute [ctor] .
op {_+_+_} : RoutingTable{X} Msg TimeInf -> Configuration [ctor] .

var R : RoutingTable{X} . var P : Oid . vars Z1 Z2 : X$Contact .
vars T1 T2 T3 T4 : Time . var T : TimeInf .

rl [RT-send] :
 < P : Peer | RT : { R + PING(Z1,Z2,T1,T2) + INF } >
=>
 PING(Z1,Z2,T1,T2) < P : Peer | RT : { R + PING(Z1,Z2,T1,T2) + T1 } > .
crl [RT-receive] :
 PING-REPLY(Z1,Z2,T1,T2) < P : Peer | RT : { R + PING(Z2,Z1,T3,T4) + T } >
=>
 < P : Peer | RT : { R + PING-REPLY(Z1,Z2,T1,T2) + T } > if T =/= INF .
```

The `add-entry` operation receives the contact to be added, the routing table and the contact of the routing table owner. The behaviour is as follows: equation `add0` deals with the case in which the bucket is not full and the new entry is added at the tail. If the bucket is full, by rule `add1` the routing table puts a `PING` message in the `RT` configuration, sets the time to the `PT` constant value, which represents the maximum time it will wait for a reply before consider the contact offline, and calls the auxiliary function `add-entry2` on the routing table. If the first contact is offline, the time value of the configuration will raise zero and rule `add2` will remove the offline contact and will add the new contact at the tail of the bucket by means of the `add-entry-aux` operation. Besides the message is removed from the configuration, and the time is set to the `INF` value. If the first contact replies, and the bucket is full and it cannot be divided (rule `add3`), the `add-entry-aux` operation moves the first contact to the tail of the bucket and discards the new contact. Finally if the bucket can be divided (rule `add4`), the entry is inserted again after splitting the bucket.

```
crl [add0] : { add-entry(Z1,R,Z2) + none + T } =>
                   { add-entry-aux(Z1,R,1,Z2,true) + none + T }
  if not full-bucket?(find-bucket(Z1,R,1)) .
crl [add1] : { add-entry(Z1,R,Z2) + none + INF } =>
   { add-entry2(Z1,R,Z2) +
     PING(Z2,first-contact(find-bucket(Z1,R,1)),RPCRemove,1) + PT }
if full-bucket?(find-bucket(Z1,R,1)) .
rl [add2] : { add-entry2(Z1,R,Z2) + M + 0 } =>
   { add-entry-aux(Z1,R,1,Z2,false) + none + INF } .
crl [add3] : { add-entry2(Z1,R,Z2) + PING-REPLY(Z,Z2,T1,0) + T } =>
   { add-entry-aux(Z1,R,1,Z2,true) + none + INF }
if B1 := find-bucket(Z1,R,1) /\ equal(Z,first-contact(B1)) /\
```

```
    full-bucket?(B1) /\ not isLastBucket?(B1,R) /\ T1 > 0 /\ T > 0 .
crl [add4] : { add-entry2(Z1,R,Z2) + PING-REPLY(Z,Z2,T1,0) + T } =>
    { add-entry(Z1,conc(R,div-bucket(last-bucket(R),Nz,NBit(Z2,Nz))),Z2) +
    none + INF }
if B1 := find-bucket(Z1,R,1) /\ equal(Z,first-contact(B1)) /\
    Nz := num-buckets(R) /\ full-bucket?(B1) /\
    isLastBucket?(B1,R) /\ T1 > 0 /\ T > 0 .
```

The Kademlia specification is written in 7 modules and it defines about 60 operations over the defined sorts. It makes use of 6 rewriting rules to simulate the routing table interaction with the node. (http://maude.sip.ucm.es/kademlia [8])

4 Kad Processes in Maude

The Kad processes responsible for the routing table maintenance are executed from periodically, instead of relying on the occurrence of an event as in the Kademlia routing table. For this reason time values are present all over the routing table specification, since contacts, bins, and the routing table must keep the time left to perform their actions.

We define the moment at which a process must occur with a constant value that is decreased as the time passes. When the value reaches zero the process is executed and the time is reset to the initial value. Time is expressed in seconds.

The Kad specification is written in 17 modules and it defines about 80 operations over the defined sorts. It makes use of 8 rewriting rules to simulate the routing table interaction with the node. (http://maude.sip.ucm.es/kademlia [8]).

4.1 The Kad Routing Table

The routing table representation in Kad differs from the one in Kademlia. Since Kad routing tables may be a complete tree up to level 4, the specification is no more a list of buckets, but a binary tree. Besides this, it is necessary to keep the time left for the processes to run.

Kad routing tables are non-empty binary trees of sort RoutingZone together with the contact information of the routing table owner and the time left to execute the consolidation process, which is performed on the whole tree.

```
sort RoutingTable .
op rt : RoutingZone Kad-Contact Time -> RoutingTable [ctor] .
```

Routing bins are defined as lists of contacts of length K.

```
sorts NeRoutingBin RoutingBin .
subsort NeRoutingBin < RoutingBin .

op empty-bin : -> RoutingBin [ctor] .
op _!_ : RoutingBin Kad-Contact -> [NeRoutingBin] [ctor] .

var T : Kad-Contact .   var B : RoutingBin .
cmb B ! T : NeRoutingBin if getSize(B) < K /\ not T in B .
```

Each contact, besides its identifier, IP address, and UDP port, keeps the time it has been active and the time remaining for checking whether it is still active. The time a contact has been active is the only time value that is increased as time passes.

```
sort Kad-Contact .
op c : BitString128 IP UDP Contact-types Time Time -> Kad-Contact [ctor] .
```

The RoutingZone sort represents a Kad routing tree. Subtrees are represented in the *Kind*, since the level and zone index values of their leaves do not correspond to a complete tree. A subtree may only be one routing bin, together with its level, zone index, and the time to populate and to remove the offline contacts. Both processes are performed in each bin independently.

```
op rb : RoutingBin Nat Nat Time Time -> [RoutingZone] [ctor] .
```

The routing zone may also be the composition of its two subtrees together with the time to populate and the time to remove offline contacts. Kad uses the time to populate in the internal tree nodes during the consolidation process to define the population time for the new node that consolidates the two old ones. We maintain the time to remove offline contacts because it is in the Kad implementation, although it is not used in our specification. There is no need in the specification to maintain the level and zone index values in the internal tree nodes.

```
op rz : [RoutingZone] [RoutingZone] Time Time -> [RoutingZone] [ctor] .
```

A membership axiom defines which of the binary trees formed with the previous constructors are Kad routing trees. In particular it checks whether the routing bin (with its level, zone index, and time values) is the root of the tree, that is, if its level and zone index are both zero or if it is a tree with more than one bin that has the correct tree structure, and the appropriate contacts in each bin.

```
mb(rb(B,0,0,T1,T2)) : RoutingZone .
cmb(rz(KR1,KR2,T1,T2)) : RoutingZone
   if is-subRT(0,0,rz(KR1,KR2,T1,T2)) /\ is-RT(rz(KR1,KR2,T1,T2)) .
```

4.2   A Kad Peer

The Kad routing table can request the node to send messages and start a looking-for process. Therefore, the routing table specification needs to consider other parts of the node. In the specification a node/peer is an object of class Peer, with four attributes: the routing table, defined in Section 4.1; the search manager, which is a list of the keys the peer is looking for; a list of the messages to send; and a list of events, that simulates the list of events in the Kad implementation and is used to go over the bins looking for the process that its time to execute.

```
class Peer | CroutingTable : RoutingTable, CSearchManager : List{vKEY},
            CMessages : MsgList, CEventMap : EventMap .
subsort Kad-Contact < Oid .
```

4.3   The Passage of Time

The passage of time is defined by the functions `delta` and `mte`. The `delta` function spreads the passage of time to all time values of the specification as follows:

```
op delta : Configuration Time -> Configuration [frozen (1)] .
op delta : RoutingTable Time -> RoutingTable .
op delta : RoutingZone Time -> RoutingZone .
op delta : RoutingBin Time -> RoutingBin .

vars CF CF' : Configuration . vars  T T1 T2 TC : Time .
var Z1 : Kad-Contact . var R : RoutingZone . var B : RoutingBin .
vars KR1 KR2 : [RoutingZone] .   var : RT : RoutingTable . var S : BitString128 .
var ip : IP . var udp : UDP . var CT : Contact-types . vars N1 N2 : Nat .

eq delta(none,T) = none .
ceq delta(CF CF', T) = delta(CF,T) delta(CF',T)
    if CF =/= none and CF' =/= none .
eq delta(< Z1 : Peer | CroutingTable : RT, ... >,TC) =
    < Z1 : Peer | CroutingTable : delta(RT,TC), ... > .
eq delta(rt(R,Z1,T2), TC) = rt(delta(R,TC),Z1,T2 monus TC) .
eq delta(rb(B,N1,N2,T1,T2),TC) = rb(delta(B,TC),N1,N2,T1 monus TC,T2 monus TC) .
eq delta(rz(KR1,KR2,T1,T2),TC) =
    rz(delta(KR1,TC),delta(KR2,TC),T1 monus TC,T2 monus TC) .
eq delta(empty-bin,TC) = empty-bin .
eq delta(B ! c(S,ip,udp,CT,T1,T2),TC)  =
    delta(B,TC) ! c(S,ip,udp,CT,T1 plus TC,T2 monus TC) .
```

while the `mte` function computes the time that passes in each step as the minimum of all the time values in the system.

```
op mte : Configuration -> TimeInf [frozen (1)] .
eq mte(none) = INF .
ceq mte(CF CF') = min(mte(CF), mte(CF')) if CF =/= none and CF' =/= none .
eq mte(< Z1 : Peer | CroutingTable : RT ,...>) = minTime(RT) .

op minTime : RoutingTable -> Time .
op minTime : RoutingZone -> Time .
eq minTime(rt(R,Z2,T2)) = min(minTime(R),T2) .
eq minTime(rb(B,N1,N2,T1,T2)) = min(T1,T2) .
eq minTime(rz(KR1,KR2,T1,T2)) = min(minTime(KR1),minTime(KR2)) .
```

The time values of the internal nodes that appear in the last equation are not taken into account in the minimum time computation, because they are not directly associated with any process.

4.4   The Populate Process

The population process is done in each routing bin when the time defined by the constant **POPULATE-TIME-BIG** in each routing bin reaches zero and the number of contacts in the bin is less than 20% of the bin size, or it is a bin that can be split. It is also required to be this the next bin in the event map attribute of the node. The process adds a random contact to the search process; and resets the population time of the bin.

```
crl [populate1] :
 < Z : Peer | CroutingTable : rt(R1,Z2,T2), CSearchManager : L ,
            CEventMap : (< N1 ; N2 > EM) , ...>  =>
 < Z : Peer |
   CroutingTable : rt(changePop(R1,N1,N2,POPULATE-TIME-BIG),Z2,T2) ,
   CSearchManager : append(L, randomKey(N1,N2)) ,
   CEventMap : (EM < N1 ; N2 >) , ... >
if (N1 =/= 0 or N2 =/= 0) /\ timePop(R1,N1,N2) /\ (N2 < KK or N1 < KBASE
  or  getNumContacts(getRZ(R1,< N1 ; N2 >)) <= (K * 2) quo 10 ) .
```

where:

- changePop changes the population time of the bin to the given value.
- append(L, randomKey(N1,N2)) appends a new look-up for a random key value to the search manager.
- timePop checks whether the time to populate in the bin is set to zero.
- getNumContacts obtains the number of contacts of a bin.
- getRZ gets the routing zone of a level and zone index.

4.5   The Remove Offline Contacts Process

The remove offline contacts process is done in each routing bin when the time defined by the constant NEXT-REM in each routing bin reaches zero.

Kad classifies contacts in five groups: type 0: very good contact, known for at least two hours; types 1 and 2: less good contact; type 3: newly inserted contact; type 4: the contact has not responded and will be removed in the next process. In addition each contact keeps two time values: the creation time, which is the time to determine how long the contact has been known, and the expire time, which is the time to check if the contact is still alive.

The process checks the first contact of the bin. If its expire time is not zero, the process removes type 4 contacts from the bin; push to the bottom the first contact (by the operation changeRem1); and puts a new remove time in the bin.

```
crl [del-dead1] :
 < Z : Peer | CroutingTable : rt(R1,Z2,T2), CEventMap : < N1 ; N2 > EM ,. >
=>  < Z : Peer | CroutingTable : rt(changeRem1(R1,N1,N2),Z2,T2),
            CEventMap : EM < N1 ; N2 > ,...>
if (N1 =/= 0 or N2 =/= 0) /\ timeRem(R1,N1,N2) == 1 .

eq [cr11] : changeRem1(rb(B1,N3,N4,T1,T2),N1,N2) =
   rb(pushToBottom(first-contact(rem4(B1)),rem4(B1)),N3,N4,T1,NEXT-REM) .
ceq [cr12] : changeRem1(rz(KR1,KR2,T5,T6),N1,N2) =
   rz(changeRem1(KR1,sd(N1,1),N2),KR2,T5,T6)
if 2 ^ sd(N1,1) > N2 .
ceq [cr13] : changeRem1(rz(KR1,KR2,T5,T6),N1,N2) =
   rz(KR1,changeRem1(KR2,sd(N1,1),sd(N2,2 ^ sd(N1,1))),T5,T6)
if 2 ^ sd(N1,1) <= N2 .
```

where the rem4 operation removes all contacts with type 4 in the bin.

If the expire time of the first contact in the bin is zero, the process: removes type 4 contacts from the bin; increases the first contact type; puts the expire time of the first contact to 2 minutes (OFFLINE-CHECK constant) by means of the changeRem2 operation; and sends a HELLO-REQ message to the first contact. If the first contact answers, it will be moved to the tail of the bin, in other case it will be declared of type 4 and removed from the routing table in 2 minutes.

```
crl [del-dead2] :
  < Z : Peer | CroutingTable : rt(R1,Z2,T2), CMessages : LM ,
       CEventMap : (< N1 ; N2 > EM) ,... >
=> < Z : Peer | CroutingTable : rt(changeRem2(R1,N1,N2),Z2,T2) ,
       CMessages : append(LM, HELLO-REQ(Z2,first-contact-bucket(R1,N1,N2),T2,1)) ,
       CEventMap : EM < N1 ; N2 >,...>
if (N1 =/= 0 or N2 =/= 0) /\ timeRem(R1,N1,N2) == 2 .

eq [cr21] : changeRem2(rb(B1,N3,N4,T1,T2),N1,N2) =
     rb(change-first(rem4(B1),OFFLINE-CHECK),N3,N4,T1,NEXT-REM) .
op changeRem2 : RoutingZone Nat Nat -> RoutingZone .
ceq [cr22] : changeRem2(rz(KR1,KR2,T5,T6),N1,N2) =
   rz(changeRem2(KR1,sd(N1,1),N2),KR2,T5,T6)
if 2 ^ sd(N1,1) > N2 .
ceq [cr23] : changeRem2(rz(KR1,KR2,T5,T6),N1,N2) =
   rz(KR1,changeRem2(KR2,sd(N1,1),sd(N2,2 ^ sd(N1,1))),T5,T6)
if 2 ^ sd(N1,1) <= N2 .
```

4.6   The Consolidation Process

The consolidation process is done in the routing tree when its time defined by the constant CONSOLIDATE-TIME of the routing table, reaches zero.

```
rl [consolidate1] :
  < Z : Peer | CroutingTable : rt(R,Z2,0), ... > =>
  < Z : Peer | CroutingTable : rt(consolidate(R),Z2,CONSOLIDATE-TIME),.. > .
```

The consolidation of a routing tree, done by the consolidate function is a recursive process. The first equation: con1 defines the case of a subtree with one bin, which cannot be consolidated. The equation con2 defines the recursive case when at most one of the subtrees is a leaf. Equations con3 and con4 define the cases where the two subtrees are leaves. If the number of contacts of the two bins is less than half the size of the bin, bins are aggregated (con3). Otherwise no consolidation is needed for these bins (con4).

```
eq [con1] : consolidate(rb(B,N1,N2,T1,T2)) = rb(B,N1,N2,T1,T2) .
ceq [con2] : consolidate(rz(KR1,KR2,T1,T2)) =
                 rz(consolidate(KR1),consolidate(KR2),T1,T2)
if not isLeaf?(KR1) or not isLeaf?(KR2) .
ceq [con3] : consolidate(rz(rb(B1,N3,N4,T1,T2),rb(B2,N3,N6,T3,T4),T5,T6)) =
        rb(add-bins(B1,B2),sd(N3,1),N4 quo 2,POPULATE-TIME,T6)
if getNumContacts(rz(rb(B1,N3,N4,T1,T2),rb(B2,N3,N6,T3,T4),T5,T6)) <
   K quo 2 /\  N6 == N4 + 1 .
ceq [con4] : consolidate(rz(KR1,KR2,T5,T6)) = rz(KR1,KR2,T5,T6)
if isLeaf?(KR1) /\ isLeaf?(KR2) /\
   getNumContacts(rz(KR1,KR2,T5,T6)) >= K quo 2 .
```

5   Case Study

We define a routing table with 4 levels and 26 contacts over 7 routing bins (Figure 2). The maximum number of contacts in a bin is set to 10. Node IDs are abbreviated to 8 bits. Contacts are of different types, from 0 to 4 and we consider realistic time values for the different process. The specification, initial terms and results of each case study can be found in `http://maude.sip.ucm.es/kademlia`.

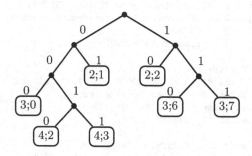

Fig. 2. Routing tree example

The owner of the routing table is supposed to have the ID 01110000. The contacts of the routing bin at level 3 and zone index 0, that are shown below, are at distance less or equal to 31 from the owner ID. This means that the three most significant bits of the contacts in the bin should agree with the ones of the routing table owner.

```
rz(rb(empty-bin !
      c(0 ; 1 ; 1 ; 1 ; 1 ; 0 ; 1 ; 0,ip,udp,type0,40000,7200) !
      c(0 ; 1 ; 1 ; 0 ; 0 ; 0 ; 1 ; 0,ip,udp,type1,3602,5400) !
      c(0 ; 1 ; 1 ; 0 ; 1 ; 0 ; 1 ; 0,ip,udp,type4,15000,60) !
      c(0 ; 1 ; 1 ; 1 ; 0 ; 0 ; 1 ; 1,ip,udp,type2,1,3600) !
      c(0 ; 1 ; 1 ; 1 ; 1 ; 1 ; 0 ; 1,ip,udp,type3,0,1) !
      c(0 ; 1 ; 1 ; 0 ; 1 ; 0 ; 0 ; 1,ip,udp,type3,0,1),3,0,3000,48)
```

There are two contacts of type 3 and one contact for types 0, 1, 2, and 4. The last two contacts have just joined the network since their creation time is set to 0 and their expire time is set to 1 second. The expire time of the first four contacts varies from 1 minute to 2 hours. The population time of the bin is set to 50 minutes and its time to remove offline contacts to 48 seconds.

Now, we can explore, for example, the shortest and the longest time it takes to reach a desired state using the **find earliest** and **find latest** commands of real-time Maude.

The syntax of these commands is as follows:

```
(find earliest initState =>* searchPattern [such that cond] .)
(find latest initState =>* searchPattern [such that cond] with no time limit .)
(find latest initState => searchPattern [such that cond] in time timeLimit .)
```

where:

- *initState* is a real-time system of sort `GlobalSystem`. Values are obtained by enclosing a system, in our specification a configuration of objects and messages, between brackets.
- *searchPattern* is the description of the global system we are looking for.

For example, we are interested in the maximum time a type 4 contact will be in the table. Thus, we obtain the time it takes to remove all type 4 contacts from the initial configuration. The property is important since having offline contacts in the table prevent new contacts to access it, limiting the chances of finding information.

We execute the command:

```
(find earliest {
    < c(0 ; 1 ; 1 ; 1 ; 0 ; 0 ; 0 ; 0,ip,udp,type0,68760,5400) : Peer |
      CroutingTable : RT1,  CSearchManager : nil ,
      CEventMap : (< 3 ; 0 > < 4 ; 2 > < 4 ; 3 > < 2 ; 1 > < 2 ; 2 >
                   < 3 ; 6 > < 3 ; 7 > < 0 ; 0 > , CMessages : nilMsgL > }
    =>* C:Configuration} such that (num-type4(C:Configuration) == 0) .)
```

where the `num-type4` operation computes the number of contacts of type 4 in a routing table.

The `find earliest` command explores in parallel all possible executions from the given configuration until one of them reaches a configuration that fulfills the given conditions. This means, that all the actions that can take place in the given configuration will be explored. In particular, type 4 contacts are removed from a bin when the time to execute the remove offline process is reached.

The result includes the reached configuration and the time it takes to obtain it. As we are interested only in the time value, we do not show the value of the node attributes.

```
Result:
{< c(0 ; 1 ; 1 ; 1 ; 0 ; 0 ; 0 ; 0,ip,udp,type0,68760,5400): Peer |
    CEventMap : ..., CMessages : ..., CroutingTable : ...>} in time 48
```

Notice that, in this example, we have changed the `OFFLINE-CHECK` constant value to a big number to prevent the remove offline contacts process from generating new type 4 contacts.

6   Conclusions and Ongoing Work

We have developed a formal specification of the Kademlia and the Kad routing tables that can be used by other system developers to use or update their DHTs. The specification includes a notion of time that allows for triggering events and detecting messages timeouts.

The specification will allow us to verify properties of the protocol. In particular properties that include time aspects, like the ones mentioned in the introduction. In this paper we have shown how to obtain the time it takes to remove offline contacts from a table. Similar properties, that will ease us the study of the behavior of the routing table are, for example to obtain the time it will take to execute the next populate process, or which will be the next bin to be populate, under different conditions.

As future work we plan to complete the study of the behavior of the routing table as time goes on, specially on the aspects related with the routing table correctness and the order in which the processes take place. We would also like to integrate the routing table specifications in a distributed system specification using the Maude sockets, and introduce the strategy language of Maude to control the execution of Kad processes.

References

1. Bakhshi, R., Gurov, D.: Verification of peer-to-peer algorithms: A case study. In: Combined Proceedings of the Second International Workshop on Coordination and Organization, CoOrg 2006, and the Second International Workshop on Methods and Tools for Coordinating Concurrent, Distributed and Mobile Systems, MTCoord 2006. Electronic Notes in Theoretical Computer Science, vol. 181, pp. 35–47. Elsevier (2007)
2. Clavel, M., Durán, F., Eker, S., Lincoln, P., Martí-Oliet, N., Meseguer, J., Talcott, C.: All About Maude. LNCS, vol. 4350. Springer, Heidelberg (2007)
3. Lu, T., Merz, S., Weidenbach, C.: Towards Verification of the Pastry Protocol Using TLA$^+$. In: Bruni, R., Dingel, J. (eds.) FMOODS/FORTE 2011. LNCS, vol. 6722, pp. 244–258. Springer, Heidelberg (2011)
4. Maymounkov, P., Mazières, D.: Kademlia: A peer-to-peer information system based on the XOR metric. In: Druschel, P., Kaashoek, M.F., Rowstron, A. (eds.) IPTPS 2002. LNCS, vol. 2429, pp. 53–65. Springer, Heidelberg (2002)
5. Mysicka, D.: Reverse Engineering of eMule. An analysis of the implementation of Kademlia in eMule. Semester thesis, Dept. of Computer Science, Distributed Computing group, ETH Zurich (2006)
6. Ölveczky, P.C., Meseguer, J.: Semantics and pragmatics of Real-Time Maude. Higher-Order and Symbolic Computation 20, 161–196 (2007)
7. Pita, I.: A formal specification of the Kademlia distributed hash table. In: Gulías, V.M., Silva, J., Villanueva, A. (eds.) Proceedings of the 10 Spanish Workshop on Programming Languages, PROLE 2010, pp. 223–234. Ibergarceta Publicaciones (2010), http://www.maude.sip.ucm.es/kademlia (Informal publication–Work in progress)
8. Pita, I., Fernández Camacho, M.I.: Formal Specification of the Kademlia Routing Table and the Kad Routing Table in Maude. Technical Report 1/2013. Dept. Sistemas Informáticos y Computación Universidad Complutense de Madrid (January 2013), http://www.maude.sip.ucm.es/kademlia
9. Pita, I., Riesco, A.: Specifying and Analyzing the Kademlia Protocol in Maude. In: 9th International Workshop on Rewriting Logic and its Applications, WRLA 2012 (2012)

10. Ratnasamy, S., Francis, P., Handley, M., Karp, R., Shenker, S.: A scalable content-addressable network. In: ACM SIGCOMM Computer Communication Review - Proceedings of the 2001 SIGCOMM Conference, vol. 31, pp. 161–172 (October 2001)
11. Rowstron, A., Druschel, P.: Pastry: Scalable, decentralized object location, and routing for large-scale peer-to-peer systems. In: Guerraoui, R. (ed.) Middleware 2001. LNCS, vol. 2218, pp. 329–350. Springer, Heidelberg (2001)
12. Stoica, I., Morris, R., Karger, D., Kaashoek, M.F., Balakrishnan, H.: Chord: A scalable peer-to-peer lookup service for internet applications. ACM SIGCOMM Computer Communication Review 31, 149–160 (2001)

A Generic Program Slicing Technique
Based on Language Definitions[*]

Adrián Riesco[1], Irina Mǎriuca Asǎvoae[2], and Mihail Asǎvoae[2]

[1] Universidad Complutense de Madrid, Spain
ariesco@fdi.ucm.es
[2] Alexandru Ioan Cuza University, Romania
{mariuca.asavoae,mihail.asavoae}@info.uiac.ro

Abstract. A formal executable semantics of a programming language
has the necessary information to develop program debugging and reasoning techniques. In this paper we choose such a particular technique
called program slicing and we introduce a generic algorithm which extracts a set of side-effects inducing constructs, directly from the formal
executable semantics of a programming language. These constructs are
further used to infer program slices, for given programs and specified
slicing criteria. Our proposed approach improves on the parametrization
of the language tools development because changes in the formal semantics are automatically carried out in the slicing procedure. We use the
rewriting logic and the Maude system to implement a prototype and to
test our technique.

Keywords: slicing, semantics, Maude, debugging.

1 Introduction

The intrinsic complexity of a modern software system imposes the need for spe-
cialized techniques and tool support, both targeting the system design and analysis aspects. It is often the case that these two aspects of the software development
are inter-dependent. On the one hand, abstraction techniques are widely used
solutions to reduce, in a systematic way, the size of the system under consideration. However this abstraction-based simplification process is usually dependent
on quality and performance-driven refinements which, in turn, would benefit
from tool support. On the other hand, the development of useful tool support
requires sound techniques to ensure the correctness of the produced results. One
possible solution to integrate techniques and tools development is to use a formal and executable framework such as rewriting logic [12]. Thus, a rewriting
logic general methodology for design and analysis of complex software systems
could and should rely on a formal executable programming language semantics
to ground the development of both abstractions and tools.

[*] Research supported by MICINN Spanish project *DESAFIOS10* (TIN2009-14599-
C03-01) and Comunidad de Madrid program *PROMETIDOS* (S2009/TIC-1465).

N. Martí-Oliet and M. Palomino (Eds.): WADT 2012, LNCS 7841, pp. 248–264, 2013.

A formal executable semantics of programming languages provides a rigorous mechanism to execute programs and in extenso, to implicitly or explicitly have access to all the program executions. The rewriting logic implementation—the Maude system [6]—comes with reachability and fully-fledged LTL model checking tool support. Thus, the notion of execution could be extended from program execution to analysis tool execution (over the same particular program). In these two settings, it is often important to simplify the executions with respect to certain criteria. One such simplification is called slicing [21] and, when applied on programs, it defines safe program fragments with respect to a specified set of variables.

In this paper we investigate, from a semantics-based perspective, the interdependent relationship between the program slicing general technique and its afferent tool—the program slicer. Modifications (i.e. extensions or abstractions) at the level of the programming language, and which are carried out at the level of the program, should be automatically reflected in the program slicing tool support. Therefore, we propose a static technique for program slicing which is based on a meta-level analysis of the formal executable semantics of programming languages. Our program slicing builds on the formal executable semantics of the language of interest, given as a rewriting logic theory, and on the source program.

Our procedure for program slicing consists of the following two steps: (1) a generic analysis of the formal executable semantics, followed by (2) a data dependency analysis of the input program. Step (1) is a fixpoint computation of the set of the smallest language constructs that may issue side-effects. Step (2) uses the resulting set from step (1) to extract safe program slices based on a specified set of variables of interest. Though our slicing technique is general, we exemplify it on the Maude specification of the classical WHILE language, named WhileL, augmented with a side-effect assignment and read/write statements. Next we present a quick, high-level overview of the semantics-based slicing methodology, grounded on the design of a formal executable semantics.

We use the standard approach to specify the WhileL programming language in rewriting logic. First, we define the (abstract) syntax, followed by the language configuration (i.e. the necessary semantic entities to define the behavior) and the language semantics (i.e. equations and rewrite rules).

We motivate our program slicing approach, starting with an alternative view on the language semantics. We elaborate on both structural and functional aspects of this view. Structurally, the formal definition of the WhileL language consists of three layers. At the top level there are the pure syntactic language constructs (i.e. the syntax), at the bottom there is the language state, while the middle layer contains the semantics equations and rewrite rules. This arrangement is important for the functionality of the definition. When we execute a program through these layers, its statements are decomposed into smaller syntactic constructs (i.e. found at the top layer) which, through transformations in the middle layer could result into state updates. In this way, a program execution establishes connections between the syntactic constructs and state updates, in other words which constructs yield side-effects. Step (1) of our program slicing method covers these meta-executions of the semantics to identify the set of syntactic constructs which results in state updates. This coverage employs unification [4] and an adaptation of the backward chaining technique [17]. During the language semantics analysis, the algorithm unfolds the middle layer (i.e. the semantics rewrite rules) into a special tree, applying labels to the visited nodes. This label-based classification is used to identify the set of side-effect constructs and to prune the unfolding tree. Step (2) takes the term representation of the input program together with the results from the previous step, represented as subterms, and does program slicing through term slicing.

Let us consider, throughout this paper, the WhileL program, say P, in Fig. 1 (left) on which we exemplify a standard slicing. We start with P and a set of variables of interest V. For example, V is {p} in Fig. 1 (middle) and V is {s} in the same figure (right). We identify and label the contexts containing variables of interest and add into the set V the other variables appearing in the current context. We run this step until V stabilizes. At the end, the program slice is represented by the skeleton term containing all the labeled contexts. Computing the slice of P with respect to variable p, in Fig. 1 (middle), identifies in the first iteration, the two assignment instructions to p. The second assignment adds the variable i to the V and the second iteration of the algorithm identifies the three input/output instructions. A third iteration considers the variable j, which contributes to the partially computed slice with its corresponding read instruction.

```
read i ; read j ;          read i ; read j ;      read i ; read j ;
s := 0 ; p := 1 ;          p := 1 ;               s := 0 ;
while not (i == 0) do {
    write (i - j) ;        write (i - j) ;        write (i - j) ;
    s := s + i ;                                  s := s + i ;
    p := p * i ;           p := p * i ;
    read i ;               read i ;               read i ;
}
```

Fig. 1. A WhileL program (left) and program slices, w.r.t. variable p (middle) and w.r.t. variable s (right)

The rewriting logic definitions of programming languages support program executability, and at the same time, provide all the necessary information to build analysis tools. In this paper we propose a generic algorithm based on a meta-level analysis of the language semantics, which extracts useful information (i.e. side-effect constructs) for the program slicing procedure. The actual program slice computation is through term slicing.

The rest of the paper is organized as follows: Section 2 presents the related work. Section 3 introduces Maude and presents an example that will be used throughout the rest of the paper. Section 4 describes the slicing algorithm and the main theoretical results, while Section 5 shows our Maude prototype of the technique and outlines its implementation. Finally, Section 6 concludes and presents some subjects of future work.

2 Related Work

Program slicing is a general and well-founded technique that has a wide range of applications in program parallelization [18], debugging [1,16], testing [9] and analysis [11,15]. It was introduced in [21] where, for a given program, is used to compute executable fragments of programs, statically (i.e. without taking into consideration the program input). Strictly from the computation perspective of program slices, the work in [21] produces backward slices, while our approach applies the set of side-effect language constructs, obtained from the formal semantics, to produce forward slices. Informally, this kind of slices, introduced in [10], represents program fragments which are affected by a particular program statement, with respect to a set of variables of interest. With respect to the general problem of program slicing, we refer the reader to the work in [19] for a comprehensive survey on program slicing techniques, with an emphasis on the distinctions between forward and backward slicing approaches and between static and dynamic slicing methods.

Our proposed approach relies on a formal executable semantics definition, specified as a rewriting logic theory, over which we apply semantics-based reasoning techniques to extract side-effect language constructs. Next we elaborate on the works in [8,2,7], all of which use formal language definitions as support for developing program slicing methods.

The approach in [8] applies slicing to languages specified as unconditional term rewriting systems. It relates the dynamic dependences tracking with reduction sequences and, applying successive transformations on the original language semantics, it gathers the necessary dependency relations via rewriting. The resulting slice is defined as a context contained in the initial term representation of the program, a context being a subset of connected subterms. Our approach handles the formal semantics at a meta-level, without executing its rewrite rules.

The recent work in [2] proposes a first slicing technique of rewriting logic computations. It takes as input an execution trace—the result of executing Maude model checker tools—and computes dependency relations using a backward tracing mechanism. Both this work and its sequent extension to conditional term rewriting systems, in [3], perform dynamic slicing by executing the semantics for an initial given state. In comparison, we propose a static approach that is centered around the rewriting logic theory of the language definition. Moreover, our main target application is not counterexamples or execution traces of model checkers, but programs executed by the particular semantics.

Our two step slicing algorithm resembles the approach in [7], where an algorithm mechanically extracts slices from an intermediate representation of the language semantics definition. The algorithm relies on a well-defined transformation between a programming language semantics and this common representation. This transformation is non-trivial and language dependent. The approach in [7] also generalizes the notions of static and dynamic slices to that of constrained slices. What we propose is to eliminate the translation step to the intermediate representation and to work directly on the language semantics.

Finally, the work in [14] presents an approach to generate test cases similar to the one presented here in the sense that both use the semantics of programming languages formally specified to extract information about programs written in this languages. In this case, the semantic rules are used to instantiate the state of the variables used by the given program by using narrowing; in this way, it is possible to compute the values of the variables required to traverse all the statements in the program, the so called coverage.

3 Preliminaries

We present in this section the Maude system [6] by means of an example.

3.1 Maude

Maude modules are executable rewriting logic specifications. Rewriting logic [13] is a logic of change very suitable for the specification of concurrent systems. It is parameterized by an underlying equational logic, for which Maude uses membership equational logic (*MEL*) [5], which, in addition to equations, allows one to state membership axioms characterizing the elements of a sort. Rewriting logic extends *MEL* by adding rewrite rules.

Maude functional modules [6, Chap. 4], introduced with syntax `fmod ...` `endfm`, are executable membership equational specifications that allow the definition of sorts (by means of keyword `sort(s)`); subsort relations between sorts (`subsort`); operators (`op`) for building values of these sorts, giving the sorts of their arguments and result, and which may have attributes such as being associative (`assoc`) or commutative (`comm`), for example; memberships (`mb`) asserting that a term has a sort; and equations (`eq`) identifying terms. Both memberships and equations can be conditional (`cmb` and `ceq`). Maude system modules [6, Chap. 6], introduced with syntax `mod ... endm`, are executable rewrite theories. A system module can contain all the declarations of a functional module and, in addition, declarations for rules (`rl`) and conditional rules (`crl`).

We present Maude syntax for functional modules specifying the natural numbers in the `MY-NAT` module. It defines the sorts `NzNat` and `Nat`, stating by means of a subsort declaration that any term with sort `NzNat` also has sort `Nat`:

```
fmod MY-NAT is
  sort NzNat Nat .        subsort NzNat < Nat .
```

The constructors for terms of these sorts, the constant 0 and the successor operator, are defined as follows:

```
op 0 : -> Nat [ctor] .          op s : Nat -> NzNat [ctor] .
```

We can also define addition between natural numbers. First, we define the operator _+_, where the underscores are placeholders and that has attributes stating that it is commutative and associative. Then we specify its behavior by means of equations. These equations can be conditional, as shown in add1, where we check that the first argument is 0, and can use patterns on the left-hand side, as shown in add2:

```
vars N M : Nat .

op _+_ : Nat Nat -> Nat [comm assoc] .
ceq [add1] : N + M = N if N == 0 .
eq [add2] : s(N) + M = s(N + M) .
endfm
```

The syntax for system modules is presented together with the semantics of the WhileL language, that we will use for our slicing example. For this semantics, assume we have defined in EVALUATION-EXP-EVAL the syntax of a language with the empty instruction skip, assignment, assignment with addition (_+=_), increment operator (_++), If statement, While loop, composition of instructions, and Read and Write functions via a read/write buffer, all of them of sort Com;[1] some simple operations over expressions and Boolean expressions such as addition and equality; and a state, of sort ST, mapping variables to values. Using this module we specify the evaluation semantics of this language in EVALUATION-SEMANTICS, that first defines a Program as a triple of a term of sort Com, a state for the variables, and a state for the read/write buffer:

```
(mod EVALUATION-SEMANTICS is
  pr EVALUATION-EXP-EVAL .
  op <_,_,_> : Com ENV RWBUF -> Statement .
```

The rule AsR is in charge of the semantics of the assignment. It first evaluates the assigned expression (note that we use another operator <_,_> to evaluate expressions which does not require the read/write buffer) obtaining its value v and a new state for variables st', and then updates this new state by introducing the new value for the variable:

```
crl [AsR] : < X := e, st, rwb > => < skip, st'[v / X], rwb >
  if < e, st > => < v, st' > .
```

This update is in charge of the upd equation, that removes the variable from the state (if the variable is not in the state it remains unchanged) with the remove function and then introduces the new value:

```
eq [upd] : ro [V / X] = remove(ro, X) X = V .
```

The rule Inc1 also uses this update operator to increase the value of X in the state. The new value is computed by first obtaining the value v of X and then adding 1 using the auxiliary Ap function, that applies the given operation (addition in this case) to the values:

---

[1] The assignment with addition, the increment operator, and the read/write buffer extend the specification of the language given in [20].

```
crl [Inc1] : < X ++, st, rwb > => < skip, st[ Ap(+., v, 1) / X ], rwb >
  if < X, st > => < v, st > .
```

Similarly, the rule SdE describes the behavior of the += assignment. It first computes the value of X in the given state to obtain the value v and then evaluates the expression e to obtain its final value v'; these values are added with Ap:

```
crl [SdE] : < X += e, st, rwb > => < skip, st''[ Ap(+.,v,v') / X ], rwb >
  if < X, st > => < v, st' > /\
     < e, st' > => < v', st'' > .
```

The rule WriteR introduces a new value in the read/write buffer after evaluating it:

```
crl [WriteR] : < Write e, st, rwb > => < skip, st', insert(v, rwb) >
  if < e, st > => < v, st' > .
```

where insert just introduces the value at the end of the buffer. Reading is performed by using the rule ReadR1. It tries to extract the next value from the buffer and, if it is not the err value, updates the state with it:

```
crl [ReadR1] : < Read X, st, rwb > => < skip, st[v / X], rwb' >
  if (v, rwb') := extract(rwb) /\
     v =/= err .
```

4   Semantics Based Program Slicing

In this section we discuss the semantics-based program slicing. This approach consists of two steps, namely the language semantics specification analysis and the program slicing as term rewriting. We introduce the algorithm for the semantics based analysis and present its execution on the WhileL language case study. We end this section with the results of the second step of the program slicing algorithm, applied on the example program in Fig.1 (left).

We consider S a specification of a program language semantics given in rewriting logic where we make the distinction between the languages syntax and the program state, via their different sorts in S. For example in the semantics of the WhileL language described in Section 3 the language syntax is given by the sort Com, while the program state is formed by sorts ENV and RWBUF.

We consider the following (standard) transformations over S. First, all equations are directed from left to right such that they become rules. Also, matching in conditions $v := e$ become $e \Rightarrow v$. Then, we assume a unique label R identifying each rule as follows:

$$[R] : l \Rightarrow r \text{ if } l_1 \Rightarrow r_1 \wedge \ldots \wedge l_n \Rightarrow r_n$$

with $n \in \mathbb{N}$. Finally, we transform any rule $[R]$ into a labeled Horn clause:

$$[R] \ l :- l_1; \ldots; l_n; r.$$

We denote $\bar{S}$ the specification S after these transformations. Note that $\bar{S}$ does not contain the r_i terms of the rules in S. We comment on this later when describing the first slicing step algorithm.

Let t be a term with variables. We use the notation $R :: t$ for "an instance of t on non-variable position can be reduced by the rule R", i.e., exists an unifier θ and a subterm s of t such that $s\theta = l\theta$.

Definition 1. *A hypernode for a valid term t, denoted as* $\boxed{\forall R \in \bar{S}, R :: t}$, *is a list* $\boxed{R_1 \to \ldots \to R_m}$ *of distinct rules in $\bar{S}$, with $m \in \mathbb{N}$, such that $R_i :: t$ for all $1 \leq i \leq m$. We define an inspection tree $i\mathcal{T}$ as a tree of hypernodes where the children of a hypernode* $\boxed{R_1 \to \ldots \to R_m}$ *are defined as:*

$$children\left(\boxed{R_1 \to \ldots \to R_m}\right) = \{successors(R_i) \mid 1 \leq i \leq m\}$$

where the successors *of a Horn clause $[R]$ $l :- t_1; \ldots; t_n$. are defined as:*

$$successors(R) = \boxed{\forall R_1 \in \bar{S}, R_1 :: t_1} \longrightarrow \ldots \longrightarrow \boxed{\forall R_n \in \bar{S}, R_n :: t_n}$$

Example 1. We give next a generic example of an inspection tree, considering $R_1, \ldots, R_7$ as all rule labels in $\bar{S}$.

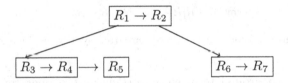

The shape of this tree is induced by the rules $R_1, \ldots, R_7$ as follows: assume there exist t and two unifiers θ_1 and θ_2 such that $t\theta_1 = l_{R_1}\theta_1$ and $t\theta_2 = l_{R_2}\theta_2$, and that no other rule in $\bar{S}$ can reduce t. Hence, $\boxed{\forall R \in \bar{S}, R :: t}$ is $\boxed{R_1 \to R_2}$. Moreover, assume that $[R_1]$ $l_{R_1} :- t_{1,1}; t_{2,1}.$ and $[R_2]$ $l_{R_2} :- t_{1,2}.$ and assume there exist θ_i, with $3 \leq i \leq 7$ such that $t_{1,1}\theta_3 = l_{R_3}\theta_3$ and $t_{1,1}\theta_4 = l_{R_4}\theta_4$, $t_{2,1}\theta_5 = l_{R_5}\theta_5$, while $t_{1,2}\theta_6 = l_{R_6}\theta_6$ and $t_{1,2}\theta_7 = l_{R_7}\theta_7$. Hence, the down arrows starting in the rules R_1 and R_2 point towards $successors(R_1)$ and $successors(R_2)$, respectively (again, under the assumption that no other rules in $\bar{S}$ can reduce $t_{1,1}, t_{1,2}$, or $t_{2,1}$). Namely, $\boxed{\forall R \in \bar{S}, R :: t_{1,1}}$ is $\boxed{R_3 \to R_4}$, $\boxed{\forall R \in \bar{S}, R :: t_{2,1}}$ is $\boxed{R_5}$, and $\boxed{\forall R \in \bar{S}, R :: t_{1,2}}$ is $\boxed{R_6 \to R_7}$. Finally, the tree is completely unfolded under the assumption that $successors(R_i) = \emptyset, 3 \leq i \leq 7$.

The algorithm in Fig. 2 computes the set of basic syntactic language constructs which *may* produce side-effects, by inspecting the conditions and the right-hand side of each rewrite rule in the definition. For this inspection, we rely on unification [4] and a backward chaining technique [17]. The algorithm unfolds the semantics rewrite rules into the inspection tree $i\mathcal{T}$ such that the final tree contains a superset of all rules which can be called during a rewrite execution "**rew**

Input: The language specification $\bar{S}$, the valid term $runPgm(X : LS, Y : PS)$,
and the set A of side-effect-sources (pre-computed based on PS).

Output: The basic syntactic constructs (non-recursive operators of sort LS)
which induce side effects in the program state (of sorts PS).

$maySE:=A;\quad noSE:=\emptyset;\quad rMix:=\emptyset;\quad lBorder:=$empty stack;

$i\mathcal{T} := \boxed{\forall R \in \bar{S}, R :: runPgm(X : LS, Y : PS)}$;

$lBorder:=insert(\text{select } R \text{ from } \forall R \in \bar{S}, R :: runPgm(X : LS, Y : PS));$

while $lBorder \neq$ empty stack **do**
   $R_{curr}:=top(lBorder);$
   if $R_{curr} \in maySE \cup noSE$
   then $backtrack(i\mathcal{T}, lBorder);$
   else if $R_{curr} \in lBorder - top(lBorder)$ or $R_{curr} \in rMix$
   then $rMix+=R_{curr}; backtrack(i\mathcal{T}, lBorder);$
   else if $newSuccessors(R_{curr}) \neq \emptyset$
   then $i\mathcal{T}+=successors(R_{curr}); lBorder:=insert(\text{select } R \text{ from } successors(R_{curr}));$
   else if $successors(R_{curr}) \cap maySE \neq \emptyset$
   then $maySE+=R_{curr}; backtrack(i\mathcal{T}, lBorder);$
   else $noSE+=R_{curr}; backtrack(i\mathcal{T}, lBorder);$
od

return $\{s \in LS \mid s \text{ subterm of } l, [R]\, l:-t_1;\dots;t_n, R \in root(i\mathcal{T}) \cap (maySE - rMix)\}$

Fig. 2. The algorithm for detecting basic program syntax producing side effects

rP", where rP is a ground term with $runPgm$ as top operator. More to the
point, we assume that there exists in S an operator $runPgm$ which is used for
executing programs based on the language semantics specification S. Note that
usually $runPgm$ contains as arguments the program term and the initial program state, i.e., $runPgm$ is defined over LS (language syntax) and PS (program
state) sorts. In other words, the inspection tree is an over-approximated result
of a *rule reachability problem*.

We call t a *valid term* if its subterms and its variables meet a set of constraints
Cns w.r.t. the possible ground instances of t. Typically we use a valid program
term, formed only with the language syntax operators from LS, and valid program state term, formed only with the constructors of the sort PS. Note that
for a valid term t some unifications are refuted, hence some rules are deleted
from $\boxed{\forall R \in \bar{S}, R :: t}$. The root of $i\mathcal{T}$ is the hypernode obtained from the input
valid term $runPgm(X : LS, Y : PS)$. The rules in $i\mathcal{T}$'s hypernodes are either
Labeled or *Unlabeled*, where $Labeled = lBorder \cup maySE \cup noSE \cup rMix$. $lBorder$
is a stack which maintains the path currently unfolded in the inspection tree.
$maySE$ and $noSE$ are two disjoint sets of rules which cover the *already traversed*
side of the inspection tree (i.e., the rules in the left half-plane determined by
the $lBorder$, labeled border, path). The $maySE$ set contains rules which *may contain side-effects*, i.e., at least a rewrite starting in an $maySE$ rule will reach the

application of a rule which modifies the state of the program. We label as *noSE* the rules for which there is no side-effect propagation, hence $maySE \cap noSE = \emptyset$. The *rMix* label is introduced for detecting and pruning recursive rules, i.e., rules that may (indirectly) call themselves which means that at some point they appear twice in the *lBorder* stack.

The $backtrack(i\mathcal{T}, lBorder)$ subroutine is an augmented backtracking procedure (that goes up in *lBorder* as long as it cannot go right via the horizontal arrows in $i\mathcal{T}$). The augmentation consists in the fact that when *backtrack* removes the top of the *lBorder* stack, it also labels that rule as *maySE* or *noSE*. Namely, a rule is labeled *maySE* if, when removed from the top of the stack, it contains at least a successor rule labeled *maySE*. Otherwise, i.e., when no successor rule is labeled *maySE*, the rule is labeled *noSE* upon removal from the stack. The labeling in the *backtrack* subroutine induces the fact that at the end of the algorithm $rMix \subset maySE \cup noSE$. Also, $newSuccessors(R) = children(R) \backslash Labeled$. Finally, the algorithm returns the language syntax subterms of the terms l in the rules R from $i\mathcal{T}$'s root that are in *maySE* but not *rMix*. In other words, the result of the algorithm identifies the language syntactic constructs (subterms of sort LS) that are basic and may produce side-effects, i.e., the rules giving its semantics in S are non-recursive (not in *rMix*) and in *maySE*.

Our algorithm allows label propagation under certain conditions, with a labeled node summarizing the information of its corresponding subtree. The results of the language semantics analysis are used as contexts in the second slicing step to infer a safe program slice. Note that the termination of the algorithm in Fig. 2 is ensured by the fact that the specification has a finite number of rules, and that any rule in $i\mathcal{T}$ that was already *Labeled* is not unfolded anymore. Notice also that the algorithm is independent of the order of the labels in a hypernode. In fact, we consider the labels as a list just for ease the presentation, but hypernodes can be just considered sets of labels, since the algorithm relies on the *existence* of a rule label generating side effects to work.

Finally, recall that upon transformation of S to $\bar{S}$ we do not consider the right hand-side of the rules' conditions (i.e., the terms r_i in the rule R have disappeared from $\bar{S}$). We allow this because the algorithm in Fig. 2 computes an *over-approximation* of the rule reachability problem. Hence, by not considering the terms r_i in the semantic rules we include in the inspection tree a superset of the reachable rules. As another consequence, the *maySE* set is also over-approximated namely, we infer more potential side-effect syntactic constructs. However, observe that the actual side-effect rules are guaranteed to appear in the *maySE* set, again, because the algorithm relies on the *existence* of a *maySE* rule descendant to induce the side-effect character on a parent rule. Nevertheless, we could consider the terms r_i as well by augmenting the algorithm with additional pruning of the inspection tree. We leave this augmentation as future work.

Example 2. We exemplify the running of the algorithm in Fig. 2 over the semantics of the WhileL language, with the set of side effect sources formed by the rules for the operators `remove`, `insert`, and `extract`. We recall the definition of these operators:

```
op remove : ENV Variable -> ENV .
op insert : Value RWBUF -> RWBUF .
op extract : RWBUF -> PairValueRWBUF .

eq [rmv1] : remove(mt, X) = mt .
eq [rmv2] : remove(X = V ro, X')
          = if X == X' then ro else X = V remove(ro, X') fi .
eq [ins]  : insert(V, buf) = buf V .
eq [ex1]  : extract(nb) = err .
eq [ex2]  : extract(V buf) = (V, buf) .
```

The set A of rules that produce modifications to the program state is formed by the rules labeled `rmv2`, `ins`, `ex2`, which are the only statements (either equations or rules) that modify the `ENV` or the `RWBUF` without using auxiliary functions. Also, `Com` is the language syntax sort LS, `ENV` and `RWBUF` give the program state sorts PS, while the "root" term is given by the operator `op <_,_,_> : Com ENV RWBUF -> Statement .`, i.e., $runPgm$ is `< C:Com, E:ENV, B:RWBUF >`.

So, after the initial assignments before the loop, the variables in the algorithm are assigned as follows:

$maySE = \{\texttt{rmv2, ins, ex2}\}$, $noSE = \emptyset$, $rMix = \emptyset$,

$$i\mathcal{T} = \boxed{\forall R \in \bar{\mathcal{S}},\ R ::\ \texttt{< C , E , B >}}$$
$$= \boxed{\texttt{AsR}\rightarrow \texttt{Inc1}\rightarrow \texttt{Inc2}\rightarrow \texttt{SdE}\rightarrow \texttt{IfR1}\rightarrow \dots \rightarrow \texttt{WriteR}\rightarrow \texttt{ReadR1}\rightarrow\texttt{ReadR2}}$$

$lBorder = \texttt{AsR}$.

During the first iteration of the loop, the variable R_{curr} is `AsR`. We recall that `AsR` represents the following Horn clause:

```
[AsR]   < X := e, st, rwb > :- < e, st > ; < skip, st'[v / X], rwb > .
```

So l_1 is the term `< e, st >`, with variables `e:Exp` and `st:ENV`, while r is the term `< skip, st'[v / X], rwb >`, with variables `st':ENV`, `X:Var`, `v:Num`, and `rwb:RWBUF`. Consequently, $successors(\texttt{AsR})$ is given by the following list of hypernodes $\boxed{\forall R \in \bar{\mathcal{S}}, R :: l_1} \longrightarrow \boxed{\forall R \in \bar{\mathcal{S}}, R :: r}$, which evaluates to the following hypernodes: $\boxed{\texttt{VarR}\rightarrow\texttt{OpR}}$, and respectively $\boxed{\texttt{upd}}$. Note that the rule `rmv2` cannot be applied over the subterm `st` of l_1 because `st` is a valid term, formed only with the constructors of `ENV`. Hence, the loop executes the branch with the condition $newSuccessors(R_{curr}) \neq \emptyset$, and the iteration tree $i\mathcal{T}$ becomes:

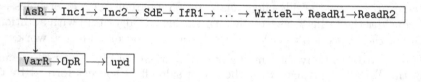

Note that the color code in $i\mathcal{T}$ signifies that $lBorder$ is the stack `VarR`, `AsR` (i.e., red for $lBorder$). The second iteration of the loop makes R_{curr} the Horn clause:

$$[\text{VarR}] \quad < \text{X, st} > :- \quad < \text{st(X), st} > \text{ .}$$

So the same branch of the loop as before is executed and $i\mathcal{T}$ becomes:

The next iteration of the loop finds that lkup is recursive, where

$$[\text{lkup}] \quad (\text{X = V ro})(\text{X'}) :- \text{ if X == X' then V else ro(X') fi .}$$

because lkup $\in$ *successors*(lkup) = $\boxed{\text{lkup}}$ (as ro(X') unifies with lkup), so *lBorder* becomes the stack lkup, lkup, VarR, AsR. Hence, in the following iteration of the loop, the top lkup becomes an element of *rMix* and, upon backtracking, the next lkup becomes an element of *noSE* (since it does not have a *maySE* successor rule). For the same reason *backtrack*($i\mathcal{T}$, VarR, AsR) makes VarR an element of *noSE*, and *lBorder* becomes OpR, AsR.

The next few iterations of the loop make OpR an element of both *rMix* and *noSE* in the same way as lkup. We skip over these steps and explain from the iteration with the *lBorder* stack containing upd, AsR.

Since *successors*(upd)=$\boxed{\text{rmv1}\rightarrow\text{rmv2}}$ which is a hypernode that contains a *maySE* rule, i.e., rmv2, then the rule upd becomes an element of *maySE*, via the last branch in the algorithm. Also, during *backtrack*($i\mathcal{T}$, upd, AsR) also AsR becomes an element of *maySE*, because of its *maySE* successor rule in the hypernode $\boxed{\text{upd}}$. After these steps, the inspection tree has the following structure:

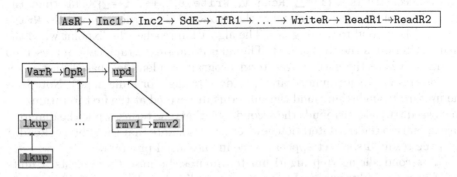

After the algorithm in Fig. 2 is applied to the WhileL language, we identify the following set of syntactic constructs that may produce side-effects: the assignment statement _:=_, the input/output statements Read_ and Write_ and the two special addition-based statements _++ and _+=_. Note that if we consider only the program environment as the side effect source (i.e., the set A is $\{\texttt{rmv2}\}$) the algorithm produces all the previous side-effect syntactic constructs, besides Write_. The reason is that WriteR rule was previously labeled as *maySE* only due to the ins rule that appeared among its descendants. Since now $A = \{\texttt{rmv2}\}$, then ins is not *maySE*, so WriteR is labeled *noSE*.

The set of basic side-effect syntactic constructs together with the term representation of the program and the slicing criterion define the input state for the second step of our slicing algorithm which computes a slice of the program.

Definition 2. *We say that a program term p produces side-effects over a variable v w.r.t. a side-effect set SE if the top operator of p is in SE and the variable v is a designated subterm of p. We define a valid slice of a program w.r.t. a slicing criterion V (i.e., a set of program variables) as a superset of program subterms that (indirectly) produce side-effects over some variables in V.*

The second slicing step is a fixpoint iteration which applies the *current* slicing criterion over the program term in order to discover new subterms of the program that use the slicing criterion, i.e., the program subterm produces side-effects over *some* variables in the slicing criterion. When a new subterm is discovered, the slicing criterion is updated by adding the variables producing the side-effects (e.g., the variables in the second argument of _:=_). We iterate this until the slicing criterion remains unchanged, so no new subterms can be discovered and added to the result, i.e., the slice set.

For example, the iterations of the second slicing step for the program in Fig. 1, the side-effect constructs obtained in Example 2, and the slicing criterion $\{\texttt{p}\}$ are listed in Fig. 3. Namely, the set $\{\texttt{p}\}$ is applied on the set of side-effect syntactic constructs to produce the slice subterms used on the term program: p:=_, Read p, Write p, p++, p+=_. Only the subterm p:=_ is matched in the program, on the term assignment for p. This iteration results are shown in the first row, in Fig. 3. Because of the matched assignment p := p *. i, the slicing criterion becomes $\{\texttt{p}, \texttt{i}\}$. Under the new slicing criterion, the set of side-effect syntactic constructs is $\{\texttt{i:=\_, Read i, Write i, i++, i+=\_}\}$. This time, the term slicing of the program matches the two Read i instructions and the Write (i-.j) (the second row in Fig. 3). The algorithm reaches the fixpoint when the slicing criterion is the set $\{\texttt{p}, \texttt{i}, \texttt{j}\}$. The final iteration is graphically represented in Fig. 4, where the slice on the term program is identified based on the set of subterms inside surrounded area (inside a triangle or a diamond). Note that the upward triangles surround the subterms discovered at the first iteration, the downward triangle surrounds the second iteration new terms, while the diamond corresponds to the third iteration new terms. Also, note that the subterm Read i from the resulted slice set appears twice in the initial program.

The second slicing step algorithm terminates, because there exists a finite set of program subterms, and it produces a valid slice, because it exhaustively saturates the slicing criterion. Moreover, the result is a minimal slice w.r.t. the set of side-effect syntactic constructs given as argument. However, the obtained slice is not minimal mainly because the set of side-effect syntactic constructs obtained in the first slicing step is already an over-approximation.

Iteration	Slicing variables	Computed slice (identified subterms)
1 △	p	$p := 1, p := p * . i$
2 ▽	p, i	$\text{Read } i, p := 1, \text{Write } (i - . j), p := p * . i$
3 ◇	p, i, j	$\text{Read } i, \text{Read } j, p := 1, \text{Write } (i - . j), p := p * . i$
4 .	p, i, j	$\text{Read } i, \text{Read } j, p := 1, \text{Write } (i - . j), p := p * . i$

Fig. 3. Program slicing as term slicing - the fixpoint iterations

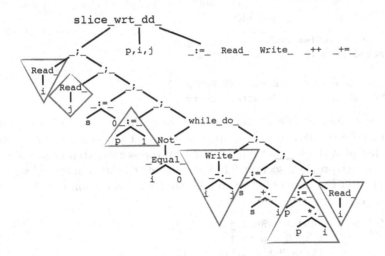

Fig. 4. Program slicing as term slicing—the result subterm

Based on the obtained slice set we can add structure to the slicing result by identifying the "skeleton" term that contains the slice set. For example, we can decide to keep the composed statements that contain subterms in the slice set. As such, the while_do_ statement from the example program has subterms in the slice set (e.g., Read i) so we add it to the "structured" slice. The resulting program representing the "structured" slice is:

```
Read i;
Read j;
p := 1;
While _ do {
  Write(i -. j);
  p := p *. i;
  Read i; }
```

5   System Description

We present in this section our Maude prototype and some details about its implementation.

5.1   Slicing Session

The tool is started by loading the `slicing.maude` file available at `http://maude.sip.ucm.es/slicing`. It starts an input/output loop where modules and commands can be introduced. Once the module in Section 3 has been introduced, we have to introduce the sort of variables and the sorts where we want to detect side effects. With this information, the tool can compute the rules and the functions generating side effects:

```
Maude> (set variables sort Var .)
Var selected as sort for the variables.

Maude> (set side-effect sorts ENV RWBUF .)
ENV RWBUF selected as side effect sorts.
```

The slicing command follows the notation shown in the previous sections. The keyword `slice` is followed by the program we want to debug, the keyword `wrt` and the list of variables that will be used by the second step of the algorithm described in the previous section. The tool outputs the label of the rules inducing side effects and the name of the variables that have been computed during the slicing stage:

```
Maude> (slice (< Read i ; Read j ;
               s := 0 ; p := 1 ;
               While Not Equal(i, 0) Do
                 Write (i -. j) ; s := s +. i ;
                 p := p *. i ; Read i,
             X:ENV, Y:RWBUF >) wrt p .)

The rules causing side effects are: AsR Inc1 ReadR1 SdE WriteR
The variables obtained by the slicing process are: p i j
```

5.2   Implementation Details

Exploiting the fact that rewriting logic is reflective, a key distinguishing feature of Maude is its systematic and efficient use of reflection through its predefined `META-LEVEL` module [6, Chapter 14], a feature that makes Maude remarkably extensible and that allows many advanced metaprogramming and metalanguage applications. This powerful feature allows access to metalevel entities such as specifications or computations as usual data. In this way, we can manipulate the modules introduced by the user, develop the slicing process, and implement the input/output interactions in Maude itself.

More specifically, our tool traverses the rules in the module indicated by the user to find the rules that modify the state of the terms modified by the side effects (ENV and RWBUF in our example). With these rules and using the predefined unification commands available in Maude our prototype can generate and traverse the hypernodes, thus computing the first step of the algorithm in the previous section. For the second part, it checks the left-hand sides of these rules, discarding the information of the side effects terms to focus on the instructions. It then uses this information to traverse the initial term (containing the program we are analyzing) checking the terms that appear in each of these terms. If any of the variables given as slicing criterion are used, then the rest of the variables appearing in this term are added to the current set and the process is repeated until the fixpoint is reached.

6 Concluding Remarks and Ongoing Work

We presented a two-phased technique for program slicing based on the formal executable semantics of a WHILE programming language, given as a rewriting logic theory. The first phase, called language semantics specification analysis considered an exhaustive inspection of the language semantics to extract a set of side-effect language constructs. The second phase was to perform program slicing as term slicing, using the previously computed set of primitives, the input (term representation of the) program and the slicing criterion. Both the formal definition of our language and the semantics-based slicing technique are implemented and tested in the Maude system.

We plan to extend this work on the following several directions. First, we incrementally analyze the impact on adding various side-effect constructs to the WhileL language, such as pointers, exceptions or file-manipulation operations. Second, we address other types of programming languages and their specific side-effect constructs. For example, if we consider assembly languages, it happens that various arithmetic instructions visibly modify a particular register value, and invisibly affect subsequent conditions in the program, via modifications of special arithmetic flags (i.e. overflow, sign, etc). Third, we improve the algorithm in the first step in our slicing technique in order to obtain a better (smaller) over-approximation of the produced set of side-effect syntactic constructs. We believe that pursuing these directions would further improve the current semantics-based program slicing technique and produce a useful design and analysis tool for language developers.

References

1. Agrawal, H., DeMillo, R.A., Spafford, E.H.: Debugging with dynamic slicing and backtracking. Software - Practice and Experience 23(6), 589–616 (1993)
2. Alpuente, M., Ballis, D., Espert, J., Romero, D.: Backward trace slicing for rewriting logic theories. In: Bjørner, N., Sofronie-Stokkermans, V. (eds.) CADE 2011. LNCS (LNAI), vol. 6803, pp. 34–48. Springer, Heidelberg (2011)

3. Alpuente, M., Ballis, D., Frechina, F., Romero, D.: Backward trace slicing for conditional rewrite theories. In: Bjørner, N., Voronkov, A. (eds.) LPAR-18. LNCS, vol. 7180, pp. 62–76. Springer, Heidelberg (2012)
4. Baader, F., Snyder, W.: Unification theory. In: Robinson, A., Voronkov, A. (eds.) Handbook of Automated Reasoning, pp. 445–532. Elsevier (2001)
5. Bouhoula, A., Jouannaud, J.-P., Meseguer, J.: Specification and proof in membership equational logic. Theoretical Computer Science 236, 35–132 (2000)
6. Clavel, M., Durán, F., Eker, S., Lincoln, P., Martí-Oliet, N., Meseguer, J., Talcott, C.: All About Maude. LNCS, vol. 4350. Springer, Heidelberg (2007)
7. Field, J., Ramalingam, G., Tip, F.: Parametric program slicing. In: Proc. of the 22nd ACM SIGPLAN-SIGACT Symposium on Principles of Programming Languages, POPL 1995, pp. 379–392. ACM Press (1995)
8. Field, J., Tip, F.: Dynamic dependence in term rewriting systems and its application to program slicing. Information & Software Technology 40(11-12), 609–636 (1998)
9. Harman, M., Danicic, S.: Using program slicing to simplify testing. Journal of Software Testing, Verification and Reliability 5, 143–162 (1995)
10. Horwitz, S., Reps, T.W., Binkley, D.: Interprocedural slicing using dependence graphs. ACM Transactions on Programming Languages Systems 12(1), 26–60 (1990)
11. Jhala, R., Majumdar, R.: Path slicing. In: Proc. of the 2005 ACM SIGPLAN Conference on Programming Language Design and Implementation, PLDI 2005, pp. 38–47. ACM Press (2005)
12. Martí-Oliet, N., Meseguer, J.: Rewriting logic: roadmap and bibliography. Theoretical Computer Science 285(2), 121–154 (2002)
13. Meseguer, J.: Conditional rewriting logic as a unified model of concurrency. Theoretical Computer Science 96(1), 73–155 (1992)
14. Riesco, A.: Using semantics specified in Maude to generate test cases. In: Roychoudhury, A., D'Souza, M. (eds.) ICTAC 2012. LNCS, vol. 7521, pp. 90–104. Springer, Heidelberg (2012)
15. Sandberg, C., Ermedahl, A., Gustafsson, J., Lisper, B.: Faster WCET flow analysis by program slicing. In: Proc. of the 2006 ACM SIGPLAN Conference on Languages, Compilers and Tools for Embedded Systems, LCTES 2006, pp. 103–112. ACM Press (2006)
16. Silva, J., Chitil, O.: Combining algorithmic debugging and program slicing. In: Proc. of the 8th ACM-SIGPLAN International Symposium on Principles and Practice of Declarative Programming, PPDP 2006, pp. 157–166. ACM Press (2006)
17. Sterling, L., Shapiro, E.Y.: The Art of Prolog - Advanced Programming Techniques. MIT Press (1986)
18. Tian, C., Feng, M., Gupta, R.: Speculative parallelization using state separation and multiple value prediction. In: Proc. of the 2010 International Symposium on Memory Management, ISMM 2010, pp. 63–72. ACM Press (2010)
19. Tip, F.: A survey of program slicing techniques. Journal of Programming Languages 3(3), 121–189 (1995)
20. Verdejo, A., Martí-Oliet, N.: Executable structural operational semantics in Maude. Journal of Logic and Algebraic Programming 67, 226–293 (2006)
21. Weiser, M.: Program slicing. In: Proc. of the 5th International Conference on Software Engineering, ICSE 1981, pp. 439–449. IEEE Press (1981)

Distances between Processes: A Pure Algebraic Approach⋆

David Romero Hernández and David de Frutos Escrig

Dpto. Sistemas Informáticos y Computación
Facultad CC. Matemáticas, Universidad Complutense de Madrid, Spain
dromeroh@pdi.ucm.es, defrutos@sip.ucm.es

Abstract. Recently, we have presented operational and denotational definitions for distances between processes corresponding to any semantics in the ltbt-spectrum. In this paper, we develop a general algebraic framework to define distances between terms from any arbitrary signature. We apply this framework obtaining a new algebraic characterization of our previous distances. Moreover, we prove the generality of our approach developing an algebraic characterization of the distances based on the (bi)simulation game by other authors.

1 Introduction

In order to define an (abstract) semantics for processes, we need just to define an adequate equivalence relation, $\equiv$, relating the processes in some universe, *Proc*. Then, the values of this semantics are just the corresponding equivalence classes, and two processes have the same semantics if and only if they are equivalent. Once we have fixed such a semantics we can compare two processes, but the output of this comparison is just a Boolean value. In particular, when two processes are not equivalent, we do not have a general way to measure "how far away" they are from being equivalent.

There are several papers which introduce several distances between processes based on the (bi)simulation game –see e.g. [3, 5]–. Even before, Ying and Wirsing studied approximate bisimilarity, following similar but simpler ideas [12, 11]. Our work started by considering those distance games, where one essentially plays the (bi)simulation game, but with the "defender" having the possibility of replying a move of the "attacker" without matching exactly his move. In such a case, he should pay to the attacker some quantity, depending on the mismatch (distance) between the two involved actions.

It is well-known the use of equivalence relations to formalize the notion of implementation: a process implements some specification (given by another process) when they are equivalent w.r.t. the adequate semantics. But, if we follow this approach, we have no flexibility at all: our process has to satisfy in a precise

---

⋆ Partially supported by the Spanish projects TESIS (TIN2009-14312-C02-01), DE-SAFIOS10 (TIN2009-14599-C03-01) and PROMETIDOS S2009/TIC-1465.

N. Martí-Oliet and M. Palomino (Eds.): WADT 2012, LNCS 7841, pp. 265–282, 2013.

way all the constraints imposed by the specification, or it will not be a correct implementation. Instead, in the real world, we often find other more flexible quality requirements, where the specification establishes the ideal behavior of the system, but some (limited) deviations from it are allowed, without invalidating the adequateness of the implementation.

We need a notion of distance between processes to make precise how far away two processes are from being equivalent w.r.t. some given semantics. It is true that metrics have been used for a long time to formalize the semantics of infinite processes, by means of (the limit of) those of their finite approximations. But these metrics were just a very particular case that only cared for "the first" disagreement between the compared processes. Instead, now we look for more general distances, which moreover should be applicable to any syntactic process algebra (i.e., to any signature Σ) and any semantics (based, for instance, on the axiomatization of the desired semantics).

We have already introduced the basic operational ideas of our approach in [8]. It is true that the most flexible way to capture a semantics for processes, $\mathcal{L}$, is based on the use of an adequate preorder $\sqsubseteq_{\mathcal{L}}$, and not just an equivalence relation $\equiv_{\mathcal{L}}$.

However, we prefer to start our presentation in Section 2 using just the better known *equational calculus*. Next, in Section 3 we will see that a simple modification of the proof system defining the classical equational calculus (see e.g. [7]), produces a general algebraic framework to define distances between processes w.r.t. any semantics. In particular, in Section 4 we study in detail the case of the bisimulation semantics. Later, in Section 5, we will see how we can easily generalize all our algebraic presentation to the inequational framework. Moreover, we define our distances for the rest of the semantics in the ltbt-spectrum. To show the flexibility of our approach, in Section 6 we see how the classic distances based on the (bi)simulation game, can be also defined in our algebraic framework. We conclude with our conclusions and some future work.

2   Preliminaries

A careful presentation of the equational calculus with application to the (testing) semantics of processes can be found in [7]. Next, we will only remind the definitions of the main concepts needed to develop that theory.

Definition 1.   *1. A signature, Σ, is a set of formal functional symbols. Each functional symbol has associated a natural number which is its arity. We use Σ_n to denote the set of symbols in Σ of arity n.*

2. If Σ is a signature, a Σ-algebra is a pair $\langle A, \Sigma_A \rangle$ where A is a set, the domain or support of the algebra, and Σ_A is a set of internal functions. A Σ-algebra is simply an interpretation of each operation in Σ, respecting of course its arity.

3. There is a particularly important Σ-algebra, called the term algebra for Σ, and denoted by T_Σ, whose support is the set of terms freely generated by Σ.

Some particular collections of Σ-algebras can be singled out by means of equations. An equation is determined by a pair of terms which may contain variables. We consider an (arbitrary) set of variables, X, and the set of terms with variables $T_\Sigma(X)$. $T_\Sigma(X)$ can be obtained by extending the signature Σ adding these variables with null arity. In fact $T_\Sigma(X)$ is just an algebra, where $\Sigma(X)$ is the classic notation for the signature $\Sigma \cup X$ which add to Σ each $x \in X$ as a new function symbol of arity 0.

Given an equation $t \equiv t'$ with $t, t' \in T_\Sigma(X)$, we say that a Σ-algebra $\langle A, \Sigma_A \rangle$ satisfies it, when the values of both t and t' under any valuation, which assigns values in A to the variables $x \subset X$, are the same. Given a set of equations E, $\mathcal{C}(E)$, is the class of Σ-algebras satisfying the equations E. The initial algebra of $\mathcal{C}(E)$ can be presented as a quotient algebra $T_\Sigma/\equiv_E$ for some particular congruence $\equiv_E$. We can obtain this congruence by means of the equational deduction system $\mathbf{DED}(E)$ in Fig. 1 whereby the equations in E may be used to derive statements of the form $t \equiv t'$, with $t, t' \in T_\Sigma$.

1. Reflexivity $\dfrac{}{t \equiv t}$ 2. Symmetry $\dfrac{t \equiv t'}{t' \equiv t}$ 3. Transitivity $\dfrac{t \equiv t' \quad t' \equiv t''}{t \equiv t''}$

4. Substitution $\dfrac{t_1 \equiv t'_1, \ldots, t_k \equiv t'_k}{f(t_1, \ldots, t_k) \equiv f(t'_1, \ldots, t'_k)}$ for every $f \in \Sigma$ of arity k.

5. Instantiation $\dfrac{t \equiv t'}{t\rho \equiv t'\rho}$ for every substitution ρ.

6. Equations $\dfrac{}{t \equiv t'}$ for every equation $\langle t, t' \rangle \in E$

Fig. 1. The proof system $\mathbf{DED}(E)$ in [7]

We write $\vdash_E t \equiv t'$ if we can derive $t \equiv t'$; and then, we say that $t \equiv t'$ is a theorem of $\mathbf{DED}(E)$. Obviously, we can see each derivable theorem as the pair of a relation $\equiv_E$, which due to 1-3 is an equivalence relation, and as a result of 4 a Σ-congruence.

As we have said, the idea in this paper is to define distance between processes. Then, we need to extend the concept of Σ-algebra with a notion of distance. This distance allows us to measure how far away is a process p of being equivalent to another process q.

In [8] we have considered as processes the terms generated by the free $(\mathbf{0}, Act, +)$-algebra, which correspond to the classic domain of $BCCSP(Act)$ processes.

Definition 2. *Given a set of actions Act, the set $BCCSP(Act)$ of processes is that defined by the BNF-grammar: $p ::= \mathbf{0} \mid ap \mid p + q$. The very well known operational semantics of BCCSP [10, 4] is defined by:*

$$(1)\ \frac{}{ap \xrightarrow{a} p} \qquad (2)\ \frac{p \xrightarrow{a} p'}{p + q \xrightarrow{a} p'} \qquad (3)\ \frac{q \xrightarrow{a} q'}{p + q \xrightarrow{a} q'}$$

Based on this operational semantics we can define all the semantics in the ltbt-spectrum [10]. In particular, we have studied the case of the finest of these semantics, that is the bisimulation semantics.

Definition 3 (Bisimulation). *A bisimulation between processes is a relation B such that, whenever pBq, we have:*

- *for every $a \in Act$, if $p \xrightarrow{a} p'$ there exists some q', with $q \xrightarrow{a} q'$ and $p'Bq'$.*
- *for every $a \in Act$, if $q \xrightarrow{a} q'$ there exists some p', with $p \xrightarrow{a} p'$ and $p'Bq'$.*

We have defined distances between processes corresponding to any semantics defined by a preorder $\sqsubseteq_{\mathcal{L}}$, this covering in particular the case of equivalence relations, such as bisimulation . Both are defined in an operational way based on transformations between processes, and in a denotational way, using SOS-rules. Next, we recall this second definition

Definition 4 ([8]). *Given a semantics $\mathcal{L}$, defined by a preorder $\sqsubseteq_{\mathcal{L}}$, coarser than bisimulation, we say that a process q is at distance at most $m \in \mathbb{N}$ of being better than some other p, w.r.t. the semantics $\mathcal{L}$ and the distance between actions $\overline{d}$, and then we write $d_{\overline{d}}^{\mathcal{L}}(p, q) \leq n$, if we can infer $p \sqsubseteq_n^{\mathcal{L}} q$, by applying the following rules:*

$$(1)\ \frac{p \sqsubseteq_{\mathcal{L}} q}{p \sqsubseteq_n^{\mathcal{L}} q} \quad (2)\ \frac{p \sqsubseteq_n^{\mathcal{L}} q}{ap \sqsubseteq_{n+\overline{d}(b,a)}^{\mathcal{L}} bq} \quad (3)\ \frac{p \sqsubseteq_n^{\mathcal{L}} p'}{p+q \sqsubseteq_n^{\mathcal{L}} p'+q} \quad (4)\ \frac{p \sqsubseteq_n^{\mathcal{L}} q \quad q \sqsubseteq_{n'}^{\mathcal{L}} r}{p \sqsubseteq_{n+n'}^{\mathcal{L}} r}$$

For instance, for the bisimulation semantics, we have $\sim$ in the place of $\sqsubseteq_{\mathcal{L}}$ and we simply write $=_d$ for the obtained collection of distance relations, that in this case are all symmetric.

The rest of the paper is devoted to the development of a pure algebraic framework that allows us to define those distances in an algebraic way. Once we have it, we could use all the algebraic techniques and general results from the area, on the study of those distance relations.

3   From Algebraic Semantics to Algebraic Distances

We will see in this section how the basic concepts appearing in the definition of algebraic semantics can be adequately modified in a natural way to obtain an algebraic theory of distances between processes. It can be used to get algebraic characterizations of all the distances previously presented in [8].

We start with the notion of Σ-algebra with distances, whose definition will be a resetting of the definition of quotient Σ-algebra.

Definition 5. *Let $\mathcal{D}$ be an adequate domain for distance values (e.g. $\mathbb{N}$, $\mathbb{R}^+$, $\mathbb{Q}^+$) and Σ a (classic) signature. A $(\mathcal{D}, \Sigma)$-algebra is a Σ-algebra $\langle \mathcal{A}, \Sigma_{\mathcal{A}} \rangle$ and a collection of relations $\langle \equiv_d, d \in \mathcal{D} \rangle$, $\equiv_d \subseteq \mathcal{D} \times \mathcal{D}$, such that:*

1. *$a \equiv_0 a$, for all $a \in \mathcal{A}$.*
2. *$a \equiv_d b \Leftrightarrow b \equiv_d a$, for all $a, b \in \mathcal{A}$ and all $d \in \mathcal{D}$.*

3. $d \leq d'$, $a \equiv_d b \Rightarrow a \equiv_{d'} b$, for all $a, b \in \mathcal{A}$ and all $d, d' \in \mathcal{D}$.

4. $(a \equiv_d b \wedge b \equiv_{d'} c) \Rightarrow a \equiv_{d+d'} c$, for all $a, b, c \in \mathcal{A}$ and all $d, d' \in \mathcal{D}$.

5. $f \in \Sigma$, $ar(f) = k$, $a_i \equiv_{d_i} b_i$ for all $i \in 1..k \Rightarrow f(\bar{a}) \equiv_{\widetilde{f}(d_1,\dots,d_k)} f(\bar{b})$,

   where $\widetilde{f}$ is a function associated to f that combines the distances between the components of its arguments and satisfies $\widetilde{f}(\bar{0}) = 0$.

Remark 1. In the previous definition we have only directly stated reflexivity of $\equiv_0$ in (1). However, applying these rules we can infer

$$(1') \ a \equiv_d a \ \text{for all} \ a \in \mathcal{A} \ \text{and all} \ d \in \mathcal{D},$$

by combining (1) and (3). Another possibility is to take (1') instead of (1), and then by combining it with (4), we get the monotonicity rule (3), which therefore could be removed.

Let us discuss and justify one by one all the ingredients of Def. 5. We have introduced a collection of relations $\equiv_d$, that intuitively describe the balls of radius d of the topology induced by our distance notion. The classical properties of distances correspond to 1-4. Note how the transitivity of equivalence relations is substituted by the triangular inequality 4.

We have said that we are generalizing the notion of quotient algebra, and not just that of (plain) algebra, because we allow that $\equiv_0$ will be any Σ-congruence, and not just the equality relation. Note that in that particular case the triangular inequality becomes transitivity, because $0 + 0 = 0$, and then $\equiv_0$ has to be first an equivalence relation, but also a Σ-congruence, by applying 5.

We have preferred to be quite general w.r.t. the constraints imposed to the combination functions $\widetilde{f}$. Certainly $\widetilde{f} = +$ for all $f \in \Sigma$, will be the most interesting case. In Section 6 we will see how taking $\widetilde{f} = max$, we obtain the algebraic characterization of other noticeable distances.

Now we can get our deduction systems for distances, $\mathbf{dDED}(E_\mathcal{D})$, by resetting the clauses that define $\mathbf{DED}(E)$, again in a very natural way.

Definition 6. *A deduction system for distances, $dDED(E_\mathcal{D})$, between terms in $\mathcal{T}_\Sigma(X)$, is a collection of rules including:*

1. *Reflexivity:* $t \equiv_d t$.
2. *Symmetry:* $t \equiv_d t' \Rightarrow t' \equiv_d t$.
3. *Triangular transitivity:* $t \equiv_d t'$, $t' \equiv_{d'} t'' \Rightarrow t \equiv_{d+d'} t''$.
4. *Substitution:* $t_1 \equiv_{d_1} t'_1, \dots, t_k \equiv_{d_k} t'_k \Rightarrow f(t_1, \dots, t_k) \equiv_{\widetilde{f}(d_1,\dots,d_k)} f(t'_1, \dots, t'_k)$

   for every $f \in \Sigma$ of arity k and the corresponding $\widetilde{f}$ composing distances.
5. *Instantiation:* $t \equiv_d t' \Rightarrow t\rho \equiv_d t'\rho$, for every substitution ρ.
6. *A set of distance equations $E_\mathcal{D} = \{t_i \equiv_{d_i} t'_i \mid i \in I\}$.*

Since variables are useful to get compact (possibly finite) axiomatizations, they are typically used when defining deduction systems. We have followed the same idea when defining $\mathbf{dDED}(E_\mathcal{D})$, that gives us distance pairs not only between closed processes, but also between open processes. Of course, we expect that any such derivable pair will reflect a universal information, which is formalized by the instantiation rule.

The roles of reflexivity, symmetry and that of the triangular transitivity, were already commented when defining the algebras with distances. Moreover, the substitution rule states the homomorphic character of the obtained distances. Of course, we have a different deduction system for each collection of composing functions $\langle \widetilde{f} | \ f \in \Sigma \rangle$, however, we prefer to maintain this small abuse of notation.

Finally, we have again adopted quite a general point of view when defining $\mathbf{dDED}(E_\mathcal{D})$ based on an arbitrary set of distance equations $E_\mathcal{D}$. But, once more it is interesting to discuss which are the most natural sets of equations, in which we are specially interested. The role of $\mathbf{DED}(E)$ is to generate the induced set of derived equations from the set E. When we start from the axiomatization of any semantics (e.g. the bisimulation semantics), the related closed processes are exactly those having the same semantics (e.g. those bisimilar).

As explained above, $\equiv_0$ just reflects the quotient algebra on top of which we will define our distance relations. As a matter of fact the following result tells us that, when E does only contain equations on $\equiv_0$, we just obtain a system totally equivalent to an ordinary deduction system.

Proposition 1. *If $E_\mathcal{D}$ is a system that only contains equations on $\equiv_0$, then the system $\mathbf{dDED}(E_\mathcal{D})$ is "essentially" equivalent to $\mathbf{DED}(E)$, where $E = \{t \equiv t' \mid t \equiv_0 t' \in E_\mathcal{D}\}$. This means that $\vdash_{E_\mathcal{D}} t \equiv_d t' \Leftrightarrow \vdash_E t \equiv t'$, $\forall d \in \mathcal{D}$.*

Proof. This is an immediate consequence of the following facts: (1) The triangular transitivity becomes plain transitivity when $d_1 + d_2 = 0$; (2) The impossibly to infer facts about $\equiv_0$ using other ones $\equiv_d$, with $d \neq 0$. □

Therefore, in order to have a useful deduction system for distances we need to start with a collection of equations $E_\mathcal{D}$ containing a set of non-trivial non-zero distance axioms $t_0 \equiv_d t_0'$ with $d > 0$. This subset is the "algebraic basis" on top of which $E_\mathcal{D}$ will derive the induced distance pairs in $\vdash_{E_\mathcal{D}} t \equiv_d t'$.

Proposition 2. *Given a system of distance equations $E_\mathcal{D}$, if we define $E_\mathcal{D}^0 = \{t \equiv_d t' \mid d = 0\}$ and we consider the set of ordinary equations $E = \{t \equiv t' \mid t \equiv_0 t' \in E_\mathcal{D}^0\}$, then we can see the family of distance relations induced by $\vdash_{E_\mathcal{D}}$ as a family of distance relations between the equivalence classes induced by $\vdash_E$,*

$$[t] \equiv_d [t'] ::= \vdash_{E_\mathcal{D}} t \equiv_d t'$$

Proof. We only need to apply the triangular transitivity rule. □

In the following section we apply this algebraic approach, defining a distance for the bisimulation semantics.

4 An Algebraic Distance for Bisimulation

As stated above, to define our processes, we use the signature including the choice operation with arity 2, and the prefix operators $a \cdot \in Act$ with arity 1, together with the constant null, $\mathbf{0}$. We expand this signature including variables in a set X to obtain the $BCCSP(Act, X)$ syntactic algebra. Then, the corresponding compositional approach to the definition of distances between processes in $BCCSP(Act, X)$ includes the rules:

- If we have $p \equiv_d p'$ and $q \equiv_d q'$, then $p + q \equiv_d p' + q'$.
- If we have $p \equiv_d q$, then $ap \equiv_d aq$.

Moreover, the equations characterizing the bisimulation axioms are turned into distance equations getting:

$$(B1)\ x + y \equiv_0 y + x \qquad\qquad (B2)\ x + x \equiv_0 x$$

$$(B3)\ (x + y) + z \equiv_0 x + (y + z) \qquad\qquad (B4)\ z + \mathbf{0} \equiv_0 z$$

Finally, we need to add the collection of equations that will work as seed for the computation of distances in an algebraic way. The idea is that we want to pay a tax for each punctual change. Since we are working under a function $\bar{d}$ defining the distance between actions in Act, we introduce the family of axioms with $a, b \in Act$, which can be considered as a single axiom scheme if we see a and b as generic action: $ax \equiv_{d(a,b)} bx$. Putting everything together we obtain the following algebraic definition of the bisimulation distance.

Definition 7. *Given a function distance $\bar{d}$ between the actions in Act, producing values in $\mathcal{D}$, and two processes $p, q \in BCCSP(Act, X)$, we can say that p is at most $d \geq 0$ far away of being bisimilar to q, if and only if $p \equiv_d q$ can be derived using the set of rules:*

1. *$p \equiv_d p$ for all $d \in \mathcal{D}$ and for all $p \in Proc$.*
2. *$p \equiv_d p' \Rightarrow p' \equiv_d p$ for all $d \in \mathcal{D}$ and for all $p, p' \in Proc$.*
3. *$p \equiv_{d_1} p'$ and $p' \equiv_{d_2} p'' \Rightarrow p \equiv_{d_1+d_2} p''$ for all $d_1, d_2 \in \mathcal{D}$ and $p, p', p'' \in Proc$.*
4. *(i) $p \equiv_{d_1} p'$, $q \equiv_{d_2} q' \Rightarrow p + q \equiv_{d_1+d_2} p' + q'$.*
   (ii) $p \equiv_d q \Rightarrow ap \equiv_d aq$ for all $a \in Act$, $d, d_1, d_2 \in \mathcal{D}$ and $p, p', q, q' \in Proc$.
5. *$p \equiv_d p' \Rightarrow p\rho \equiv_d p'\rho$, for every substitution ρ.*
6. *(i) $ax \equiv_{\bar{d}(a,b)} bx$ for all $a, b \in Act$.*   *(ii) $x + y \equiv_0 y + x$.*
   (iii) $x + x \equiv_0 x$.   *(iv) $(x+y)+z \equiv_0 x+(y+z)$.*   *(v) $z + \mathbf{0} \equiv_0 z$.*

Remark 2. 1. The definition above only considers finite terms in $BCCSP(Act, X)$, but we can extend the application of these rules to infinite processes. However, this extension will only produce distances for the case of pairs of processes that are bisimilar up to the change of finitely many actions (e.g., $a^\omega \equiv_{d(a,b)} ba^\omega$). In Section 7 we will discuss how we can get other more interesting distances in the infinite case.

2. Once we use addition as the composition function of distance for the choice operator, we could substitute rule $4(i)$ in Def. 7 by the simpler rule $p \equiv_d p' \Rightarrow p + q \equiv_d p' + q$. We immediately obtain the original rule by combining this with the triangular transitivity rule 3.

By combining that simplified rule with rule $4(ii)$ is easy to prove that for any linear context $\mathcal{C}(x)$ we have the preservation rule

$$p \equiv_d p' \;\Rightarrow\; \mathcal{C}(p) \equiv_d \mathcal{C}(p').$$

This will not be true for any arbitrary non-linear $\mathcal{C}$, where in principle if x appears k times in $\mathcal{C}$, then we will have $\mathcal{C}(p) \equiv_{k*d} \mathcal{C}(p')$ but not $\mathcal{C}(p) \equiv_d \mathcal{C}(p')$. Obviously, this is an important difference to what happens in $\mathbf{DED}(E)$. There we can modify the global substitution rule 4 in Fig. 1 by a local substitution, where only an argument of f is substituted. It is possibly to do it without obtaining nothing new due to the presence of the transitivity rule.

Example 1. For instance, let us take $Act = \{a, b, c\}$ and define $\overline{d}(a, b) = 1$, $\overline{d}(a, c) = 2$ and $\overline{d}(b, c) = 1$. Now we will show that $ab\mathbf{0} + bb\mathbf{0} \equiv_3 ac\mathbf{0} + cc\mathbf{0}$:

$$b\mathbf{0} \equiv_1 c\mathbf{0} \;(Def.7.5,\,7.6) \;\Rightarrow\; \left\{ \begin{array}{c} ab\mathbf{0} \equiv_1 ac\mathbf{0} \;\;(Def.7.4) \\ \wedge \\ bb\mathbf{0} \equiv_1 bc\mathbf{0} \;\;(Def.7.4) \\ \wedge \\ bc\mathbf{0} \equiv_1 cc\mathbf{0} \;\;(by\;Def.7.5,\,7.6) \end{array} \right\} \overset{(Def.7.3)}{\Rightarrow} bb\mathbf{0} \equiv_2 cc\mathbf{0}$$

and finally applying Def. 7.4 we can conclude $ab\mathbf{0} + bb\mathbf{0} \equiv_3 ac\mathbf{0} + cc\mathbf{0}$.

We developed in [8] our operational and our denotational approaches to the definition of distance relations without considering variables. However we can easily extend both of them to cover the full set $BCCSP(Act, X)$. In particular, for the denotational approach that we are using here, we can extend Def. 4 by first extending the preorder $\sqsubseteq_{\mathcal{L}}$ to open terms in the classic way, and simply applying the rest of the rules also to these open terms.

Now it is immediate to check that if we can derive $p(\overline{x}) =_d q(\overline{x})$, using this extension of Def. 4, where the variables in $\overline{x}$ are those appearing in either p or q, then we can also derive any instance $p(\overline{r}) =_d q(\overline{r})$. Here we have used the classical notation $p(\overline{r})$, to denote the application of the instantiation of each variable x_i in $p(\overline{x})$ by the corresponding term, r_i.

Lemma 1. *If we can derive $p(\overline{x}) =_d q(\overline{x})$, then we can also derive any instance of it, $p(\overline{r}) =_d q(\overline{r})$.*

Proof. By induction on the derivations of $p(\overline{x}) =_d q(\overline{x})$

1. $p(\overline{x}) \sim q(\overline{x}) \Rightarrow p(\overline{r}) \sim q(\overline{r})$, by definition of bisimulation.
2. and 3. The application of (2) and (3) in Def. 4, is immediate because instantiation satisfies the homomorphic rules:

$$(ap(\overline{r})) = a(p(\overline{r})) \;\text{ and }\; (p + q)(\overline{r}) = p(\overline{r}) + q(\overline{r}).$$

4. Finally, by a direct application of (4) in Def. 4, from $p(\overline{r}) =_n s(\overline{r})$ and $s(\overline{r}) =_{n'} q(\overline{r})$ it produces $p(\overline{r}) =_{n+n'} q(\overline{r})$. $\square$

Next we prove the equivalence between this algebraic definition of the distance relations and the extension of the denotational definition in Def. 4.

Theorem 1. *For all $p, q \in BCCSP(Act, X)$ we have $p \equiv_d q \Leftrightarrow p =_d q$*

Proof. $\Leftarrow$ | We want to show that if $p =_d q$ then $p \equiv_d q$. We use induction over the depth of derivations.

1. $\dfrac{p \sim q}{p =_d q}$. If $p \sim q$, then we can prove it using the set of axioms for bisimulation that are mimicked by the last four axioms in Def. 7.6. So that we have $p \equiv_0 q$, and then by Def. 7.1 we have $q \equiv_d q$, and finally applying Def. 7.3 we get $p \equiv_d q$ as we wanted to show.

2. $\dfrac{p =_d q}{ap =_{d+\overline{d}(b,a)} bq}$. By the i.h. we have $p \equiv_d q$. We use the equation $ax \equiv_{\overline{d}(a,b)} bx$, Def. 7.1 and 7.5 to get $ap \equiv_{\overline{d}(a,b)} bp$. Moreover, applying Def. 7.4 (ii) and the i.h. we get $bp \equiv_d bq$; and finally by the triangular transitivity rule (7.3) we can conclude $ap \equiv_{d+\overline{d}(a,b)} bq$.

3. $\dfrac{p =_d p'}{p + q =_d p' + q}$. We have, by the i.h., that $p \equiv_d p'$. Trivially $q \equiv_0 q$ using Def. 7.1. Then, we can conclude applying 7.4 (i) that $p + q \equiv_d p' + q$.

4. $\dfrac{p =_d r \quad r =_{d'} q}{p =_{d+d'} q}$. Immediate, using the i.h. and Def. 7.3.

$\Rightarrow$ | By induction on the derivation of $p \equiv_d q$.

1. $p \equiv_d p$. Obviously, we have $p \sim p$ and then by Def. 4.1 we get $p =_d p$.
2. $p \equiv_d p' \Rightarrow p' \equiv_d p$. By the i.h. we get $p =_d p'$, and it is immediate to check, by induction on the derivation of $p =_d p'$ that we can generate a symmetric derivation concluding $p' =_d p$. We use that $\sim$ is an equivalence relation in Def. 4.1, and $\overline{d}$ is a symmetric relation in Def. 4.2.
3. $p \equiv_{d_1} p'$ and $p' \equiv_{d_2} p'' \Rightarrow p \equiv_{d_1+d_2} p''$. By the i.h. we get $p =_{d_1} p'$, and $p' =_{d_2} p''$, and applying Def. 4.4 we conclude $p =_{d_1+d_2} p''$.
4. (i)$p \equiv_{d_1} p'$, $q \equiv_{d_2} q' \Rightarrow p + q \equiv_{d_1+d_2} p' + q'$. By the i.h. we get $p =_{d_1} p'$ and $q =_{d_2} q'$ using Def. 4.3 and 4.4 we get $p + q =_{d_1+d_2} p' + q'$.
   (ii)$p \equiv_d q \Rightarrow ap \equiv_d aq$. Immediate, applying the i.h. and Def. 4.2.
5. $p \equiv_d p' \Rightarrow p\rho \equiv_d p'\rho$, for every substitution ρ. Trivially, applying Lemma 1.
6. (i) $ax \equiv_{\overline{d}(a,b)} bx$, applying 4.2 and 4.1.
   (ii)-(iv). Immediate, applying 4.1. $\square$

Remark 3. Certainly, we could also remove variables from the algebraic presentation simply expanding every axiom, including instead of it, all its closed instances. Obviously the instance rules will not be necessary after that. But, of course the role of variables in any algebraic presentation is crucial in order to obtain finite axiomatizations where we reflect by a single action an infinity of facts.

5   Algebraic Distances for Other Semantics

Once we have studied in detail the algebraic definition of the distance for the case of bisimulation, we will briefly discuss the case of the rest of the semantics which are collected in the ltbt-spectrum [10]. These semantics are not induced by an equivalence relation but by a preorder. Hennessy also presented in [7] a theory of Σ-po algebras, $\langle A, \leq_A, \Sigma_A \rangle$, which are endowed with a partial order. Now for each f in Σ of arity k, there is a monotonic function $f_A \colon A^k \to A$. You can find in [7] all the details about this theory of ordered algebras which smoothly extends that of plain algebras. In particular, Hennessy purposes the proof system $\mathbf{DED}(I)$, in Fig. 2, where I is now a set of inequations.

1. Reflexivity $\dfrac{}{t \leq t}$

2. Transitivity $\dfrac{t \leq t' \quad t' \leq t''}{t \leq t''}$

3. Substitution $\dfrac{t_1 \leq t'_1, \ldots, t_k \leq t'_k}{f(t_1, \ldots, t_k) \leq f(t'_1, \ldots, t'_k)}$, for every $f \in \Sigma$ of arity k.

4. Instantiation $\dfrac{t \leq t'}{t\rho \leq t'\rho}$, for every substitution ρ.

5. Equations $\dfrac{}{t \leq t'}$, for every inequation $t \leq t' \in I$

Fig. 2. The proof system $\mathbf{DED}(I)$

As in Section 4 we can adapt this theory of ordered algebras, obtaining an algebraic theory which allows us to measure the distance between processes w.r.t. a given semantics $\mathcal{L}$ (in)equationally defined by axioms on an order $\subseteq^{\mathcal{L}}$. So, we define the following ordered deduction system with distances $\mathbf{dDED}(I)$.

Definition 8. *Given a semantics $\mathcal{L}$ algebraically defined by means of an axiomatization I on $\leq^{\mathcal{L}}$, and a distance $\overline{d}$ over the set of actions Act, we will say that a process p is at most $d \geq 0$ far away of being better than another process q, w.r.t. the preorder $\leq^{\mathcal{L}}$, if and only if we have $\vdash_{\mathbf{dDED}(I)} p \leq_d^{\mathcal{L}} q$. $\mathbf{dDED}(I)$ is the following deduction system:*

1. $p \leq_d^{\mathcal{L}} p$, *for all $d \in \mathcal{D}$.*
2. $p \leq_{d_1}^{\mathcal{L}} p'$ *and* $p' \leq_{d_2}^{\mathcal{L}} p'' \Rightarrow p \leq_{d_1+d_2}^{\mathcal{L}} p''$.
3. *(i)* $p \leq_{d_1}^{\mathcal{L}} p'$ $q \leq_{d_2}^{\mathcal{L}} q' \Rightarrow p + q \leq_{d_1+d_2}^{\mathcal{L}} p' + q'$.
   (ii) $p \leq_d^{\mathcal{L}} q \Rightarrow ap \leq_d^{\mathcal{L}} aq$, *for all $a \in Act$.*
4. $p \leq_d^{\mathcal{L}} p'$ *then* $p\rho \leq_d^{\mathcal{L}} p'\rho$, *for every substitution ρ.*
5. $ax \leq_{\overline{d}(a,b)}^{\mathcal{L}} bx$.
   $t \leq_d^{\mathcal{L}} t'$, *for every inequation $t \leq^{\mathcal{L}} t' \in I$.*

Remark 4. 1. By the way, the only difference between the proof systems **DED**(E) and **DED**(I) is that in this second case we are defining a preorder and not an equivalence relation, therefore symmetry is lost. The same happens in **dDED**(I) by means of which we measure "how far away" a process is to be better than another process w.r.t. the corresponding order $\subseteq^{\mathcal{L}}$.

2. Once we are defining an order and not an equivalence it is natural to consider asymmetric distances d, where $d(b, a)$ denotes what we have to add to b to obtain (at least) a. For instance if $a = 1€$ and $b = 2€$ we could have $d(a, b) = 1$ but $d(b, a) = 0$. We have developed our operational distances in [8] including this generalization, and it can be also introduced here simply changing the symmetric distance $\overline{d}$ by an asymmetric distance d.

It is easy to translate the results in Section 4 to this more general framework. In particular, we can prove that each denotational distance, obtained by application of Def.4, is equivalent to the corresponding algebraic distance, obtained by application of Def. 8.

6   Characterizing the Bisimulation Distance Game

The classic approaches to the definition of distances between processes are based on valued versions of the (bi)simulation game [9, 6, 2].

Definition 9. *(Bisimulation game) Given two LTSs, L_1 and L_2, we call configurations the pairs (p, q), with $p \in L_1$ and $q \in L_2$. The bisimulation game is played by two players: the attacker $\mathbb{A}$ and the defender $\mathbb{D}$. The initial configuration of the game deciding if $p_0 \sim q_0$, is just the pair (p_0, q_0). A round of the game, when the current configuration is (p, q), proceeds as follows:*

1. *$\mathbb{A}$ chooses either p or q.*
2. *Assuming it chooses p, he next executes a transition in L_1: $p \overset{a}{\rightarrow} p'$.*
3. *$\mathbb{D}$ must choose a transition with the same label at the other side of the board, i.e., it must execute an a-move in L_2: $q \overset{a}{\rightarrow} q'$. If $\mathbb{A}$ plays at L_2, then $\mathbb{D}$ replies at L_1.*
4. *The game proceeds in the same way from the new configuration (p', q').*

The bisimulation game can be turned into the "classical" bisimulation distance game [1], by allowing to reply any a-move by some b-move with $b \neq a$. However, in this case the defender should pay $\overline{d}(b, a)$ to the attacker for the mismatch. The value of the game, $V(p, q)$, provides the "classical" bisimulation distance between p and q, $d_\sim(p, q)$.

In order to illustrate the generality of our algebraic approach to the definition of distances between processes, next we present an algebraic characterization of that bisimulation game distance.

Definition 10. *We define the algebraic bisimulation game collection of relations, $\{\equiv_d^g, d \in \mathcal{D}\}$, as the set of tuples $p \equiv_d^g q$ which can be derived by applying the following set of rules:*

1. $p \equiv_d^g p$ for all $d \in \mathcal{D}$ and $p \in Proc$.

2. $p \equiv_d^g p' \Rightarrow p' \equiv_d^g p$ for all $d \in \mathcal{D}$ and $p, p' \in Proc$.

3. $p \equiv_{d_1}^g p'$ and $p' \equiv_{d_2}^g p'' \Rightarrow p \equiv_{d_1+d_2}^g p''$ for all $d_1, d_2 \in \mathcal{D}$ and $p, p', p'' \in Proc$.

4. (i) $p \equiv_{d_1}^g p', q \equiv_{d_2}^g q' \Rightarrow p+q \equiv_{max\{d_1,d_2\}}^g p' + q'$.

   (ii) $p \equiv_d^g q \Rightarrow ap \equiv_d^g aq$ for all $a \in Act$, $d, d_1, d_2 \in \mathcal{D}$ and $p, p', q, q' \in Proc$.

5. $p \equiv_d^g p' \Rightarrow p\rho \equiv_d^g p'\rho$, for every substitution ρ.

6. (i) $ax \equiv_{\overline{d}(a,b)}^g bx$ for all $a, b \in Act$.   (ii) $x + y \equiv_0^g y + x$.

   (iii) $x + x \equiv_0^g x$.   (iv) $(x+y) + z \equiv_0^g x + (y+z)$.   (v) $z + \mathbf{0} \equiv_0^g z$.

Remark 5. Note that the definition of the algebraic bisimulation distance by means of the collection $\{\equiv_d, d \in \mathcal{D}\}$ in Def. 7, and that of the algebraic bisimulation game distance by means of $\{\equiv_d^g, d \in \mathcal{D}\}$, are nearly the same. We only modify the composition rule 4 (i), where the composition function $\widetilde{f}$–see Def. 5.5– was initially $+$ and now it becomes max.

As we did for our bisimulation distance, there is still a third equivalent denotational characterization of the bisimulation game distance.

Definition 11. We define the denotational bisimulation game collection of relations, $\{\equiv_d^g, d \in \mathcal{D}\}$, as the set of tuples $p \equiv_d^g q$ which can be derived by applying the following set of rules:

(1) $\dfrac{p \sim q}{p \equiv_n^g q}$ (2) $\dfrac{p \equiv_d^g q}{ap \equiv_{d+\overline{d}(b,a)}^g bq}$ (3) $\dfrac{p \equiv_{d_1}^g p' \quad q \equiv_{d_2}^g q'}{p+q \equiv_{max\{d_1,d_2\}}^g p' + q'}$

We have proved in [8] that the relations defined by Def. 11 remain the same if we add the transitivity rule

(4) $\dfrac{p \equiv_{d_1}^g q \quad q \equiv_{d_2}^g r}{p \equiv_{d_1+d_2}^g r}$

so further in this paper we consider Def. 11 including this rule. Then, it is easy to get the following theorem based on the proof of Th. 1

Theorem 2. $p \equiv_d^g q \Leftrightarrow p \equiv_d^g q$.

Proof. Immediate, just substituting $+$ by max in the reasoning related to the application of Def. 4.3 and Def. 7.4.

$\square$

Remark 6. Although rule (3) in Def. 4 has not $+$, we can see applying transitivity that this rule is equivalent to

(3') $\dfrac{p \equiv_{n_1}^{\mathcal{L}} p' \quad q \equiv_{n_2}^{\mathcal{L}} q'}{p+q \equiv_{n_1+n_2}^{\mathcal{L}} p' + q'}$

which produce a simpler proof than the given in Th. 1 when we use 4.3.

Next, we see that the original definition of the distances by means of the distance bisimulation game is also equivalent to these. We start proving two lemmas that provide us some properties of the denotational characterization and of the values of the distance bisimulation game.

Lemma 2 (Prefix lemma). *Given two processes $P = ap$ and $Q = bq$, we have $ap =_d^g bq$ if and only if there exists some d' such that $d = d' + \overline{d}(a, b)$ with $p =_{d'}^g q$.*

Proof. $\Leftarrow |$ Immediate, since if we have Def. 11.2 to $p =_{d'}^g q$ we get the result.
$\Rightarrow |$ We will prove a more general result. It says that $\sum_{i \in I} a_i p_i =_d^g \sum_{j \in J} b_j q_j$ implies $(\forall i \in I \exists j \in J \quad p_i =_{d-\overline{d}(a_i, b_j)}^g q_j)$. So, the result of this theorem will be just the particular case of this result when we have only one element in the sum. We use induction over the derivation of $p =_d^g q$ taking $p = \sum_{i \in I} a_i p_i$ and $q = \sum_{j \in J} b_j q_j$.

1. $\dfrac{p \sim q}{p =_d^g q}$. If $p \sim q$ then we have for each $i \in I$ such that $a_i p_i \xrightarrow{a_i} p_i$ there

   exists some $j \in J$ with $b_j q_j \xrightarrow{b_j} q_j$ where $b_j = a_i$ and $p_i \sim q_j$. Then we have $p_i =_d^g q_j$ and using the rule (2) we can conclude that $a_i p_i =_0^g b_j q_j$ as we wanted to show, since $\overline{d}(a_i, b_j) = 0$.

2. $\dfrac{p =_{d-\overline{d}(b,a)}^g q}{ap =_{d-\overline{d}(b,a)+\overline{d}(b,a)}^g bq}$. Then we have a single summand at both sides, and the premise of the last set of the derivations exactly expresses the thesis to be proved.

3. If we have $p = p' + p''$, $q = q' + q''$ and $\dfrac{p' =_{d'}^g q' \quad p'' =_{d''}^g q''}{p =_{d=max\{d',d''\}}^g q}$. We will have $p' = \sum_{i \in I'} a_i p_i$, $p'' = \sum_{i \in I''} a_i p_i$, $q' = \sum_{j \in J'} b_j q_j$, and $q'' = \sum_{j \in J''} b_j q_j$ with $I = I' \cup I''$ and $J = J' \cup J''$. Then, by applying the i.h. we have $\forall i \in I'$ $\exists j \in J'$ with $p_i =_{d'-\overline{d}(a_i, b_j)}^g q_j$ and $\forall i \in I''$ $\exists j \in J''$ with $p_i =_{d''-\overline{d}(a_i, b_j)}^g q_j$. As $d = max\{d', d''\}$ we immediately conclude the result applying the triangular transitivity (rule (3)).

4. $\dfrac{p =_d^g r \quad r =_{d'}^g q}{p =_{d+d'}^g q}$. We take $r = \sum_{k \in K} c_k r_k$ and $r_k =_{d'-\overline{d}(c_k, b_j)}^g q_j$, so applying the i.h. we get $\forall i \in I$ $\exists k \in K$ $p_i =_{d-\overline{d}(a_i, c_k)}^g r_k$ (and $\forall k \in K$ $\exists j \in J$ $r_k =_{d'-\overline{d}(c_k, b_j)}^g q_j$). Then, applying the triangular inequality $(\overline{d}(a_i, b_j) \leq \overline{d}(a_i, c_k) + \overline{d}(c_k, b_j))$ we get that $\forall i \in I$ $\exists j \in J$ such that $p_i =_{d+d'-\overline{d}(a_i, b_j)}^g q_j$. $\square$

Corollary 1. $d(\alpha, \beta) = \sum \overline{d}(a_i, b_i)$ *where* $\alpha = a_1 \ldots a_n$ *and* $\beta = b_1 \ldots b_n$. *This means that we have* $\alpha =_d^g \beta$ *with* $d = \sum_{i=1}^n \overline{d}(a_i, b_i)$, *and for all* $d' < d$ *we have* $\alpha \neq_{d'}^g \beta$.

Proof. That $\alpha =_d^g \beta$ is immediate by iterated application of the triangular transitivity rule. We prove the second negative result by induction on n.

$n = 0|$ We have $d = 0$ and then the result is trivial.

$n > 0|$ Applying the Prefix Lemma –Lemma 2– we should have $\alpha' =^{g}_{d'-\overline{d}(a_1,b_1)} \beta'$ with $\alpha' = a_2 \ldots a_n$ and $\beta' = b_2 \ldots b_n$. But, if $d' < \sum_{i=1}^{n} \overline{d}(a_i, b_i)$ then we have $d' - \overline{d}(a_1, b_1) < \sum_{i=2}^{n} \overline{d}(a_i, b_i)$ contradicting the i.h. for the shorter sequences α' and β' □

Lemma 3. $V(p,q) \leq V(p,r) + V(r,q)$ for all processes p, q and r.

Proof. We use induction over the depth of processes.

For all $i \in I$ there exists some $k \in K$ with $V(p_i, r_k) = V(p,r) - \overline{d}(a_i, c_k)$. For all $k \in K$ there exists some $j \in J$ with $V(r_k, q_j) = V(r,q) - \overline{d}(c_k, b_j)$. Combining both, we obtain:

For all $i \in I$ there exists some $j \in J$ and $k \in K$ with
$$V(p_i, r_k) + V(r_k, q_j) = V(p,r) + V(r,q) - \overline{d}(a_i, c_k) - \overline{d}(c_k, b_j).$$

and then by applying the i.h. and the triangular inequality for d, we obtain:

$$\forall i \in I \; \exists j \in J \text{ with } V(p_i, q_j) \leq V(p,r) + V(r,q) - \overline{d}(a_i, b_j).$$

In a symmetric way, we prove that For all $j \in J$ there exists some $i \in I$ with $V(p_i, q_j) \leq V(p,r) + V(r,q) - \overline{d}(a_i, b_j)$. Applying the definition of the value of the distance game for bisimulation we conclude: $V(p,q) \leq V(p,r) + V(r,q)$.   □

The value of the bisimulation game between two processes p and q, $V(p,q)$, gives us "the" distance between them $d_\sim(p,q)$. Next we will see that our denotational definition gives us all the bounds of this distance.

Theorem 3. $d_\sim(p,q) = \min_d \{p =^{g}_{d} q\}$; i.e. $p =^{g}_{d'} q$ iff $d_\sim(p,q) \leq d'$.

Proof. In order to simplify our notation we denote simply by $d(p,q)$ the distance between these two processes. We use induction on the depth of p and q.

$\supseteq |$ We have that $p =^{g}_{v} q$ and we want to check that $d(p,q) \leq v$. We prove it by induction on the derivation of $p =^{g}_{v} q$.

1. $\dfrac{p \sim q}{p =^{g}_{v} q}$. If $p \sim q$ then the value of the bisimulation game is 0, because all along a game we can reply an a-move of p by an a-move of q, and $\overline{d}(a,a) = 0$, and conversely.

2. $\dfrac{p =^{g}_{v} q}{ap =^{g}_{v+\overline{d}(b,a)} bq}$. By applying the induction hypothesis we have $d(p,q) \leq v$. Then, the definition of the bisimulation game, produces $d(ap, bq) = d(p,q) + \overline{d}(a,b) \leq v + \overline{d}(b,a)$ as we wanted to see.

3. $\dfrac{p =^{g}_{v} q \quad p' =^{g}_{v'} q'}{p + p' =^{g}_{\max\{v,v'\}} q + q'}$. By applying the induction hypothesis we have $d(p,q) \leq v$ and $d(p',q') \leq v'$. Now, any a-move on the p-side (resp. p'-side) of $p + p'$ can be replied by some b-move on the q side (resp. q'-side) of $q + q'$ guaranteeing a payment less or equal than v (resp. v'), and conversely. Thus concluding that $d(p + p', q + q') \leq \max\{v, v'\}$.

4. $\dfrac{p =^g_v r \quad r =^g_{v'} q}{p =^g_{v+v'} q}$. By applying the induction hypothesis we have $d(p,r) \leq v$

and $d(r,q) \leq v'$. Now, applying Lemma 3 we conclude $d(p,q) \leq v + v'$.

$\subseteq$ | The proof is by induction on the depth of the processes. If it is 0 the result is trivial. Therefore, let us consider decomposition of the two involved processes $p = \sum a_i p_i$ and $q = \sum b_j q_j$.

Applying the definition of the bisimulation game, from $d(p,q) = v$ we obtain that for all $i \in I$ there exists some $j_i \in J$ such that $d(p_i, q_{j_i}) + \overline{d}(a_i, b_{j_i}) \leq d(p,q)$. Reciprocally, for all $j \in J$ there exists some $i_j \in I$ such that $d(p_{i_j}, q_j) + \overline{d}(a_{i_j}, b_j) \leq d(p,q)$. Now, by applying the induction hypothesis we have

$$\forall i \in I \; p_i =_{d(p_i,q_{j_i})} q_{j_i} \text{ and } \forall j \in J \; p_{i_j} =_{d(p_{i_j},q_j)} q_j.$$

From which applying Def. 11.2 we obtain $\forall i \in I \; a_i p_i =_{d(p_i,q_{j_i})+\overline{d}(a_i,b_{j_i})} b_{j_i} q_{j_i}$ and $\forall j \in J \; a_{i_j} p_{i_j} =_{d(p_{i_j},q_j)+\overline{d}(a_{i_j},b_j)} b_j q_j$, that applying the monotonicity of the bounds computed by Def. 11 produces

$$\forall i \in I \; a_i p_i =_{d(p,q)} b_{j_i} q_{j_i} \text{ and } \forall j \in J \; a_{i_j} p_{i_j} =_{d(p,q)} b_j q_j.$$

These two can be combined applying Def. 11.3 to produce $p = \sum a_i p_i =_{d(p,q)} \sum_{i \in I} b_{j_i} q_{j_i}$ and $\sum_{j \in J} a_{i_j} p_{i_j} =_{d(p,q)} \sum b_j q_j = q$.

We only need to combine these two again using Def. 11.3 and the idempotency of bisimilarity, to conclude $p =_{d(p,q)} q$, as we wanted to show. $\square$

7   Conclusions and Future Work

We have presented an algebraic framework to define distances between processes. In particular those associated to the semantics that are axiomatizable. Although a part of our definitions and properties could be applied to arbitrary processes, most of them are based on the consideration of finite image processes. It can be syntactically represented by a (finite) term of a certain signature.

Currently we are working on the extension of our results to the infinite case. Following [7] the idea is to approximate processes by their finite approximations and compare them level by level. Then, we would state that $p \equiv_d q$ if and only if we have $p_n \equiv_d q_n \; \forall n \in \mathbb{N}$. But whenever Act is finite, or the non-null values of $\overline{d}(b,a)$ are low bounded, we could only obtain a finite distance $p \equiv_d q$ for some $d \in \mathcal{D}$, when q can be obtained from p by a finite number of applications of the rules in Def. 7; that is, whenever p and q are bisimilar up to a finite number of changes of the actions occurring in them.

In order to obtain a more general distance that also produces finite values for pairs of processes which cannot be transformed one into the other by a finite number of transformations, we would need to adopt a discounting function. The idea is that the weights of the disagreements between the two compared processes decrease with the depth they occur. This is easily formalized in our algebraic framework, simply changing our rule 4 ii) in Def. 7 by a discounted rule

$$p \equiv_d q \;\Rightarrow\; ap \equiv_{\alpha d} aq$$

where $\alpha > 1$. As a matter of fact this is another instantiation of our Def. 5. In such a case it would be immediate to check that when $\overline{d}(b, a) = 1$ and $\alpha = \frac{1}{2}$, we would have $a^\infty \equiv_2 b^\infty$.

We are also working on the definition of these distances by applying a coinductive approach that avoids the consideration of finite approximations to obtain the distances between infinite processes.

We have used the algebraic developments in [7] to base our algebraic theory on distances. We did that, not only due to the simplicity and clarity of its presentation of the theory, but also because of its detailed study of the testing semantics. We hope indeed that most, if not all, of the concepts and results on this semantics will be transferable to the distance scenario. So, we will obtain a nice theory of approximated pass of test both producing a distance for the induced semantics, and an interesting new concept to be applicable in practice.

References

[1] Černý, P., Henzinger, T.A., Radhakrishna, A.: Quantitative simulation games. In: Manna, Z., Peled, D.A. (eds.) Time for Verification. LNCS, vol. 6200, pp. 42–60. Springer, Heidelberg (2010)

[2] Černý, P., Henzinger, T.A., Radhakrishna, A.: Simulation distances. In: Gastin, P., Laroussinie, F. (eds.) CONCUR 2010. LNCS, vol. 6269, pp. 253–268. Springer, Heidelberg (2010)

[3] Chen, X., Deng, Y.: Game characterizations of process equivalences. In: Ramalingam, G. (ed.) APLAS 2008. LNCS, vol. 5356, pp. 107–121. Springer, Heidelberg (2008)

[4] de Frutos-Escrig, D., Gregorio-Rodríguez, C., Palomino, M.: On the unification of process semantics: equational semantics. ENTCS 249, 243–267 (2009)

[5] Fahrenberg, U., Legay, A., Thrane, C.R.: The quantitative linear-time–branching-time spectrum. In: FSTTCS 2011. LIPIcs, vol. 13, pp. 103–114. Schloss Dagstuhl - Leibniz-Zentrum für Informatik (2011)

[6] Fahrenberg, U., Thrane, C.R., Larsen, K.G.: Distances for weighted transition systems: Games and properties. In: QAPL 2011. EPTCS, vol. 57, pp. 134–147 (2011)

[7] Hennessy, M.: Algebraic theory of processes. MIT Press (1988)

[8] Romero Hernández, D., de Frutos Escrig, D.: Defining distances for all process semantics. In: Giese, H., Rosu, G. (eds.) FMOODS/FORTE 2012. LNCS, vol. 7273, pp. 169–185. Springer, Heidelberg (2012)

[9] Stirling, C.: Modal and temporal logics for processes. In: Moller, F., Birtwistle, G. (eds.) Logics for Concurrency. LNCS, vol. 1043, pp. 149–237. Springer, Heidelberg (1996)

[10] van Glabbeek, R.: The linear time-branching time spectrum I: the semantics of concrete, sequential processes. In: Handbook of Process Algebra, ch. 1, pp. 3–99. Elsevier (2001)

[11] Ying, M.: Topology in process calculus - approximate correctness and infinite evolution of concurrent programs. Springer (2001)

[12] Ying, M., Wirsing, M.: Approximate bisimilarity. In: Rus, T. (ed.) AMAST 2000. LNCS, vol. 1816, pp. 309–322. Springer, Heidelberg (2000)

Appendix: On the Operational Definition of the Distances between Processes

It is well known that the set of (unordered) finite trees labelled in their arcs by actions in Act: $FTree(Act)$, constitute an initial model for the theory of bisimulation. A similar result can be obtained, adding variables in X, for the set $FTree(Act, X)$ of trees which besides the constants in Act also can have variables in X labeling their leaves.

By applying Def. 7 and Prop. 2, we immediately obtain a family of distance relations on both $FTree(Act)$ and $FTree(Act, X)$. In other words these two algebras become a (D, Σ)-algebra and a $(D, \Sigma(X))$-algebra, respectively, when considering these two families of distance relations.

Next, we present in detail the corresponding operational definition, already studied in [8]. Later, we develop the proof of its equivalence with the denotational and algebraic characterizations, that were proved to be equivalent each other at Section 4.

Definition 12. *We say that an unordered tree p is at most at distance d from another tree q, w.r.t. the symmetric distance between actions $\overline{d}$, and then we write $d_{\overline{d}}(p, q) \leq d$, if and only if:*

- *(C1) $p = ap'$, $q = bp'$, and $d \geq \overline{d}(a, b)$, or*
- *(C2) $p = p' + r$, $q = q' + r$, and $d \geq d_{\overline{d}}(p', q')$, or*
- *(C3) $p = ap'$, $q = aq'$, and $d \geq d_{\overline{d}}(p', q')$, or*
- *(C4) $d \geq 0$ and q can be obtained from p by application of (B1)-(B4), or*
- *(C5) There exist r, d' and d'' s.t. $d' \geq d_{\overline{d}}(p, r)$, $d'' > d_{\overline{d}}(r, q)$ and $d \geq d' + d''$.*

Definition 13. *We define $p \leadsto_d^1 q$ if and only if*

1. *$ap' \leadsto_d^1 bp'$ and $d \geq d(a, b)$, or*
2. *$p' + r \leadsto_d^1 q' + r$ and $p' \leadsto_d^1 q'$, or*
3. *$ap' \leadsto_d^1 aq'$ and $p' \leadsto_d^1 q'$, or*
4. *$d \geq 0$ and q can be obtained from p by application of (B1)-(B4).*

Definition 14. *We define $p \leadsto_d p'$ if and only if there exist $p_1, \dots, p_n$ such that $p \leadsto_{d_1}^1 p_1 \leadsto_{d_2}^1 p_2 \leadsto_{d_3}^1 \cdots \leadsto_{d_n}^1 p_n = p'$, where $\sum d_i = d$.*

Definition 15. *We define $p \leadsto_d^* p'$ if and only if we have $p \leadsto_d^1 p'$, or there exists some p'' such that $p \leadsto_{d_1}^1 p''$ and $p'' \leadsto_{d_2}^* p'$, where $d = d_1 + d_2$.*

Definition 16. *We define $p \mid\leadsto_d p'$ if and only if we have $p \leadsto_d^1 p'$, or there exists some q such that $p \mid\leadsto_{d_1} q$ $q \mid\leadsto_{d_2} p'$, where $d_1 + d_2 = d$.*

Proposition 3. *Def. 14, 15, 16 are obviously equivalent.*

Proof. Routine well known induction.   □

Lemma 4 (Structural lemma). *If $p \leadsto_d q$ then (1) $ap \leadsto_d aq$ and (2) $p + r \leadsto_d q + r$.*

Proof. (1)| If $p \leadsto_d^1 q$ then trivially we have $ap \leadsto_d^1 aq$ applying Def. 13.3.
If $p \leadsto_d q$ then we use induction over the path length. Similarly, if we have
$p = p_0 \leadsto_{d_1}^1 p_1 \leadsto_{d_2}^1 p_2 \leadsto_{d_3}^1 \cdots \leadsto_{d_n}^1 p_n = p'$ with $d = \sum d_i$, we will get
$ap_i \leadsto_{d_{i+1}}^1 ap_{i+1}$ for all $i < n$. So that, we have

$$ap = ap_0 \leadsto_{d_1}^1 ap_1 \leadsto_{d_2}^1 ap_2 \leadsto_{d_3}^1 \cdots \leadsto_{d_n}^1 ap_n = ap' \ with \ d = \sum d_i$$

thus proving $ap \leadsto_{d_1} ap'$.

(2)| Analogous to (1). $\qquad\qquad\qquad\qquad\qquad\qquad\qquad\qquad\qquad\qquad\qquad\square$

Theorem 4. *The operational definition of distance relations,* $\leadsto_d$, *and the denotational one, defining the family* $=_d$, *are equivalent.*

Proof. We want to prove that $p \leadsto_d q \leftrightarrow p =_d q$.

$\Rightarrow$ | We use induction over the length of the path generating $p \leadsto_d q$. We also use induction on the derivation of $p \leadsto_d^1$ to prove this basic case.

1. $p \leadsto_d^1 p'$ with $p = ap''$ and $p' = bp''$ and $d \geq d(a,b)$. We only need to apply rule (2) in Def. 4 to get that $ap'' =_d bp''$.
2. $p' + r \leadsto_d^1 q' + r$ with $p \leadsto_d^1 p'$. The i.h. now produces that $p' =_d q'$, and then applying rule (3) in Def. 4 we get that $p' + r =_d q' + r$.
3. $ap \leadsto_d^1 ap'$ with $p \leadsto_d^1 p'$. One more time, the i.h. produces $p =_d p'$, and applying rule (2) in Def. 4 we get $ap =_d ap'$.
4. $p \leadsto_d^1 p'$ with $d \geq 0$ and p' can be obtained from p by application of (B1)-(B4). As $p \sim p'$ trivially we have using rule (1) in Def. 4 that $p =_d p'$.
5. $p \leadsto_d p'$ if and only if $\exists p''\ p \leadsto_{d_1} p''\ p'' \leadsto_{d_2} p'$ with $d = d_1 + d_2$. We can apply the i.h. obtaining $p =_{d_1} p''$ and $p'' =_{d_2} p'$, and now applying rule (4) in Def. 4 we get $p =_{d_1+d_2} p'$, i.e., $p =_d p'$, as we wanted to show.

$\Leftarrow$ | Now we proof $p =_d q \Rightarrow p \leadsto_d q$, by induction on the derivation of $p =_d q$.

1. $\dfrac{p \sim q}{p =_d q}$. Trivially if $p \sim q$ then $p \leadsto_d q$ with $d \geq 0$, applying Def. 13.4.

2. $\dfrac{p =_{d - \bar{d}(b,a)} q}{ap =_d bq}$. By applying the induction hypothesis we have $p \leadsto_{d - \bar{d}(a,b)} q$ and applying the structural lemma we obtain $bq \leadsto_{d - \bar{d}(a,b)} bq$, from where using Def. 13.1 and Def. 15 we can conclude that $ap \leadsto_d bq$.

3. $\dfrac{p =_d p'}{p + q =_d p' + q}$. Again, by applying the induction hypothesis we have $p \leadsto_d p'$, and applying the structural lemma we conclude $p + q \leadsto_d p' + q$.

4. $\dfrac{p =_d q\ \ q =_{d'} r}{p =_{d+d'} r}$. By the i.h. we have $p \leadsto_d q$ and $q \leadsto_{d'} r$. So, applying Def. 15 and Prop. 3 we get $p \leadsto_{d+d'} r$. $\qquad\qquad\square$

Author Index